Age Factors in Biometric Processing

Age Factors in Biometric Processing

Edited by
Michael Fairhurst

The Institution of Engineering and Technology

Published by The Institution of Engineering and Technology, London, United Kingdom

The Institution of Engineering and Technology is registered as a Charity in England & Wales (no. 211014) and Scotland (no. SC038698).

First published 2013

The Institution of Engineering and Technology
Michael Faraday House
Six Hills Way, Stevenage
Herts, SG1 2AY, United Kingdom

www.theiet.org

British Library Cataloguing in Publication Data
A catalogue record for this product is available from the British Library

ISBN 978-1-84919-502-7 (hardback)
ISBN 978-1-84919-503-4 (PDF)

Typeset in India by MPS Limited
Printed in the UK by CPI Group (UK) Ltd, Croydon

Contents

Section 1

Introduction and basic issues

Chapter 1

Ageing and biometrics: an introduction

Michael Fairhurst

1.1 Introduction

The chapters of this book which follow will discuss and analyse a wide variety of issues which describe the effects of human ageing on the way in which biometric systems are designed and deployed. This is a particularly important issue since, of course, biometrics is an area which is intrinsically concerned with the measurement of human characteristics which are subject to change as a consequence of ageing. It is appropriate, therefore, to begin by briefly examining why the ageing issue is potentially a matter of concern from a general societal perspective, before introducing the nature of the problem when biometric systems are considered. This initial chapter will identify the principal features of this general landscape, setting the scene for the more detailed studies to be reported subsequently.

A briefing paper produced by UK Parliament House of Commons Library researchers in 2010 [1] contained some interesting and very striking statistics about the ageing characteristics of the population of the United Kingdom. In that year, for example, there were 10 million people in the United Kingdom over 65 years old, but available projections predicted that figure to rise by more than another 5 million over the next 20 years or so, and likely approximately to double by 2050. If the most elderly group is specifically considered, the report showed a figure of around 3 million people over the age of 80 years, projected almost to double by around 2030, and reaching around 8 million by 2050. In other words, in 2010 1 in 6 of the UK population was over 65, but by 2050 this proportion is expected to be around 1 in 4. Figure 1.1 shows the projected increase in the UK population over the next 20 years according to age group distribution and Figure 1.2 shows specifically predictions of the future increase in those over the age of 100 years.

There appear to be two principal factors underlying these statistics. The first is that part of the cause of this rapid change is the natural shift in demographics which can be traced to the significant increase in the birth rate during the 1960s. However, another principal factor is the inherent change in life expectancy in recent years, with the cohort projection of life expectancy by 2030 now 91 for a man, and 95 for

University of Kent, UK

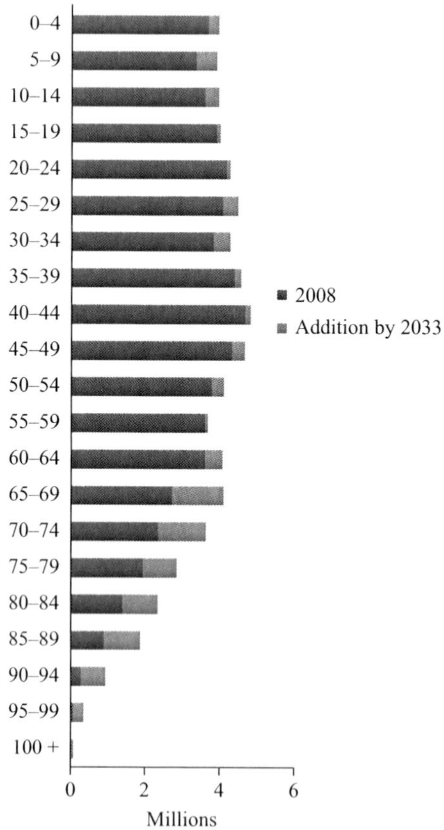

Figure 1.1 *Projected increase in the UK population 2008–2033. From Reference 1, reproduced with permission*

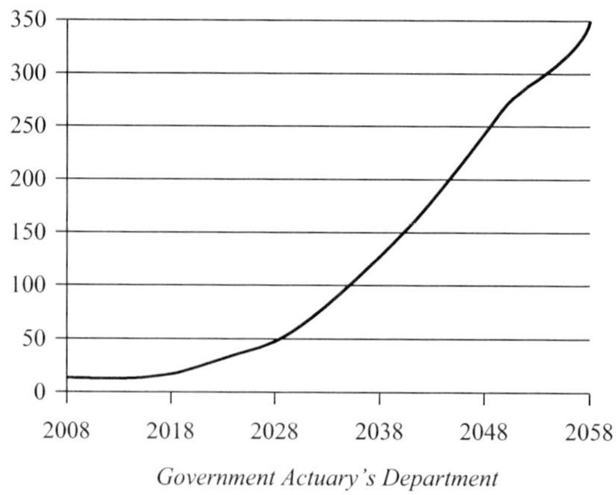

Figure 1.2 *Projected increase in number of centenarians to 2058. From Reference 1, reproduced with permission*

a woman. Such predictions have serious implications for governments, which need to understand the implications in terms of spending on, for example, social benefits such as pensions, health provision, and so on, and it is not surprising that this has led to legislation to progressively increase the age at which state pensions can be paid, as one natural consequence of what the figures reveal. Other responses to the statistics naturally include concerns about their more immediate political impact (the 'grey vote'), the distribution of wealth and spending power, the implications for work and wealth-creation and so on. And, looking more widely, similar patterns of changing demographics (and rather similar concerns about their implications) can be found across Europe and, indeed, pretty generally worldwide.

What is easily overlooked in this sort of analysis, however, is that the phenomenon of ageing not only has implications in the political or economic spheres. There are also clearly implications for technology and how it is used. On the one hand, for example, technology can bring enormous benefits to populations of all ages, and maybe especially in supporting older people, but it is also true that an increasing uptake of 'high-tech' devices and systems in everyday life has sometimes proved particularly challenging for the older generation. Thus, while attitudes to technology, and acceptance of technology-based solutions, are perhaps likely to become significantly less of a problem with succeeding generations, it is also evident that ageing will necessarily bring changes which may have an unavoidable, and perhaps less positive, impact on the relationship between technology-based systems and the capabilities of their users.

The studies which are presented in later chapters will describe and explore some of the implications of statistics such as these in one particular area of technology which is becoming increasingly taken up by society and which, because we live in a world which is becoming more and more interconnected, is seen as especially important in maintaining our safety and security, and providing greater convenience in executing the basic transactions of everyday life. This relates to fundamental questions of individual *identity* and how we can increase our confidence that those with whom we have to interact really are who they claim to be. This brings us naturally to the area of *biometrics*, which is the key focus of this book, addressing specifically the relationship between how biometric technologies are developing and how they should develop, and the sort of statistics about patterns of ageing on target populations which were presented above.

1.2 A context for understanding ageing effects

There is no escaping the self-evident fact that human beings age, and even less opportunity to avoid the effects of ageing in one way or another. Yet, while we all age identically in the sense that we all follow the same timeline, so that the ageing 'clock' ticks at the same rate for all of us, experience shows that the effect of the ageing process can be fundamentally different in different individuals. Moreover, the effects of the ageing process seem to affect different aspects of our personal characteristics in different ways. For example, our ability to reason and to analyse information generally develops and improves for a good proportion of our lives while, say, a decreasing ability for our eyes to accommodate to clear vision across all distances is a progressive phenomenon for many, if not most, people.

These effects – the ageing-related changes themselves, and the unpredictable and highly variable nature of the changes across the set of individuals in a population – can present major difficulties for activities which depend on constancy and stability for their functionality. This is especially true in situations in which there is a need for consistency to be maintained for significant periods of time, or where we typically find it practically helpful to make the assumption that this is the case. The effects of ageing tend to be incremental and gradual, but there can be no doubt that as time passes they introduce change on an increasing scale with respect to any particular time-related baseline. Not surprisingly, therefore, ageing and its effects represent an issue of considerable importance to the development of an area of technological development such as biometrics.

But it is not only this inherent volatility of our individual characteristics which raises difficult questions for both designers and users of biometric systems. We are constantly being reminded, for one reason or another, that we are an 'ageing population'. That is to say, it is a simple but important fact that in many, indeed most societies, population demographics have reached a situation where the proportion of older people in a country's population is increasing, as we have already seen. This is a statistic which raises a number of important issues – it reflects the obvious fact that life expectancy is increasing, but further investigation also shows that older people can often remain more active participants in society than has been the case in the past. Thus, both generational changes and prevailing attitudes mean that older people are perhaps much more willing to engage with newer technologies than hitherto, and this is likely only to escalate as time goes by. It is consequently more important than ever that we understand the effects of ageing on those aspects of physiology and behaviour which directly affect our interaction with systems and new technologies. In relation to biometrics in particular, since a fundamental defining characteristic of a biometric system is its requirement directly to exploit measurements derived from physiological features or behavioural activity, it is critically necessary that we understand the impact that changes associated with the ageing process are likely to have. Unless we recognise and act upon this necessity, we are likely to exclude an increasing number of people from participating in some of the most important societal developments, leading to a philosophy of exclusion rather than the prioritisation of inclusion as a practical imperative.

1.3 Biometrics and ageing

The field of biometrics, as almost every reader of this book will already know, deals with the problem of determining the identity of individuals, or of verifying a claimed identity. Yet in aiming to do this, a biometric system relies on particular measurements of physiological or behavioural characteristics of individuals – exactly those characteristics most likely to change, and perhaps often in rather unpredictable ways – as a result of the natural ageing process. While over a short time period in the operation of a system this may not be a significant issue, as the field of biometrics has matured and as the deployment of practical systems has consequently increased, both in terms of numbers of application scenarios and the scale of deployment (e.g. we are increasingly witnessing very large scale implementations involving whole national populations), questions about how the progressive ageing effects in users are likely to impact on individual or overall system performance, and how a system can be effectively managed so as to minimise the effects of ageing, become both more important and increasingly pressing.

Early research in biometrics, not surprisingly, tended to focus principally on techniques to acquire and reliably process biometric data for short-term use, or at least without much consideration of a possibly significant timescale for eventual operational lifespan. But with the field now having reached a level of performance, robustness, reliability and effectiveness which allows the successful deployment of biometrics in many, and potentially very varied, practical scenarios, it is entirely understandable that, increasingly, researchers have turned their attention to how systems can be designed for longer term application, and how they can most effectively be managed in order to maintain reliability in the face of these naturally changing characteristics of their users.

It is for these reasons that it is timely to provide, as the contributions to this book will do, a survey and an illustration of some of the best research of recent years into the subject of ageing effects and their impact on biometric systems. The contributors cover a wide range of backgrounds and technical interests, reflecting exactly one of the most fundamental features of this fascinating field, namely the diversity to be found in the study of biometrics, since a principal characteristic of the biometrics field is its inherent breadth and interdisciplinarity. It is a field which brings together researchers from many different disciplines and backgrounds, especially when considering a topic such as age and ageing, and it will be apparent that the contributors, all very well-known and respected in their respective research fields, each brings a particular perspective and viewpoint on this very important topic. Collectively, then, the contributions to be found in later chapters will provide a solid platform from which to become familiar with the nature and scope of current work and future possibilities, and a strong foundation on which to build future work.

1.4 Some practical implications for biometrics

It is important and useful to give some initial context for studies of ageing in biometric systems, and to put into perspective the issues which will be considered in more detail in the various chapters to follow. This will give us a basis for understanding better the nature of the problem, its potential severity and likely impact, and will also help us to judge the effectiveness of measures which we can take to combat any undesirable effects which the ageing process might introduce when we consider the long-term performance of biometric systems.

For example, as a starting point, let us consider the most fundamental nature of biometrics-based recognition processes by going right back to some initial requirements and definitions. Traditionally, we have defined an acceptable and useful biometric characteristic according to whether or not it can meet four general criteria, usually categorised as follows:

Universality: All individuals should possess the characteristic being used as a basis for the identification measurement.

Uniqueness: No two individuals should be the same in terms of the characteristic chosen.

Permanence: The chosen biometric characteristic should be invariant with respect to time.

Collectability: The chosen characteristic should be measurable and describable quantitatively.

While other general attributes are sometimes also invoked, the four noted above are the principal elements of the typically adopted definition of what constitutes a 'biometric'.

However, we can see that pretty well all these criteria need to be defined rather flexibly in a practical situation when we look at them in the present context. For example, consider an obvious biometric modality such as face recognition (this is a useful example because it is a very common and widely studied biometric and also one which uses a characteristic which is easily and relatively reliably adopted by humans too). We might quite readily concede that the property of *uniqueness*, for example, does indeed apply to an individual face, since we know we all look different. However, we may also, especially if we think about this a little more, want to add a proviso – which might be to say that, in biometrics, we generally have to define some particular aspect(s) of the face on the basis of which to decide upon identity – and thus uniqueness might depend on what these measurements are. We might also make a more general observation – which is that, whether or not at some deep level two faces are indeed genuinely different, we might perceive them in practice as actually so similar as to lack much confidence in trying to distinguish them. This could actually be the case for two completely randomly chosen individuals, of course, but is an increasingly difficult issue if we start to look at typical family likenesses, and even more so if we consider, say, twins in a family. And this is before we have even thought about whether or not we are dealing with identical twins, so we see how difficult it can be to demonstrate the fundamental criterion in a precise and categorical way.

In the context of this book, however, it is the property of *permanence* which is of most interest, and it does not take much reflection to see that the concept of invariability is impossible to apply to facial characteristics in any absolute sense, just as was the case for *uniqueness*. Changes in facial appearance occur constantly, whether because of facial expression, blemishes which appear and disappear, temporary or longer term changes in hairstyle, natural cyclic changes (such as facial hair and beard 'stubble'), changes perhaps in texture and colouring as a result of environmental factors and so on. Such changes are often unpredictable but are nevertheless a natural part of the way things are in the real world. Changes due to ageing are usually rather different (although the effects might be rather similar in practice), especially if we look over longer time periods, and indeed can be quite dramatic, though it is difficult to generalise because changes and patterns of change can vary hugely in different individuals. But, by way of an illustration, Figure 1.3(a,b) shows the facial image of two of the authors contributing to this book at different ages (from childhood through to adulthood), while Figure 1.4 shows a set of different images of one of these authors, all produced at the same age (of 25 years in this case). And, finally, to provide a little insight into another modality, Figure 1.5 shows examples of fingerprint images typical of those found in individuals at the ages of 23 and 73 years.

The point being made here is that while we define criteria which it is perfectly clear must be met by any biometric measure, we do, in fact, use the definitions in rather unconventional and, some might say, less than precise ways. Thus, to all intents and purposes in a practical sense we regard facial characteristics as 'unique'

Figure 1.3 (a) Facial images of the same person at ages 3 and 25; (b) a sequence of facial images of the same person at various ages

Figure 1.4 Different images of the same person all at the age of 25

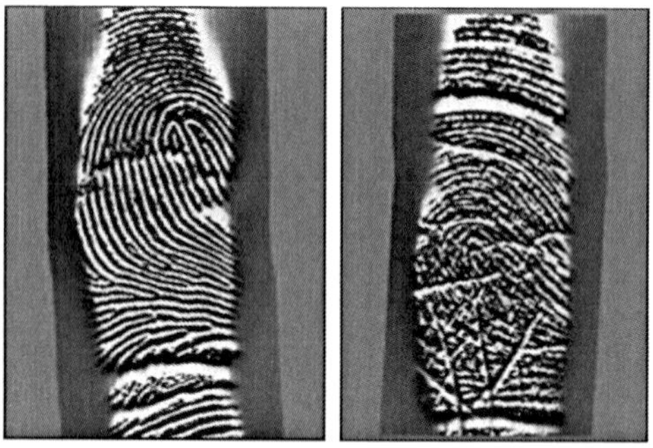

Figure 1.5 Fingerprint images typical of those found at the ages of 23 and 73

and 'permanent' (e.g. few people would dismiss facial biometrics on the basis of the present discussion), even though it is easy to see that we do not use these terms in an absolute way, and certainly not in accordance with any typical dictionary definition. In fact, one of the rather paradoxical questions which we are left with in this discussion could be set in the rather puzzling formulation 'When does "permanent" not mean permanent'?

So, we take as our starting point that we use these defining terms in a rather particular and non-rigid way. In turn, this means that we may need to ask questions about if, how, and under what circumstances age-induced changes transform a problem from one which we have become accustomed to dealing with on a routine basis to one which is, in some sense, a separate, or more important, or maybe a more difficult problem where new and more powerful techniques are required. Is it merely a question of degree or are there more varied and subtle factors involved? The contributions to this book may help us to understand this sort of issue more clearly and to be in a position to deal with it more effectively.

We could also ask some questions about the relationship between generic analysis and generic solutions and particular individual cases. Researchers in the biometrics research community are used to the idea of categorising users of a biometric system according to labels such as *lambs, sheep, goats* and *wolves*, for example, depending on the distributions of parameters which define their identity according to biometric measures, and the relationship between individual distributions and those of an overall population [2, 3]. Thus, for example, we may characterise a user as a 'goat' if (s)he has a naturally high variability in the biometrics of interest, and a knowledge of this categorisation can help us to manage a system and its user population most effectively. But it may be less clear how the effect of ageing impacts on individual user characterisation such as this, and whether there may be some useful information to be gathered by asking questions about whether certain factors might translate a user from one category to another.

There is, too, another very interesting and potentially important issue which needs to be considered. This is because we tend to use the term 'ageing' in the context of biometrics-based studies in two rather different ways. On the one hand, we can be fairly sure that biological age will have an impact on how we design or use a system (e.g. we now know that the ability of the human iris to accommodate to different levels of incident lighting decreases with age, and that this in turn can lead to issues about the availability of the features used to recognise a particular individual's iris pattern) and so a knowledge of the absolute age of an individual might be very useful and practically important information if it is available [4]. Yet it is also quite evident that a perhaps even more critical factor in determining the performance of a biometric system is always going to be the time lapse between when an individual enrolled into the system (i.e. the nature of the samples provided which are used to give the details of an individual's 'identity' to a system in the first place, by constructing a reference *template*) and when the system is actually used to make a specific identification decision by seeking to match current data against stored templates (i.e. how much change has occurred as a result of ageing in the interim).

We must therefore be aware of some sort of relationship, and perhaps even a tension, between these ideas of *biological/physical ageing* and *template ageing* in the context of biometric system performance. Both are important, and they are clearly related, but they are not quite the same thing. Perhaps the easiest way to capture the essence of this issue is simply to note that the effects of template ageing are likely to depend not just on the time lapse between enrolment and use, but that this effect will also depend to some extent on the absolute age of the subject within the relevant time frame.

There is a further complication here, related to one of the points introduced earlier. It is well understood that no biometric measure is stable in an absolute sense, since there are a variety of factors beyond the ageing issue (data collection mechanisms, interaction variations, environmental factors, noise, presentational pose and so on) which mean that no two biometric measurements are likely to be identical, even for the same person and acquired under nominally identical conditions. In other words, all biometric systems must inherently be designed to cope with variability between different individual samples – and so we become more concerned with feature *distributions* than with specific individual archetypes, and we thus try to engineer an operating environment where intra-individual sample distributions are as tightly clustered as possible while inter-individual distributions are as distinct as possible. In this sense, the effects of ageing are somewhat comparable to sample variations expected to be encountered in all normal circumstances. On the other hand, the effects of ageing are progressive and likely to induce more permanent, and often more significant and extensive data modulations, and so the question remains about whether, when and how to introduce mechanisms which are specifically designed to address the variability introduced only because of the ageing process, and to ensure that the ageing process is understood so as to be very clear about what factors are driving change in any particular system.

Of course, it is easy to observe that an obvious way to counteract the effects of ageing is to ensure that regular updating of the templates adopted is carried out, perhaps by a wholesale re-enrolling of all individuals in a user population at regular and preferably short intervals of time. But re-enrolment is likely to be a very time-consuming exercise, especially in a large population and will in most cases almost certainly turn out to be expensive [5]. Moreover, it is also likely to be viewed as inconvenient by many system users while, for all the reasons already considered, there is perhaps no generally agreed optimal plan for how often a re-enrolment exercise should ideally be carried out (this will, anyway, depend on modality and application in most cases) [5], while individually varying physiological or behavioural characteristics may make such decisions appear rather arbitrary in some circumstances. So, while we would not be sure whether to embark on re-enrolment every year, every five years, every ten years (although in some cases guidelines are beginning to emerge about these questions), even if we can establish some agreed best practice, this does not overcome the other issues about acceptability to a user population, cost factors, and so on.

It is clear, therefore, that the question of what the effects of human ageing are likely to be in relation to biometric measurements, and how this phenomenon affects the performance of a particular biometric system, is very complex, and will almost certainly depend on many very different factors.

Perhaps it is better in the end not to regard the problem of population ageing in relation to biometric system as something which is to be, as it were, *solved*, but rather more as a problem which needs to be *managed* in an appropriate, efficient and effective way. This sort of thinking allows considerably more scope for a system designer to recognise that the way in which a biometric system is configured and implemented is always something which is problem- and application-dependent. Systems can have very different requirements in terms of performance (e.g. from the perspective of error rates, related to false match rates, false non-match rates and indeed in terms of other error-based requirements too), and different operational scenarios will impose different criteria which need to be met in practice.

Thus, the key to taking account of ageing in biometric systems in an optimal way is first of all to understand as far as possible the nature of the changes which ageing is likely to cause and, second, to be aware of what tools, methods, and strategies are available to help in minimising any negative impact which such phenomena may have for a particular system and operational environment.

This book will directly pick up on these principal key issues, providing a timely source of useful information with which they can be approached, and support for those wanting to discover more about what current research is now achieving. And beyond that, it is important to remember that a consideration of the impact of ageing on a biometric system can be viewed in a rather different light too. While much of the work reported in the following chapters begins from a starting point which identifies the fact of human ageing as a problem to be addressed in real applications, we should also remember that the age of a subject can, in many applications, be considered also as an information source itself. For example, having a knowledge of an individual's age can actually provide a piece of information which can contribute to identifying that person, or at least narrowing the search space for a required identification decision. Similarly, there are many applications where being able to predict an individual's age can be an easier and more tractable problem than predicting individual identity, but can nevertheless be a very valuable outcome – in some forensic examinations, perhaps, or in a situation where constraints (e.g. restrictions on access to a physical or virtual space) are age-determined. In this book we have tried to cover some of the work currently being undertaken in these areas too – providing a more comprehensive insight into age-related issues across a spectrum of applications which use biometrics as a core technology.

1.5 Plan and structure of the book

In the following chapters, which collectively survey a wide range of aspects of research and formative thinking about the effects of ageing on biometric systems, the separate contributions are grouped in a loosely structured way to help guide the reader through a logical progression of topics.

The first section aims to give some context for the detailed subsequent discussions, while the second addresses some fundamental issues relating to individual biometric modalities. In the third section, some of these fundamentals are picked up and discussed further in the context of their implications for practical systems and implementations. Finally, the book points to some issues for the future, indicating just some of the challenges which might influence the way in which studies such as those reported here might develop.

It can be argued that there is a good deal of overlap across the sections and, indeed, that some chapters might fit just as comfortably in different sections. This is undoubtedly the case, but the aim was not to be too rigid or prescriptive in structuring the presentation of the book's material, but to provide at least some sort of direction of travel to help the reader find ready access, in a fairly logical way, to this representative selection of topics which are currently exercising the research community, to see a progression of ideas which are emerging and to sense a momentum in research effort.

Thus, Lanitis and Tsapatsoulis provide an initial review of ageing as it occurs across a range of modalities, giving a very helpful background to what follows. The chapter by Mordini and Rebera addresses some of the social issues which are often overlooked in much technical writing about biometrics, but which are enormously important in an area which is so centred on human activity. In the final chapter of the first section, Abreu and Fairhurst demonstrate how, while ageing can present a problem for the designer of a biometric system, this is balanced to some extent by the fact that a knowledge of the age of a system user can also be turned to some advantage in the right circumstances.

In the second section, we principally focus on specific individual modalities, including the major and most popular modalities currently widely used, and also ensuring that we cover both the common physiological modalities and also some behavioural modalities too. We thus have chapters which cover all the principal modalities such as facial images (from Ricanek and Mahalingham), the handwritten signature, one of the archetypal behavioural modalities (Galbally, Martinez-Diaz and Fierrez), iris patterns (Liu-Gimenez and Sanchez-Reillo), fingerprints (Uhl and Wild) and voice (Kelly and Harte). These chapters provide some valuable information about ageing issues of particular current interest across these individual modalities. This section then concludes with a chapter from Poh and Kittler, which introduces a rather different perspective and which embraces a more multi-modal view of some important and relevant issues.

The third section begins with a chapter from Partington, which brings a particularly appropriate industrial viewpoint on the topic of ageing, while the chapter from Reedman follows this up with another rather different slant on some of the relevant issues, offering us a view of ageing in relation to fingerprint analysis more from a forensics perspective, and from the point of view of a human fingerprint expert. The chapter from Guo brings us back to the face modality, but in this case looks not at how age might impact on performance in biometrics, but instead addresses the question of how we might be able to learn something about subject age starting from the face biometric itself, and the section also concludes with a discussion (from Tistarelli, Yadavi, Vatsu and Singh) of the face biometric, exploring some of the important issues relevant to short-term and long-term ageing effects in this modality.

The final section first brings together three chapters which each reflect some aspect of ageing reflecting relevant issues which might suggest some particular topics likely to be high on the research agenda in the future. Recognising an issue which consistently affects research across the whole biometrics spectrum, the chapter by Erbilek and Fairhurst focuses on the topic of research databases of biometric measurements, not so much from the point of view of exhaustively discussing relevant available databases, but instead exploring some of the ways in which databases have been used, and some of the questions this raises when studying ageing issues and their impact on biometric systems. The chapter from Sasse and Krol reminds us of a frequently neglected topic, which is that of how changes during ageing affect a range of human capabilities which have a particular bearing on the usability of biometric systems, and which brings into focus some important issues about usability and system interaction which are raised for biometric systems when human ageing processes are considered. It is very important to include this type of discussion, to ensure that we keep in sight an area of research which is pivotal when transferring biometric technologies and systems from the research environment to an operational environment. The third chapter in this section from Rathgeb, Uhl and Busch introduces us to an area of increasing research interest, discussing the effects of ageing in iris biometrics, but where the ultimate aim is deployment in biometric cryptosystems.

Fittingly, this section and thus the whole book then concludes with what might be seen as an extended 'postscript' from Peter Hawkes, one of the UK's most respected and knowledgeable senior members of the biometrics community. He has worked at the interface between the research community and the commercial sector for many years, and has seen the biometrics enterprise develop from the days of small pockets of work and a few exploratory systems to the pervasive and mature set of technologies with which we are more familiar today. It is a great privilege to have access to the benefits of his long association with the field, his expertise and his recognised wisdom to bring a book such as this, with its wide span and considerable diversity, to a conclusion.

The main intention of the book is to provide a reference source which sets out a representative sampling and overview of what research has achieved to date in relation to this most important topic, and to provide a strong base from which to continue to investigate and develop a better understanding of how the natural processes of ageing are likely to impact on the theory and practice of biometrics, how to manage these effects in practical systems, and how to think in a more focused and effective way about the future. Despite its relative maturity, the field of biometrics still has many unresolved issues, and still needs on-going high-quality research and the continued sharing of knowledge, expertise and ideas. Some evidence for this is the fact that the IET (the publishers of this book) launched in 2012 a new Journal (*IET Biometrics* [6]) to focus explicitly on the biometrics field in all its diversity. The inevitability of ageing, and the indiscriminate way in which ageing affects each and every potential user of a biometric system, means that this is a topic which should indeed be towards the very top of the research agenda, and is increasingly important with each year that passes.

All of the contributors hope that this book will be a useful reference source for researchers both in the academic sector and in industry, for users of biometric systems, for students who are studying them, and for anyone who has an interest in a fascinating area where research is able to shine a light into that space between an individual user and a powerful system which (s)he would like to access. It has long been a guiding principle of system design that a system should be able to adapt to the needs of the user, rather than the other way round. With respect to a phenomenon such as natural ageing, this is surely a principle which researchers should strive to embrace and adopt as a real priority.

References

[1] R. Cracknell. (2010). *The ageing population.* Available: http://www.parliament.uk/business/publications/research/key-issues-for-the-new-parliament/value-for-money-in-public-services/the-ageing-population/

[2] G. Doddington, W. Liggett, A. Martin, M. Przybocki, and D. Reynolds, 'SHEEP, GOATS, LAMBS and WOLVES: A statistical analysis of speaker

performance in the NIST 1998 speaker recognition evaluation', in *International conference on spoken language processing*. Sydney, Australia; 1998.

[3] N. Yager and T. Dunstone, 'The biometric menagerie', *IEEE Transactions on Pattern Analysis and Machine Intelligence,* vol. 32, pp. 220–230, 2010.

[4] M. Fairhurst and M. Erbilek, 'Analysis of physical ageing effects in iris biometrics', *Computer Vision,* vol. 5, pp. 358–366, 2011.

[5] G. Marcialis, A. Rattani, and F. Roli, 'Biometric template update: An experimental investigation on the relationship between update errors and performance degradation in face verification', in *Structural, Syntactic, and Statistical Pattern Recognition*, vol. 5342, N. da Vitoria Lobo, T. Kasparis, F. Roli, J. Kwok, M. Georgiopoulos, G. Anagnostopoulos, and M. Loog (Eds.), Berlin/Heidelberg: Springer, 2008, pp. 684–693.

[6] See www.ietdl.org/IET-BMT.

Chapter 2
Review of ageing with respect to biometrics and diverse modalities

Andreas Lanitis, Nicolas Tsapatsoulis and Anastasios Maronidis

2.1 Introduction

2.1.1 Overview of biometric systems

Biometric systems are related to the extraction and analysis of personal characteristics in an attempt to derive information and/or interpret human actions. Biometric systems aim to extract information about attributes such as identity, emotional state and age or to interpret human actions such as speech, handwritten text and behaviour.

The endless possibilities in embedding biometric modules in existing systems along with the great potential of developing contemporary applications involving biometric units have caused a dramatic rise in the market share of biometric systems. At the same time, the general topic of biometrics has attracted a considerable amount of interest by researchers who aim to develop new methods for extending the effectiveness, accuracy and scope of application of biometric systems.

Among different possible applications of biometric systems the problem of secure access is regarded as a major problem and as a result on many occasions biometric systems are associated with identity recognition and/or verification. Indeed the increased need for fast, secure and automatic access to real and electronic sites has prompted the development of biometric systems that can replace traditional identity verification methods.

The major difficulty in developing biometric systems is the variability encountered in biometric traits that causes variation in biometric features belonging to the same class, a situation known as intra-class variation. A successful biometric system should be able to isolate and use the right type of variability for a particular application while other types of variation, which impede the process of correct interpretation, should be suppressed.

Cyprus University of Technology, Cyprus

2.1.2 Ageing variation

Ageing variation can be defined as the type of variation that causes alteration to biometric characteristics of a subject captured at different time instances. When compared to other types of variation encountered in biometric characteristics, ageing variation has the following unique characteristics:

Uncontrolled: The ageing process cannot be stopped, controlled or reversed. Although cosmetic industries promote anti-ageing treatments, such initiatives do not offer a way to eliminate entirely the process of ageing. In addition anti-ageing treatments are usually directed towards the facial appearance and/or skin, hence they are not usually applicable to all biometric modalities.

Uniqueness: Ageing variation is associated with physiological changes taking place throughout the lifetime of humans. Although ageing effects for different persons display common characteristics, both the rate of ageing and the actual effects may cause different types of variation to different persons. Apart from the personal uniqueness of ageing variation, ageing can cause diverse effects for subjects belonging to different genders and ethnic origins.

Uncertainty: The process of ageing could be affected by external factors that may not be predictable. For example, the ageing process for a person who experiences a traumatic event in his/her life may influence both the rate and the type of ageing variation for that person. As a result the long-term impact of ageing is highly unpredictable.

In Lanitis (2010) a separation of ageing effects into direct and indirect is suggested. Within this context direct ageing effects are associated with the normal ageing process while indirect ageing effects are associated with unpredictable events taking place through the lifetime (e.g. illnesses and accidents) resulting in a disruption of the 'normal' ageing process. In addition to the separation of ageing effects into direct and indirect effects, it is possible to divide ageing effects into short- and long-term ones. Within this context, short-term ageing effects usually involve variations caused within a five-year interval, whereas long-term ageing effects may refer to the effects occurring throughout the lifetime of a subject.

2.1.3 Biometric systems and ageing variation

Ageing variation can have adverse effects on the performance of biometric identity authentication systems. Usually such systems are trained to recognize/verify the identity of subjects based on biometric templates created during the initial user enrolment. Such biometric templates may include features derived from any biometric modality. As the time from the initial enrolment of a user increases, ageing variation causes modification of biometric features, impeding in that way the process of identity recognition/verification. The longer the time interval between enrolment and identification, the greater the probability of failure due to the effects of ageing variation.

On the other hand, the presence of ageing variation can be beneficial to other types of biometric systems, such as biometric-based age estimation. In such cases the intensity of ageing-related modifications of biometric templates can play a key role in the development of accurate age estimation biometric systems. It should be noted though that due to the uncertainty involved, even for age estimation-related applications, indirect ageing effects can play a major role in distracting the operation of an otherwise reliable biometric age estimator.

Until recently, the problem of ageing in relation to biometric system development did not receive adequate attention when compared to other types of inter- and intra-person variation. However, for the last few years the problem of ageing in biometric systems has been formulated as a key problem that requires the development of dedicated methods. As a result of research efforts in this area methods that aim to suppress the effects of ageing based on template updating (Rattani *et al.* 2009; Uludag *et al.* 2004), quantifying the extent of ageing on certain features (Lanitis & Tsapatsoulis 2011; Ortega *et al.* 2009), and the modelling/simulation of the effects of ageing (Fu *et al.* 2010; Ramanathan *et al.* 2009) have been reported in the literature. So far, due to the availability of appropriate data sets, most of the work on ageing effects was reported in relation to facial ageing. However, it is anticipated that ageing-related research for other modalities will receive increased attention in the near future.

2.2 Ageing databases

The evolution of biometric systems relies on the availability of suitable datasets that allow the development, experimentation, benchmarking and performance evaluation of biometric systems. For this reason a large number of biometric datasets, for either individual or multimodal biometrics, have been made publicly available. However, only few datasets are appropriate for experimentation related to the ageing of biometric features. Suitable ageing datasets should have the following features:

(I) They contain samples for numerous subjects. For each subject there should be multiple templates obtained at different ages.

(II) Non-ageing related types of variability should not be present in the samples. This is a key difficulty as ageing datasets with samples taken at different ages, most probably, were captured with different equipment and standards. For example, in the case of face images, images showing an 80-year-old during his/her childhood were captured about 70 years ago using different photographic technologies than images of the same person captured today. In addition to equipment-related variation, in ageing datasets it is very possible to encounter other significant non-ageing-related variations.

(III) In order to allow the development and assessment of multimodal techniques it is desirable to have datasets containing multimodal samples of subjects captured at the same instance.

Based on the criteria set above, it is evident that the acquisition of 'perfect' ageing datasets is an almost impossible task. For this reason the availability of ageing datasets is very limited. In the remainder of this section we present the most typical ageing-related biometric datasets that are currently publicly available.

2.2.1 Face datasets

Currently the most widely used face ageing datasets are MORPH and FG-NET. The initial version of the MORPH dataset (Ricanek & Tesafaye 2006) was made available in 2003. The dataset contains facial images of numerous individuals and includes metadata, such as age, gender, ancestry, height and weight. Images from the MORPH dataset are divided into Albums. Album 1 contains 1,690 scanned photographs of 631 individuals taken between 1962 and 1998. The acquisition dates correspond to increasing ages for individuals in the database; these dates range from 46 days to 29 years after the earliest photograph. The number of images per subject varies from 1 to 6, with an average of 2 images per subject. Album 2 contains 55,608 digital images obtained from 13,673 individuals. The number of images per subject varies from 1 to 53, with an average of 4 images per subject.

The FG-NET Ageing Database (Lanitis 2008) is an image database containing face images showing a number of subjects at different ages. The database consists of 1,002 JPG greyscale and colour images from 82 subjects. The number of images per subject varies from 6 to 18. The age of the depicted subjects is in the range of 0–69 years. The images exhibit significant pose, expression and illumination variation.

The two datasets described above have several limitations. Efforts in creating new more useful datasets are currently under development. For example, VADANA (Somanath *et al.* 2011) is a face image dataset that has been constructed as a complement of the MORPH and FG-NET datasets in order to facilitate comparison and benchmarking of different algorithms on face recognition, face verification, age progression, etc. VADANA includes 43 subjects (26 males, 17 females) and 2,298 images. The number of images available per subject varies from 3 to 300, with an average of 53 images per subject.

2.2.2 Other datasets

Despite the huge effort undertaken for the last few years from various researchers and institutions to construct multimodal databases (Faundez-Zanuy *et al.* 2006; Flynn 2008) none of these datasets was specifically designed to allow for ageing-related research. However, in a few cases short-time ageing effects can be studied due to the time difference between the data collecting periods. Some, indicative, examples are listed below:

- BIOMET (Garcia-Salicetti *et al.* 2003): Includes 5 different biometric modalities (face, speech, hand, fingerprints and signature) of 91 subjects. The data were acquired in three temporally separated sessions spanning an eight month period. The age distribution of the subjects is not detailed explicitly but 10% of the subjects were students (at the time of data collection) while the age span of the other 90% of the samples is 35–60.

- BIOSECURE (Ortega-Garcia *et al.* 2010): Includes 6 biometric traits (face, speech, fingerprints, hand, iris and signature) of 667 subjects.[1] The data were collected in 2 sessions within a 3 month time span. The age distribution of the subjects is 18–25 ⇒ 42%, 25–35 ⇒ 21%, 35–50 ⇒ 21%, >50 ⇒ 17%.
- BiosecurID (Fiérrez-Aguilar *et al.* 2007): Includes 8 biometric traits (face, speech, fingerprints, hand, iris, signature, handwritting and keystroking) of 400 subjects (not all subjects provided all traits). The data were collected in four sessions within a 4 month time span. The age distribution of the subjects is 18–25 ⇒ 42%, 25–35 ⇒ 22%, 35–45 ⇒ 16%, >45 ⇒ 20%.
- BIOSEC (Fiérrez-Aguilar *et al.* 2007): Includes 4 biometric traits (face, speech, fingerprints and iris) of 250 subjects (the extended version) and was acquired under the BioSec Integrated Project. The data were collected in four sessions (the extended version, since in the first version only 2 sessions were conducted) within a 4 month time span. The age distribution of the subjects is not explicitly stated in the description of the database but includes subjects aged 18–60.
- MbioID (Dessimoz *et al.* 2007): Includes face (2D and 3D), speech, fingerprints, signature (online) and iris collected from approximately 120 subjects in 2 time-separated sessions. The time between the sessions is not explicitly stated but the authors argue consistency with MyIDEA and BIOMET databases; therefore, 1 month period between sessions is assumed. The subjects are mainly students and university staff aged 18–60.
- MyIDEA (Dumas *et al.* 2005): Includes face, speech, fingerprints, signature, handwriting and hand geometry collected from 104 subjects in 3 sessions separated by, at least, 1 month each. The age distribution of the subjects is not explicitly reported but includes subjects aged 18–60.
- POLYBIO (Antonakoudi *et al.* 2009): Includes face (still images and video), speech, fingerprints and hand geometry collected from 50 subjects. Data were collected in 2 sessions spanning a 3 month period. The age distribution of the subjects is <20 ⇒ 8%, 20–30 ⇒ 60%, 30–40 ⇒ 12%, >40 ⇒ 10%.

In addition to the above-mentioned multimodal biometric databases there are many others (see Ortega-Garcia *et al.* (2010) for more details). However, the time between collection sessions is either small or not reported at all. The same holds also for the age distribution of the subjects. These two attributes are absolutely necessary to allow experimentation concerning the age variation of biometric templates as discussed below.

As reported in Fiérrez-Aguilar *et al.* (2010) there are 29 common subjects in the BIOSEC, BiosecurID, and BIOSECURE datasets. Since the time span of data collection in these multimodal databases is 2 years, some ageing-related studies can be performed. It is important to note here that the same technology was used for the creation of the three datasets mentioned above; thus, variation caused by the sensors is limited. However, as mentioned earlier, a time period less than 5 years is not

[1] Here we report only the version recorded using an office environment (Dataset II).

enough to study long-time ageing effects. Ageing is not a linear process, and as a result, the majority of biometric modalities (e.g. speech, hand-related, iris, behavioural) are quite stable in the ages 18–50. Although there are variations in biometric modalities (see Lanitis & Tsapatsoulis 2011) in this age range, these variations are smooth and a time period of less than 5 years is not sufficient to record direct ageing effects. Contrary to other biometric modalities, mentioned above, ageing effects are quite significant in the face modality in the ages 18–50. However, as already reported there are many face datasets specifically designed for ageing studies, covering also this age range.

Obviously, the difficulty in creating multimodal datasets that span several years is high. We know that the data collection process in modalities such as fingerprints, iris and signatures is inherently annoying. Thus, the willingness of subjects to participate in data collection sections that span several years is low. Furthermore, changes in recording technology, staff and even culture impede the creation of a proper ageing multimodal database. The situation is much better for faces due to the availability of personal photographs and photographs in identity documents, and, of course, the explosion of social networks like Facebook. Creating ageing databases for speech is also easier due to the fact that speech is a natural medium for human communication.

It should be clear from the preceding discussion that the limited studies on the ageing effects for biometric modalities other than face is the unavailability of appropriate datasets. Lanitis and Tsapatsoulis (2011) proposed recently a different approach to study the ageing effects in these biometric modalities. Instead of focusing on changes on the biometric templates of the same person across several years, they tried to quantify the ageing effects in biometric templates of distinct age groups. The results taken therein show that important conclusions can be drawn. The fact that in the majority of existing multimodal biometric databases the age distribution is high allows extension of their work as well as new studies to emerge based on this approach.

2.3 Biometric modalities and ageing

In this section, a brief overview of the physiological mechanisms that cause ageing effects, the impact of ageing in different biometric modalities and attempts to deal with ageing variation are briefly described.

2.3.1 Face

The human face is probably the most prominent biometric modality conveying important information about the expression, identity, gender, ethnicity and age of a human. The appearance of a human face is subjected to the influence of many factors caused by physical laws, the emotional state and the actions of the human. Examples of such factors are exposure to the sun, weather conditions, gravity, person's habits, diet, health conditions and lifestyle. These factors cause diverse changes in both the shape and texture of the face as the age progresses.

Similar to other biometric modalities, age-related facial effects could be categorized into direct and indirect ones (Lanitis 2010). Direct ageing effects are

Figure 2.1 Images of the same person in the age of 2, 8, 16, 29 and 43 years, from left to right

caused as a consequent result of a person's age evolution. For example, during formative years, a face is predominantly subjected to shape changes in the form of bone growth, while during adulthood, face changes are mainly attributed to texture modifications (e.g. appearance of wrinkles) (Ramanathan *et al.* 2009). Particularly from infancy to puberty the cranium is stretched downward and outward. The eyes, nose, and mouth expand to fill a relatively greater area of the surface of the cranium and the area occupied by the forehead shrinks. Some of these changes may continue, but less dramatically, into adulthood. During adulthood, the decline of the face takes place. During this process, facial skin becomes slack, darker, less flexible and rougher. Lines, wrinkles, folds and blemishes gradually appear as a result of the reduced skin elasticity. Muscles and connective tissues change their elasticity and fatty deposits and bone may be lost to produce pouches in the cheeks, bags under the eyes, sagging under the chin and opening of pores[2] (Patterson *et al.* 2007). Direct ageing effects can be reasonably simulated, because they act uniformly across the human population. In Figure 2.1, a series of pictures taken from the FG-NET ageing dataset (Lanitis 2008) depict the same person at gradually progressed ages. The corresponding ages are 2, 8, 16, 29 and 43. Despite the fact that the first three images are greyscale, the ageing process described above is pronounced for both the shape and the texture. For instance, the chin clearly becomes more protrusive from the leftmost to the rightmost image. Furthermore, in the first image, the eyes lie around the middle of the head, while in the last image they lie in the upper part of the head. This implies that the head is stretched downward, while the eyes move upward. Regarding the texture, in the two last images, the loss of skin elasticity is evident while some lines appear under the eyes. Also, some wrinkles appear in the forehead of the rightmost image. Despite the often dramatic age-related facial metamorphosis, some facial characteristics are retained throughout the lifetime.

Indirect ageing effects are caused by factors which are extraneous to physical ageing. Examples pertaining to this category are diseases, usually associated with the elder population, that cause alterations in facial appearance (e.g. appearance of skin lesions, tumours and weight gain/loss). The nature of indirect effects is almost impossible to be accurately determined due to their unpredictable character.

[2] http://face-and-emotion.com/dataface/facets/aging.jsp

In recent years the topic of modelling the process of ageing in face images has received considerable attention. The interest is motivated by the multi-disciplinary applications that can be addressed once the process of facial ageing in modelled. Both age estimation and age progression are discussed explicitly in recently published survey papers (Fu *et al.* 2010; Ramanathan *et al.* 2009). In summary facial ageing-related applications are divided into facial age estimation and age progression. In facial age estimation the aim is to estimate the age of a subject by analysing information derived from face images. Within this framework, the main objective is to exploit the appearance of age-sensitive facial features in an attempt to obtain feature vectors that reflect the age of a subject. In age progression the aim is to deform the appearance of a face in order to predict what the subject will look like in the future. Within this context the aim is to retain identity-specific facial features while transforming the appearance of age-sensitive features. In both cases the main issue to be addressed is the isolation of ageing variability in face images so that it can be used according to the intended application. Both age estimation and age progression methodologies can be used in numerous real-life applications involving secure age-restricted access, age-adaptive interfaces, age-based image indexing, identification of missing persons and entertainment applications.

The main facial ageing modelling approaches employed are subspace approaches that relate a low dimensional face representation with age (see Lanitis *et al.* 2002; Geng *et al.* 2007; Ricanek & Tesafaye 2006), physical model-based approaches where the process of ageing is modelled using functions that emulate the physics of ageing variation (Ramanathan & Chellappa 2006) and approaches that operate directly on the images (Rowland & Perret 1995).

2.3.2 Fingerprints

In forensic applications and in general applications involving identity verification, fingerprints are possibly among the most frequently used biometric modality. So far, the topic of ageing in relation to fingerprints has been mainly investigated in an attempt to estimate the age of a fingerprint trace (Merkel *et al.* 2011) rather than estimating the age of the fingerprint owner. Only few researchers have investigated the effects of fingerprint ageing in relation to identity verification. Modi and colleagues (Modi & Elliott 2006, Modi *et al.* 2007) describe a study where the effects of ageing on fingerprint-based person identification are examined. Based on an experimental evaluation they claim that the quality of fingerprints in older age groups deteriorates, affecting in that way the recognition performance. According to that study ageing effects on fingerprint templates are mainly attributed to the reduction of skin elasticity encountered in older ages. Due to the reduced elasticity of fingerprints the contact between the skin and scanner during the process of fingerprint scanning is not firm affecting in that way the quality of fingerprint images of older subjects. In addition, at older ages the possibility of observing damaged fingerprints due to wearing and injuries increases. The factors stated above inflict both direct (i.e. skin elasticity, wearing) and indirect (i.e. injuries) age-related modifications on fingerprint biometric templates. In Lanitis and Tsapatsoulis (2011) an initial attempt to quantify the effects of ageing on fingerprints is described. Within this context the so-called 'ageing impact factor' (AIF), which provides an indication of the dependence of certain features to ageing variation, is estimated. According to the results colour-based feature vectors display higher AIF when compared to edge-based and texture-based fingerprint features.

Based on the results reported in the literature, there is concrete evidence that fingerprints are affected by ageing. Bearing in mind the widespread use of fingerprints in identity verification applications as well as in forensics, it is essential that the topic of fingerprint ageing effects is further investigated.

2.3.3 Hand biometrics

Hand biometric templates usually contain information about the hand geometry and/or texture information derived from the palm or upper hand side. In the case of palmprints, ageing effects are attributed to similar causes as the ones encountered with fingerprints, that is, mainly loss of elasticity, wearing and injuries. Usually palm images are acquired using scanners that require contact between the palm and the scanning surface. The loss of elasticity, caused by ageing, affects the quality and appearance of palmprint images. In addition the palm of a hand is more vulnerable to injuries; hence, there is an increased risk of observing palmprint indirect ageing effects.

Texture-based templates derived from the upper side of a hand are usually captured by cameras; hence, the loss of elasticity does not affect the data acquisition process, making this type of template more tolerant to short-term ageing variation. However, at increased ages, the loss of elasticity causes the appearance or wrinkles and other skin deformation that result in texture metamorphosis.

Age-related modifications of hand geometry are usually recorded during the growth of the skeleton that takes place during childhood and puberty (Greulich & Pyle 1999). Therefore hand geometry-based features display increased ageing variation during the growth process but after the growth process is completed the geometry of hands remains reasonably stable limiting in that way the adverse effects of ageing to hand geometry-based biometric templates. However, an indirect age-related factor that distorts hand geometry is the disease of arthritis (Kirwan 1995) that is usually encountered in the elderly. Arthritis causes damage and distortions to joints, including the joints of hand and fingers, affecting in that way the accuracy of hand geometry-based identification.

The effects of ageing on the performance of hand-based biometric systems have been investigated by few researchers so far. Uhl and Wild (2009) investigate how the age of subjects affects the performance of identity verification using fingerprints, palmprints and hand geometry. Within this context, they compare the verification performance obtained when systems were tested using samples of subjects belonging to the age groups of 3–10, 11–18 and over 19. The results show that the performance for the youngest age group is worse when fingerprints, palmprints and data from the middle, ring and little fingers are used. According to the authors this effect is mainly attributed to the fact that subjects from the age group 3–10 do not cooperate fully during the data acquisition process.

According to Lanitis and Tsapatsoulis (2011) the AIF of texture-based features derived from palms is significantly higher when compared to the same features derived from fingerprints, indicating that the effects of ageing on palmprints is more serious when compared to fingerprints. The result is also verified in age estimation experiments, where higher age estimation results are obtained when palmprint features are employed. Based on the results reported, one can conclude that in cases where the time interval between enrolment and identification is long, fingerprints may provide better identification clues when compared to palmprints.

Hand geometry is obviously more stable in terms of age variation compared to both palmprints and fingerprints. However, involvement of hand geometry in large-scale biometric systems is limited because this trait is not as distinctive as fingerprints and palmprints.

2.3.4 Voice

Speech is a biometric modality that is used more often for speech-recognition applications rather than speaker identification. In both types of applications ageing can play an important role in the deterioration of the performance of a system as the human voice is affected by direct and indirect ageing effects. Direct ageing effects are mainly caused by loss of elasticity of the lungs that limits the ability of the lungs to move air, vocal muscle atrophy, anatomical changes in the larynx and thickening of the epithelium (Linville 2004). According to several studies (see Benjamin 1997; Mueller 1997; Ramig & Ringel 1983), as a result of direct ageing effects, the pitch level, tremor, hoarseness and breathiness of speech undergo variations as the age of a subject increases. Features usually used for constructing biometric speech templates, such as frequency range and perturbation, intensity level, intensity range and speaking rate are highly dependent on the age of a user (Jain *et al.* 2004, 2008). Apart from direct ageing effects, the human voice is distorted as a result of indirect ageing that result from pathologic vocal conditions usually encountered in the elderly. Pathological age-related conditions that usually affect the voice include Parkinson's disease (Jankovic & Kapadia 2001), head and neck cancer and laryngitis (Schulz 2006).

In speech recognition-related applications, an important challenge is the ability to recognize speech despite the age of a user (Narayanan & Potamianos 2002). In most cases the problem of age-invariant speech recognition is treated based on frequent retraining. For speaker identification applications, the problem of ageing has also been highlighted by researchers working in the field (Pawlewski & Jones 2006) but so far the problem did not receive adequate attention. Only a few researchers (Gurbuz *et al.* 2000; Lamel & Gauvain 2000) investigate the use of dedicated techniques for eliminating short-term ageing effects. A possible approach to deal with long-term ageing effects in speech biometrics is to reverse the techniques followed for age estimation from speech. A corresponding literature review can be found in Dobry *et al.* (2011).

2.3.5 Behaviour

Behavioural biometric templates (Yampolskiy & Govindaraju 2008) are constructed using features collected as a result of actions performed by a subject. This broad category includes voice-based biometrics and body movement biometric templates. Here we focus our attention to body movement biometrics. Body movement biometrics utilize information about the way that certain actions are accomplished by different subjects, so that it is possible to identify subjects based on their actions. It is well known that body movements are critically affected by ageing as the muscle elasticity, muscle strength, stamina and style (Larsson *et al.* 1979; Wu 1998) are heavily linked to the age of a subject. Apart from the factors stated above, which correspond mainly to direct ageing effects, certain diseases such as Parkinson's (Jankovic & Kapadia 2001), arthritis (Kirwan 1995) and reduced sight and reduced hearing cause indirect ageing-related effects on behavioural biometric templates.

Signature is a commonly used proof of identity and intent. It can be considered as a behavioural biometric because it involves hand movement, which in online signature verification systems is critical. Signature is a personal stylized depiction of one's name, nickname or some other characters that a person writes down to prove his/her identity or to declare consent. It contains inherent handwriting habits and differs from person to person. Signature development is influenced by physical condition, family environment, education and many other factors. Handwritten signature has already been established as one of the most widespread methods of personal verification with a long history. It is commonly used in many aspects of daily life, such as validating cheques, physical entry to protected areas, financial and legal documents. Handwritten signatures are non-invasive, easy to collect, widely accepted and can be hard to forge. Thus they are considered as one of the most appropriate biometrics for identity verification.

Methods of signature verification are generally divided into two categories depending on the nature of data acquisition: offline (static) signature verification and online (dynamic) signature verification. In an offline signature verification system, data are usually acquired, e.g. by scanning, after they have been written and signatures are represented as grey-level images. On the other hand, online signature verification uses special instruments, such as digital pen or tablet, to record the dynamic information of the pen tip during signing, taking into account dynamic features, such as coordinates, azimuth and pressure of the pen tip. In this way, the signature is represented as a sequence of signal values at sampling points and provides a higher quantity of information and higher reliability than in offline approaches. Although the shape of a signature can be forged through professional training, dynamic features are much more difficult to imitate, for the signing process of online signature involves several factors: experimental psychology, neuroscience, physics, education and so on. Thus, online signature verification is more robust than offline signature verification. On the other hand, the signing process is also affected by the signer's mood, physical condition, muscular coordination and other external factors such as writing conditions, drugs, alcohol, and nervousness

(Plamondon & Srihari 2000). As a result, while online signatures have enough consistency for personal identification, there are some natural variations. No one can write two signatures identically because the body cannot carry out actions as precisely as machines while signatures of the same person may have a large intra-class variation. The latter is a challenging problem that a signature verification system must tackle to improve the performance.

As already mentioned above, online signatures depend on a subject's physical condition and muscular coordination. Both of these, especially the latter, are highly affected by ageing as indicated by both theoretical (Woollacott *et al.* 1986) and experimental (Shim *et al.* 2004) studies.

Guest (2006) investigates the effect of ageing on signature verification systems by studying features extracted from signatures from 274 volunteers belonging to 3 distinct age groups (26–40, 41–60 and over 60). Experiments where the intra- and inter-session repeatability of features extracted from signatures are conducted in an attempt to assess the accuracy of reproducing signatures. An interesting outcome from this study is that aged subjects require longer time periods to complete their signatures when compared to younger subjects; hence, movement dynamics are affected by ageing. Similarly Zaphiris and Ellis (2000) describe an experiment where the response time of computer users belonging to different age groups is evaluated. According to the results, of that study, significant differences in response time were recorded for subjects older than 57 years when compared to subjects with ages less than 36. These results are in agreement with the conclusions reported by Guest (2006) indicating that movement dynamics are affected by ageing.

2.3.6 Iris

Iris is a tissue that covers the pupil of an eye. Due to the unique colours and patterns appearing on it, it is possible to accurately identify individuals based on iris patterns. The iris is regarded to be invariant to within-class variation, presenting in that way an ideal biometric feature. The appearance of the iris is formed within a few months from an infant's birth, and is believed to remain relatively unchanged throughout a person's lifetime. For this reason iris is regarded as an ageing invariant biometric feature. However, in recent studies (Baker *et al.* 2009; Bowyer *et al.* 2007) the effects of both direct and indirect ageing effects on iris are highlighted. In a survey on iris-recognition state (Bowyer *et al.* 2007) it is stated that certain eye-related diseases such as cataract (Jaffe *et al.* 1990) and glaucoma (Rhee 2003) affect the appearance of the iris and consequently the accuracy of iris-based biometric identification systems. Usually diseases such as cataract and glaucoma appear in subjects belonging to older age groups; hence, this type of variation can be regarded as an ageing indirect one.

Recently Baker *et al.* (2009) attempted to quantify the effect of ageing on iris templates by comparing the Hamming distance between iris templates of the same subject captured within the same month against the Hamming distance between iris templates of the same subject captured with a time difference of four years. According to the results, the difference in the Hamming distances between the two cases is statistically significant, demonstrating that, despite earlier views regarding the ageing invariance of ageing patterns, iris may undergo direct ageing effects.

2.3.7 Other modalities

In this section we provide a general account of other biometric modalities in relation to ageing variation. One of the biometric modalities regarded as being totally ageing invariant is DNA (Jeffreys *et al.* 1985). The uniqueness of DNA patterns in conjunction to the fact that certain DNA sequences remain unchanged throughout the lifetime makes DNA-based identity verification one of the most accurate biometric identification methods. However, the provision of samples of human tissue needed for DNA analysis requires the cooperation of a user, and the fact that DNA analysis is a time-consuming procedure makes the use of DNA suitable only in dedicated applications, rather than using DNA as a general biometric modality.

In some circumstances biometric modalities have limited significance for identification but can be useful for age estimation applications. For example, the human skin on its own does not possess enough discriminatory power to allow the development of accurate skin-based person recognition applications. On the other hand, the human skin is vulnerable to the ageing process and other diseases. Actually, skin appearance can provide indications regarding the overall health status of a subject. In a related study (Tanaka *et al.* 2008), an image analysis system that automatically assesses integrity of the skin is presented. As the age of a person increases his/her skin loses its elasticity and as a result wrinkles are formed. In addition indirect ageing effects, in the form of skin spots and other local deformations, are encountered in the elderly. Kim *et al.* (2009) describe a skin-based age estimation system where features associated with skin wrinkles are extracted and used for training an SVM-based age estimator. Experimental results demonstrate the feasibility of achieving age-group classification based on skin information.

In some cases biometric modalities can be used, under some circumstances, for identity or age determination. Dental biometrics (Jain *et al.* 2003) usually refer to the analysis of dental radiograms in an attempt to verify the identity of a subject. The practicality of using dental biometrics is limited by the data acquisition method (x-ray) and the fact that dental-related information is highly variable. Such variability includes changes associated with the normal cranium growth that results in changes in teeth patterns. In addition the number of teeth, teeth location and teeth size are highly dependent on the age of a subject. Such direct ageing effects are also coupled with indirect ageing effects in the form of teeth replacement that usually occurs intensively in the elderly. The increased dependency of dental features on age has been exploited by several researchers for age estimation rather than identity verification applications. In Chaillet *et al.* (2004) the ages of children in the range of 2–19 years and 8 dental-related parameters (Demirjian *et al.* 1973) are related using polynomial functions. According to the authors this representation can provide an accurate basis for age estimation based on dental radiograms. It should be noted that the polynomial representation for modelling age variation only applies to direct ageing effects observed during the formative years.

Vein-based biometrics are becoming popular for the last few years especially in the Asia Pacific region, where, due to some controversy surrounding fingerprint biometrics, vein recognition has found widespread acceptance (see Im *et al.* 2001; Badawi 2006). Vein-based biometrics involve scanning the vein pattern of some parts of the hand (usually the dorsal, palm or fingers (Wu *et al.* 2010)) using a hardware module that shines near-infrared light onto it (Such 1996). The deoxygenated blood in the veins absorbs the near-infrared light, while the rest of the hand lets the light pass through (Haxthausen 1933). A sensor in the module records the image of the hand, where the veins show up as dark lines (Miura *et al.* 2004). Thermal imaging has been also used for vein pattern scanning (see Lin & Fan 2004; Cross & Smith 1995). Vein patterns are considered age-invariant (Wilson 2010) and they are, indeed, far less sensitive than other hand-based biometrics (fingerprints, palmprints, hand geometry) to ageing effects both direct and indirect. However, the most prominent vein pattern, that of the back of the hand, cannot be assumed totally age-invariant. Dorsal hand veins, especially for hand-workers, become more prominent as the age increases. As a result, and due to the imaging techniques, the vein patterns that are recorded in time-distant sessions are quite different. Furthermore, reduction of skin elasticity creates noise during imaging especially in the case of finger vein patterns. Despite this, vein patterns can be considered among the less age-sensitive of the non-invasive and easily collected biometrics.

2.4 Conclusions, summary and future research directions

Concluding this chapter we can safely state that ageing affects biometric templates due to a number of reasons which might be different for each biometric trait. As a result of ageing effects the false rejection rate usually increases and the long-term performance of a biometric system decreases.

Although there are biometric characteristics that are considered age-invariant the truth is that there are less and more age-sensitive biometrics but none of them can be safely assumed to be age-invariant (except DNA which, as already stated, cannot be considered a biometric trait for everyday use for authentication). In this chapter we have seen the physiological effects of ageing in a variety of biometric modalities. Theoretical studies investigating ageing from the perspective of medicine and cosmetic surgery can be adapted to the area of biometrics but experimental evaluation is missing. The main reason for this is the lack of appropriate databases that would allow such experimental studies to take place. With the exception of faces, where ageing databases do exist (although by no means perfect for ageing or modelling tests), databases of biometric modalities do not include samples of the same subjects taken in a time period exceeding two years (best existing situation) while the typical time-difference between recording sessions is a month or less. As a result only short-time ageing effects can be experimentally examined. For long-time ageing effects several recording sessions of the same subject should exist within a time span exceeding a decade. Ageing variation in humans is highly non-linear in very young (age range 1–18) and older (>65) ages. Despite this, the age distribution of the subjects in existing multi-biometric databases is 18–60.

Obviously, the difficulty in creating multimodal datasets that span several years is high. Not only is the willingness of subjects to participate in data collection sections that span several years low, but changes in recording technology, staff and even culture impede the creation of a proper ageing multimodal database. What are our options then? Trying to identify the ageing effects on an individual basis definitely requires the existence of properly designed ageing databases. However, quantifying the ageing effects on biometric templates is still feasible in an accumulated fashion. Lanitis and Tsapatsoulis (2011), instead of focusing on changes on the biometric templates of the same person across several years, tried to quantify the ageing effects in biometric templates of distinct age groups by defining and introducing the so-called 'ageing impact factor' (AIF). This approach revealed that significant knowledge about the ageing effects on the biometric templates can be acquired by using existing biometric datasets. Biometric databases in which the age distribution of the subjects spans the expected human lifetime would be extremely important for such an approach. Unfortunately, as stated earlier, very few databases include subjects of very young (less than 18 years old) and older (over 65 years old) ages. Nevertheless, in this chapter we provide a relatively extended report on the multimodal databases that could be used for ageing effects quantification on biometric templates using age groups.

The other action line to account for the ageing effects in biometric templates is template adaptation and updating. Although the corresponding research area is new, there exist several studies dealing with it (see Rattani *et al.* (2009) for a quite recent study).

It is anticipated that the topic of ageing in relation to biometrics will be among the most important topics to be studied within the next few years in the biometrics research community. The main reasons for this trend are the increased need for more secure and efficient biometric systems to be used in conjunction with the development of advanced human machine interaction applications. Bearing in mind that in a digital world biometric templates of young people are commonly encountered, it is necessary that such templates are age-invariant in order to expand the maximum period of validity of such templates. The only way to tackle this problem is to focus the attention of the research community on issues related to biometric template ageing so that truly age-invariant templates are produced. During the last ten years, despite the fact that the existing publicly available face ageing datasets are not ideal, a significant number of research papers have been published in the area of facial age estimation and facial age progression. It is certain that the rest of the biometric templates will follow the same path so that in the near future a plethora of novel approaches and techniques, which will offer viable solutions to the problem of ageing, will be developed.

References

Antonakoudi A., Kounoudes A., Theodosiou Z. (2009): 'POLYBIO multibiometrics database: Contents, description and interfacing platform'. *AIAI Workshops*, vol. 475 CEUR-WS.org, 2009, pp. 150–157.

Badawi A. M. (2006): 'Hand vein biometric verification prototype: A testing performance and patterns similarity'. *Proceedings of IPVC*, 2006, pp. 3–9.

Baker S. E., Bowyer K. W., Flynn P. J. (2009): 'Empirical evidence for correct iris match score degradation with increased time-lapse between gallery and probe matches'. *Proceedings of the 3rd IAPR/IEEE International Conference on Biometrics*, LNCS 5558, pp. 1170–1179.

Benjamin B. J. (1997): 'Speech production of normally aging adults'. *Seminars in Speech and Language*, 18(2), 135–141.

Bowyer K. W., Hollingsworth K., Flynn P. J. (2007): 'Image understanding for iris biometrics: A survey'. *Computer Vision and Image Understanding*, 110, 281–307.

Chaillet N., Nystrom M., Kataja M., Demirjian A. (2004): 'Dental maturity curves in Finnish children: Demirjian's method revisited and polynomial functions for age estimation'. *Journal of Forensic Sciences*, vol. 49, issue 6, pp. 1324–1331.

Cross J. M., Smith C. L. (1995): 'Thermographic imaging of the subcutaneous vascular network of the back of the hand for biometric identification'. *Proceedings of the 29th International Carnahan Conference on Security Technology*. Surrey, UK; pp. 20–35.

Demirjian A., Goldstein H., Tanner J. M. (1973): 'A new system of dental age assessment'. *Human Biology*, vol. 45, pp. 211–227.

Dessimoz D., Richiardi J., Champod C., Drygajlo A. (2007): Multimodal biometrics for identity documents (MBioID). *Forensic Science International*, vol. 167, pp. 154–159.

Dobry G., Hecht R. M., Avigal M., Zigel Y. (2011): 'Supervector dimension reduction for efficient speaker age estimation based on the acoustic speech signal. *IEEE Transactions on Audio, Speech, and Language Processing*, vol. 19, issue 7, pp. 1975–1985.

Dumas D., Hennebert J., Humm A., Ingold R., Petrovska D., Pugin C., Von Rotz D. (2005): *MyIdea – Sensors Specifications and Acquisition Protocol*. Research Report DIUF-RR 2005.01. Department of Informatics, University of Fribourg.

Faundez-Zanuy M., Fiérrez-Aguilar J., Ortega-Garcia J., Gonzalez-Rodriguez J. (2006): 'Multimodal biometric databases: An overview'. *IEEE Aerospace and Electronic Magazine*, vol. 21, issue 8, pp. 29–37.

Fiérrez-Aguilar J., Ortega-Garcia J., Torre-Toledano D., Gonzalez-Rodriguez J. (2007): 'BioSec baseline corpus: a multimodal biometric database'. *Pattern Recognition*, vol. 40, issue 4, pp. 1389–1392.

Fiérrez-Aguilar J., Galbally J., Ortega-Garcia J., Freire, M. R., Alonso-Fernandez F., Ramos D., ... Garrido-Salas J. (2010): 'BiosecurID: A multimodal biometric database'. *Pattern Analysis and Applications*, vol. 2010, pp. 235–246.

Flynn P. (2008): 'Biometric databases'. In Jain A. K., Flynn P., Ross A. A. (Eds.), *Handbook of Biometrics*, Springer; pp. 529–548.

Fu Y., Guo G., Huang T. S. (2010): 'Age synthesis and estimation via faces: A survey'. *IEEE Transactions on Pattern Analysis and Machine Intelligence*, vol. 32, no.11, pp. 1955–1976.

Garcia-Salicetti S., Beumier C., Chollet G., Dorizzi B., les Jardins J.-L., Lunter J., Ni Y., . . . Petrovska-Delacrétaz D. (2003): 'BIOMET: A multimodal person authentication database including face, voice, fingerprint, hand and signature modalities'. *Lecture Notes in Computer Science*, vol. 2688/2003, pp. 845–853.

Geng X., Zhou Z. H., Smith-Miles K. (2007): 'Automatic age estimation based on facial aging patterns'. *IEEE Transactions of Pattern Analysis and Machine Intelligence*, vol. 29, issue 12, pp. 2234–2240.

Greulich W. W., Pyle S. I. (1999): *Radiographic Atlas of Skeletal Development of the Hand and Wrist*. 2nd edn. Stanford University Press, Palo Alto, US.

Guest R. (2006): 'Age dependency in handwritten dynamic signature verification systems'. *Pattern Recognition Letters*, vol. 27, issue 10, pp. 1098–1104.

Gurbuz S., Gowdyl J. N., Tufekci Z. (2000): 'Speech spectrogram based model adaptation for speaker identification'. *Proceedings of the IEEE Southeastcon*, Nashville, Tennessee, USA; pp. 110–115.

Haxthausen H. (1933): 'Infrared photography of subcutaneous veins'. *British Journal of Dermatology*, vol. 45, pp. 506–511.

Im S.-K., Park H.-M., Kim Y.-W., Han S.-C., Kim S.-W., Kang C.-H., Chung C.-K. (2001): 'A Biometric identification system by extracting hand vein patterns'. *Journal of the Korean Physical Society*, vol. 38, issue 3, pp. 268–272.

Jaffe N. S., Jaffe M. S., Jaffe G. F. (1990): *Cataract Surgery and Its Complications*. 5th edn. St. Louis, CV Mosby, pp. 385–411.

Jain A., Chen H., Minut S. (2003): 'Dental biometrics: Human identification using dental radiographs'. *Proceedings of the Audio- and Video-Based Biometric Person Authentication*, LNCS 2688. Guildford, UK; pp. 429–437.

Jain A., Ross A., Prabhakar S. (2004): 'An introduction to biometric recognition'. *IEEE Transactions on Circuits and Systems for Video Technology*, Special Issue on Image- and Video-Based Biometrics, vol. 14, issue 1, pp. 4–20.

Jain A., Patrick F., Ross A. (Eds.) (2008): *Handbook of Biometrics*, Springer, New York.

Jankovic J., Kapadia A. S. (2001): 'Functional decline in Parkinson disease'. *Archives of Neurology*, vol. 58, issue 10, pp. 1611–1615.

Jeffreys A. J., Wilson V., Thein S. L. (1985): 'Individual-specific "fingerprints" of human DNA'. *Nature*, vol. 316, issue 6023, pp. 76–79.

Kim K., Choi Y. H., Hwang E. (2009): 'Wrinkle feature-based skin age estimation scheme'. *Proceedings of the ICME 2009*. Cancun, Mexico; pp. 1222–1225.

Kirwan J. R. (1995): 'The effect of glucocorticoids on joint destruction in rheumatoid arthritis'. *New England Journal of Medicine*, vol. 333, pp. 142–147.

Lamel L. F., Gauvain J. L. (2000): 'Speaker verification over the telephone'. *Speech Communication*, vol. 31, issue 2–3, pp. 141–154.

Lanitis A. (2008): 'Comparative evaluation of automatic age progression methodologies'. *EURASIP Journal on Advances in Signal Processing*, Article ID 239480, 2008.

Lanitis A. (2010): 'A survey of the effects of aging on biometric identity verification'. *International Journal of Biometrics*, vol. 2, issue 1, pp. 34–52.

Lanitis A., Taylor C. J., Cootes T. F. (2002): 'Toward automatic simulation of aging effects on face images'. *IEEE Transactions of Pattern Analysis and Machine Intelligence*, vol. 24, issue 4, pp. 442–455.

Lanitis A., Tsapatsoulis N. (2011): 'Quantitative evaluation of the effects of aging on biometric templates'. *IET Computer Vision Journal*. Special Issue on Future Trends in Biometric Processing, vol. 5, issue 6, pp. 228–347.

Larsson L., Grimby G., Karlsson J. (1979): 'Muscle strength and speed of movement in relation to age and muscle morphology'. *Journal of Applied Physiology*, vol. 46, issue 3, pp. 451–456.

Lin C. L., Fan K. C. (2004): 'Biometric verification using thermal images of palm-dorsa vein patterns'. *IEEE Transactions on Circuits Systems for Video Technology*, vol. 14, issue 2, pp. 199–213.

Linville S. E. (2004): 'The aging voice'. *ASHA Leader* (available online at http://www.asha.org/Publications/leader/2004/041019/041019e.htm, last accessed August 2012).

Merkel R., Dittmann J., Vielhauer C. (2011): 'How contact pressure, contact time, smearing and oil/skin lotion influence the aging of latent fingerprint traces: First results for the binary pixel feature using a CWL sensor'. *Proceedings of the 2011 IEEE International Workshop on Information Forensics and Security (WIFS)*. Mabu Thermas, Brazil; pp.1–6.

Miura N., Nagasaka A., Miyatake T. (2004): 'Feature extraction of finger-vein patterns based on repeated line tracking and its application to personal identification'. *Machine Vision and Applications*, vol. 15, issue 4, pp. 194–203.

Modi S. K., Elliott S. J. (2006): 'Impact of image quality on performance: Age comparison of young and elderly fingerprints'. *Proceedings of the 6th International Conference on Recent Advances in Soft Computing*, pp. 10–12.

Modi S. K., Elliott S. J., Whetsone J., Hakil K. (2007): 'Impact of age groups on fingerprint recognition performance'. *Proceedings of the IEEE Workshop on Automatic Identification Advanced Technologies*, pp. 19–23.

Mueller P. B. (1997): 'The aging voice'. *Seminars in Speech and Language*, vol. 18, issue 2, pp. 159–168.

Narayanan S., Potamianos A. (2002): 'Creating conversational interfaces for children'. *IEEE Transactions on Speech and Audio Processing*, vol. 10, issue 2, pp. 65–78.

Ortega, M., Brodo, L., Bicego, M., Tistarelli, M. (2009): 'Measuring changes in face appearance through aging'. *Computer Vision and Pattern Recognition Workshops*. Miami, Florida, USA; pp. 107–113.

Ortega-Garcia J., Fierrez J., Alonso-Fernandez F., Galbally J., Freire M. R., Gonzalez-Rodriguez J., Gonzalez-Rodriguez J., . . . Savran A. (2010): 'The multi-scenario multi-environment BioSecure Multimodal Database (BMDB)'. *IEEE Transactions on Pattern Analysis and Machine Intelligence*, vol. 32, issue 6, pp. 1097–1111.

Patterson E., Sethuram A., Albert M., Ricanek K., King M. (2007): 'Aspects of age variation in facial morphology affecting biometrics'. *Proceedings of the 1st IEEE International Conference on Biometrics: Theory, Applications, and Systems*, pp. 1–6.

Pawlewski M., Jones J. (2006): 'Speaker verification: Part 1'. *Biometric Technology Today*, vol. 14, issue 6, pp. 9–11.

Plamondon R., Srihari S. N. (2000): On-line and off-line handwriting recognition: A comprehensive survey. *IEEE Transactions on Pattern Analysis and Machine Intelligence*, vol. 22, issue 1, pp. 63–84.

Ramanathan N., Chellappa R. (2006): 'Modeling age progression in young faces'. *Proceedings of the IEEE Conference on Computer Vision and Pattern Recognition*. New York, NY, USA; pp. 387–394.

Ramanathan N., Chellappa R., Biswas S. (2009): 'Computational methods for modeling facial aging: A survey'. *Journal of Visual Languages and Computing*, vol. 20, issue 3, pp. 131–144.

Ramig L. A., Ringel R. L. (1983): 'Effects of physiological aging on selected acoustic characteristics of voice'. *Journal of Speech and Hearing Research*, vol. 26, pp. 22–30.

Rattani A., Freni B., Marcialis G., Roli F. (2009): 'Template update methods in adaptive biometric systems: A critical review'. *Advances in Biometrics*, vol. LNCS 5558/2009, pp. 847–856.

Rhee D. J. (2003): *Glaucoma: Color Atlas & Synopsis of Clinical Ophthalmology*. McGraw-Hill Professional, USA.

Ricanek K., Tesafaye T. (2006): 'MORPH: A longitudinal image database of normal adult age-progression'. *Proceedings of the 7th IEEE International Conference on Automatic Face and Gesture Recognition*. Southampton, UK; pp. 341–345.

Rowland D. A., Perrett D. I. (1995): 'Manipulating facial appearance through shape and color'. *IEEE Computer Graphics and Applications*, vol. 15, issue 5, pp. 70–76.

Schulz R. (2006): *The Encyclopedia of Aging: A Comprehensive Resource in Gerontology and Geriatrics*, 4th edn., Springer, New York.

Shim J. K., Lay B. S., Zatsiorsky V. M., Latash M. L. (2004): 'Age-related changes in finger coordination in static prehension tasks'. *Journal of Applied Physiology*, vol. 97, issue 1, pp. 213–224.

Somanath G., Rohith M. V., Kambhamettu C. (2011): 'VADANA: A dense dataset for facial image analysis'. *Proceedings of the 2011 International Conference on Computer Vision Workshops (ICCV Workshops)*. Barcelona, Spain; pp. 2175–2182.

Such O. (1996): 'Near infrared imaging of hemodynamics'. *Proceedings of the 18th International Conference of the IEEE/EMBS*. Amsterdam, Netherlands; vol. 5, pp. 2105–2108.

Tanaka H., Nakagami G., Sanada H., Sari Y., Kobayashi H., Kishi K., Konya C., Tadaka E. (2008): 'Quantitative evaluation of elderly skin based on digital image analysis'. *Skin Research and Technology*, vol. 14, issue 2, pp. 192–200.

Uhl A., Wild P. (2009): 'Comparing verification performance of kids and adults for fingerprint, palmprint, hand-geometry and digitprint biometrics'. *Proceedings of the 3rd IEEE International Conference on Biometrics: Theory, Application, and Systems (BTAS 2009)*. Washington , USA; pp. 1–6.

Uludag U., Ross A., Jain A. (2004): 'Biometric template selection and update: a case study in Fingerprints'. *Pattern Recognition*, vol. 31, issue 7, pp. 1533–1542.

Wilson C. (2010): *Vein Pattern Recognition: A Privacy-Enhancing Biometric*. 1st edn., CRC Press, New York, ISBN: 978-1439821374.

Woollacott M. H., Shumway-Cook A., Nashner L. M. (1986): 'Aging and posture control: Changes in sensory organization and muscular coordination'. *International Journal of Aging and Human Development*, vol. 23, issue 2, pp. 97–114.

Wu G. (1998): 'Age-related differences in body segmental movement during perturbed stance in humans'. *Clinical Biomechanics*, vol. 13, issue 4–5, pp. 300–307.

Wu X., Gao E., Tang Y., Wang K. (2010): A novel biometric system based on hand vein. *Proceedings of the 5th International Conference on Frontier of Computer Science and Technology (FCST)*. Changchun, China; pp. 522–526.

Yampolskiy R. V., Govindaraju V. (2008): 'Behavioural biometrics: A survey and classification'. *International Journal of Biometrics*, vol. 1, issue 1, pp. 81–113.

Zaphiris P., Ellis R. D. (2000): 'Mathematical modeling of age differences in hierarchical information systems'. *Proceedings of the ACM Conference on Universal Usability*, pp. 157–158.

Chapter 3

Biometrics and ageing: social and ethical considerations

Andrew P. Rebera and Emilio Mordini

3.1 Introduction

That the ageing process poses a problem for biometrics is well-understood. No biometric is 100% permanent: people's biometrics change over time. Hence the technical challenge is to develop techniques whereby an individual may be identifiable by his or her biometrics, throughout his or her lifetime, despite this mutability [1].

Such at least is the challenge from the purely technical perspective. But reflection upon wider societal and ethical considerations necessarily broadens our view. Ageing and the role of older people in society raise difficult questions concerning *inclusion, discrimination, inter-generational justice*, and so forth. Biometrics cannot stand apart from those issues. There is first an ethical imperative to ensure that technological developments do not aggravate or reinforce existing problems. On the face of it, biometrics does not fare well here. For instance, one existing problem is that older people are, in various ways, subject to exclusion and discrimination in our society. Hence that biometrics identifies older people as 'problematic' – even if this is simply a *façon de parler* – is unfortunate and likely to exacerbate rather than ameliorate the problem. A second more pragmatic imperative is imposed by the fact that society is rapidly ageing. Between 2004 and 2050, the proportion of the total population aged 65 years or older will grow by a massive 77% [2], and while median age (globally) in 2000 was only 26, it is predicted to be 44 by the year 2100 [3]. Older people constitute the fastest-growing demographic: if biometrics is to be at the forefront of mass identification schemes in the future, it must be able to efficiently identify them. Failure to meet this challenge will render biometrics societally and ethically unacceptable, and practically and economically unviable.

In this chapter we document and discuss the major societal and ethical issues raised by biometrics in an ageing society. There are various well-known problems associated with biometrics – privacy concerns, for example – which certainly affect older people, but which in fact pertain to all users, regardless of age. And there are various ethical and societal issues arising in relation to biometrics and older

CSSC – Centre for Science, Society and Citizenship, Italy

people – issues of inclusion and exclusion, for example – which also arise with other non-biometric technologies. We will discuss these kinds of issues in section 3.3.1. On the other hand, we will also discuss the claim that biometrics, in its current guise, unintentionally deepens and embeds the *problematisation of ageing and of older people*. We argue in section 3.3.2 that biometrics, in its approach to technical issues such as template ageing and perhaps even in its terminology, implicitly casts the ageing process and older people as *problems, difficulties*, or *obstructions* to be overcome.[1] Such a view is unhelpful at the societal level, and biometrics' association with such a view can only impede its acceptance among the general public. Biometrics offers advantages and possibilities for older people, hence developing its societal acceptability is very much in the interests of all parties. We argue that the problematisation of ageing – which can appear to spring organically from the natural development of biometrics as a discipline – is in no way essential to biometrics, and that such societal and ethical issues as arise at present can, in principle, be addressed. In section 3.4 we sketch a biometric system for border control based on the principle that there are no 'normal' or 'standard' users of a biometric system, and hence that no age group should be more or less problematic for biometrics.

We begin in section 3.2 by elucidating the way in which ageing and older people come to be seen as problematic for biometrics.

3.2 The nature of the problem

All personal identification systems involve two distinct phases. In the first ('primary identification'), a person is linked to a primary identifier which singles her out as an individual, different from all others. In all known human societies the most important primary identifier is a word, or a combination of words (e.g. given name plus family name). However, there are many fewer names than persons and, as a consequence, other primary identifiers are often necessary to unambiguously indicate a person. Today the main primary identifiers used together with names are gender, and place and date of birth.[2] The attribution of these primary identifiers is usually carried out at birth, provided by an adult (often the parents) who 'recognises' the baby by giving her a personal name according to the name system of that society. The personal name is then certified by state officers who also register (at least) gender, and place and date of birth. From this moment until death, the state becomes the guarantor that there is a unique person, male or female, with that name, born in that place, on that date.[3] In particular circumstances, it can happen that a person changes her primary identifiers, and so becomes a

[1] We are, no doubt, guilty of those same terminological slips ourselves. In this chapter, when we speak of, for example, the 'quality' of a biometric feature, we do not mean to imply by that term a value judgement.

[2] In some countries other information (e.g. race or ethnicity) is also registered at birth.

[3] One of the main consequences of this primary identification is that a person becomes legally entitled to certain rights and duties.

'new' person. This process could imply the permanent erasure of previous primary identifiers (e.g. rectification for gender change, or new names for people in witness protection programmes) or it can just create a new, parallel, primary identity (e.g. codes used to anonymise individuals in a database[4]).

The second phase of a personal identification system ('secondary identification') is actually a process of re-identification. It aims to attribute an existing personal name to an unknown individual, or to verify whether a person is who she claims to be. Secondary identification is the linking of known primary identifiers with an actual person whose identity is to be ascertained. Usually this is achieved by using one or more tokens. Tokens have traditionally been either small objects that can be presented (e.g. seals, rings, ID cards), knowledge known only by the individual or other trusted parties (e.g. secret lore, passwords, answers to security questions), or physical augmentations (e.g. scarring, mutilation, tattooing).

With biometrics, however, one may – in effect – be one's own token. As conceptualised in contemporary biometry,[5] the body is a source of uniquely identifying, machine-readable, tokens or 'templates' which can be stored and later compared with other templates generated from the body. If a later template matches an earlier stored template, an individual is authenticated (re-identified); if later and earlier templates do not match, the individual's identity is not confirmed.[6]

3.2.1 The vulnerabilities of tokens

One of the main advantages of biometrics is thought to be that – in certain respects at least – it does not require tokens susceptible to common human failings (which are often particularly associated with older people). Physical tokens can be lost, stolen, or broken; mental tokens can be forgotten. Biometric features cannot be forgotten or lost, since wherever one goes, there they go too. In this, biometrics has the edge over more traditional forms of identification. However, biometric features are far from perfect in any of these capacities. Vulnerabilities remain. We divide them into two groups:

3.2.1.1 Group 1 vulnerabilities

Group 1 vulnerabilities include the chances of a token being (i) forgotten; (ii) lost; or (iii) broken.

(i) *Can biometrics be forgotten?* Unlike physical and mental tokens, which can be easily forgotten, one cannot forget (say) one's fingerprint. Yet the shift from first to second generation biometrics [5] entails more widespread use of activity-based and behavioural biometrics, and at least some of these can be forgotten (*written signature* being an obvious example).

[4] One speaks of 'anonymisation' when it is impossible for anyone to connect two primary identifiers, and of 'pseudo-anonymisation' when it is still possible – under certain circumstances – to connect two identifiers (e.g. a personal name with a code).

[5] As opposed to, say, the anthropometry of Galton, or the *Bertillonage* of Bertillon. On the conceptualisation of the 'biometric body', see Reference 4.

[6] For brevity we describe only a case of 1:1 authentication ('verification'), leaving 1: n ($n > 1$) authentication ('identification') aside.

(ii) *Can biometrics be lost?* Fingerprints, to continue with that modality, can be lost temporarily (due to 'pruning' when wet, for example [6]), or permanently, through ageing or otherwise-induced wear and tear [7, 8]. They may also be lost intentionally. The 1930s gangster John Dillinger is said to have attempted to burn off his fingerprints with acid, and some asylum seekers are known to have attempted to burn off their fingerprints in order not to be traceable to the first European state in which they arrived [9]. It should also be noted that when biometric data is stored on a chip carried by the individual subject (rather than in a database of templates), the chip may be as easily lost as any other physical token.

(iii) *Can biometrics be broken?* Although 'broken' is not the most natural word, biometrics can cease to function optimally. Image quality degradation can occur for many reasons (e.g. sub-optimal scanner functioning, infelicitous lighting conditions, and other 'noise'), but with respect to an individual's biometric itself, it has been shown that accidents and certain medical treatments (e.g. burns, some cancer treatments [10]), wear and tear (e.g. older people, builders, bakers [7]), and temporary alterations (e.g. in wet conditions [6]) can all have a deleterious effect on fingerprint recognition.

3.2.1.2 Group 2 vulnerabilities

Group 2 vulnerabilities include the chances of a token being (i) stolen or (ii) faked.

(i) *Can biometrics be stolen?* Biometrics cannot be stolen in the usual sense. However, they may be *faked* (on which see directly below) or *hijacked* in coercion attacks (in which an individual is forced to use a biometric feature against their will[7]). Actual theft of biometrics is rather grizzly, as in Reference 12.

(ii) *Can biometrics be faked?* Biometrics are vulnerable to 'spoofing' attacks in which one individual attempts to pass himself off as another individual by using a fake biometric.[8] Examples of fake biometrics include silicon or gummy finger [13] (for fingerprint); photograph, video, or 3D mask [14] (face); printed iris, video playback, fake eye, or printed contact lens [15] (iris); impersonation or speech synthesis [16] (voice); and printed image [17] (vein pattern). Proposed countermeasures include liveness detection, multi-modality, and the development of modalities inherently robust to spoofing attacks.[9]

Our focus must then be on the Group 1 vulnerabilities (*forgotten, lost*, and *broken*); for these match the technical problem posed by ageing to biometrics.[10] As a subject ages,

[7] Cf. [11], p. 214.

[8] The European Commission funded project *Tabula Rasa* (https://www.tabularasa-euproject.org/) addresses the urgent need to develop standards for examining the problem of spoofing, as well as investigating potential countermeasures.

[9] For discussion of the societal and ethical issues arising from the use of these measures see References 18 and 19.

[10] To the best of our knowledge, there is no research into the question of whether the age and condition of a biometric significantly affects the ease with which it may be stolen or spoofed (Group 2 vulnerabilities). It may well be that age factors impact upon security when, for example, a sensor's acceptance threshold is widened in order to accommodate older people with lower quality biometrics. But if and how the age of a biometric affects its vulnerability to spoofing is not yet, as far as we are aware, properly researched.

his biometrics tend to become either broken (i.e. they cease to function optimally with respect to sensors) or lost (i.e. they cease to be amenable to sensors at all).[11]

3.2.2 Ageing and age as 'problems' for biometrics

3.2.2.1 The ageing problem

The challenge for biometrics is, then, to come to terms with the fact that ageing-induced bodily and behavioural changes reduce system performance by increasing 'within-person' variations, i.e. variations between features captured at an earlier time t_1 and a later time t_2. Lanitis [1] writes that 'with the wide spread of biometric authentication systems, sooner or later users will be faced with performance degradation of their systems as a result of aging' (p. 35).

Note that this is conceived as a problem for the future ('sooner or later', but in any case not *now*). The issue is 'increasingly important' (p. 35) and 'needs serious consideration by the biometrics community' (p. 47), but it is 'deterioration in the long-term performance of biometric authentication systems' (p. 1) that wants minimising. Biometrics in its automated, digital form is a relatively young technology: the other end of the long-term has not yet arrived. That the problem of 'template ageing' cannot be avoided is, on this view, simply a matter of the impressive dissemination of biometric systems: being now more widely spread, the influence of the challenges biometric systems face will be correspondingly greater.

Template ageing is assumed to be a problem affecting all modalities. However, despite more interest in recent years, there is still research to be done to establish the extent of the problem for each modality, and to develop practicable solutions.

Fingerprints are known to deteriorate significantly with age, with ridge structures and minutiae losing definition. Modi and Elliot [8] note that ageing sees a reduction in collagen, with skin becoming loose and dry. The skin being less firm and less elastic, the contact between the finger and the sensors can be sub-optimal. This can be worsened by conditions more common among older people, such as arthritis. Studies [8, 20] suggest that age-related deterioration of fingerprints results in reduced performance of fingerprint identification systems [1].

The modality most obviously altered by ageing is the face. Faces change over time in many ways: soft tissue modifications (loss of skin elasticity, sagging, weakening of ligaments, drooping of brows, etc.); hard tissue modifications ('bony' changes, alteration of skull dimensions, movement of teeth and jaws, etc.); muscle atrophy; fat accumulation; forward and downward movement of the nose; elongation and thinning of lips; alteration in ear dimensions; and so on [21]. Diseases common among the elderly may also result in facial changes [1].

It was, for a time, supposed by some that iris represented a relatively stable and permanent modality. This view was represented in a number of US patents for iris systems [22]. It has been suggested that diseases such as cataract and glaucoma alter the appearance of the iris, leading to reduced performance by iris recognition

[11] For present purposes the third vulnerability, *forgotten*, is sufficiently similar to *lost* that we may treat them as one.

systems [1]. Moreover it has recently been argued [22] that there is 'clear and consistent evidence' (p. 1) of template ageing for iris [23, 24].[12]

Four approaches are commonly cited in response to the problem of template ageing:

1. *Updating templates*. Periodically updating templates appears to be a slightly inconvenient and inefficient proposal – although it is, in effect, the system currently used in most countries for renewal of passports and national ID cards.
2. *Simulation of the effects of ageing*. A more advanced proposal is to develop ways of updating templates without the cooperation or presence of the subject, by simulating the effects of ageing. This requires detailed understanding (currently lacking) of the effects of ageing, and the use of 'age-progression methodologies' to produce accurately age-progressed biometric templates [25, 26]. This can be combined with techniques in adaptive biometrics [27, 28].
3. *Use of age-invariant biometric features*. The development of age-invariant features takes two forms [1]. First, researchers may be able to isolate and utilise age-invariant features of existing modalities (e.g. some parts of the face may be more stable than others). Second, it may be that emerging biometric modalities (such as hand-vein pattern) are more age-invariant than existing common modalities.
4. *Multi-modality*. The combination of several modalities may enable more accurate and reliable identification by allowing an element of tailoring whereby modalities can be matched to individuals to accommodate their needs (e.g. if fingerprint is unsuitable for an older person, they may be able to use, say, iris instead).

The template ageing problem is a problem of *ageing*, rather than of *age*. The problem concerns the ageing-caused variation between a given biometric feature at t_1 and t_2. Here t_1 and t_2 may represent a subject at age 35 and 75, but they might equally represent a subject at age 6 and 16. The issue is one of change and variation, hence of ageing rather than age itself.[13]

Ageing, in the guise of template-ageing, is a serious problem for biometrics. But to focus only on this would be an error. Biometrics faces further age-related challenges stemming from demographic shifts.

3.2.2.2 The age problem

To the many difficult questions posed by population growth,[14] biometrics has the potential to provide at least some answers. Weak or non-existent identity documents – around one in every three births goes unregistered across the globe [31] – restrict

[12] Other biometrics, including *hand* (palmprint or hand geometry), *voice*, and a variety of *behavioural* modalities based on bodily movement are surveyed in Reference 1, and age-related performance reductions found.

[13] The ageing/age distinction we are drawing is witnessed in the exemption of under-12s from the requirement to provide fingerprints for European Union passports (cf. Regulation (EC) No. 444/2009 (28 May 2009), amending Council Regulation (EC) No. 2252/2004 (13 December 2004)).

[14] From over 7 billion in 2011 [29], the United Nations predicts that by 2050 global population will reach somewhere between 8 and 11 billion [30].

access to health care, welfare, rights, and opportunities. Appropriately managed [32], biometrics could offer societally acceptable and technologically viable solutions, as we are beginning to see in India and in other low- and medium-income countries [33].

In addition to population growth, we are witnessing dramatic demographic shifts. Older population growth – i.e. growth of the population of people aged 65 and over – is considerably more marked than overall population growth (i.e. across all age groups), and it is foreseen that this trend will continue for at least the next quarter of a century [34]. By 2050, it is estimated that around one third of the European population will be aged 65 or over [35].[15] The 'oldest old', i.e. those aged over 80, is the most rapidly growing age group of all [34].

As the proportion of the population aged over 65 increases, the number of people likely to pose a problem of some kind for biometric identification systems increases. Older people tend to have weaker biometric features. Age itself can be problematic insofar as the quality of an image capturable from an older person is likely to be inferior to that of a younger person. Thus the fact that biometric features are not permanent poses problems not only of *variation* but also of *quality*. This was illustrated by, for example, Hong Kong's development of an e-Channel border crossing system using fingerprint-based smart ID cards. A small percentage of people have trouble clearing the e-Channels due to the quality of their fingerprints [37]. Now if the numbers are small, the problem is perhaps not so serious, since old-style alternatives to e-Channels are available. But there are two clear problems. First, as Mordini and Ashton [38] note, 'fall-back alternatives [. . .] are often time-consuming to pursue and can create some perception of stigma at 'failing' the system' (p. 272). Second, as the numbers using a system get larger – as they must if biometrics are to be rolled out on a larger scale in an ageing society – so even small percentage values for false acceptance, false rejection, failure to enrol, and failure to capture rates translate into large numbers of people affected. A false rejection rate of just 0.5% among, say, 250,000 travellers passing through an airport works out as 1,250 people inconvenienced.[16]

3.3 Social and ethical factors

We have distinguished two problematic aspects of ageing for biometrics: problems of ageing, i.e. problems deriving from within-person variation brought on by the ageing process, and problems of age, i.e. problems deriving from the increased incidence of lower quality biometric features among older populations. This section begins with the latter, addressing the societal and ethical issues raised by problems of age.

[15] In China, the 'dependency ratio' (ratio of number of people aged 65 or over to the number of people of working age (15–64)) will increase by a factor of 4 by 2050 [36].

[16] Not to mention people indirectly inconvenienced by the inefficient running of the system (e.g. queues and hold-ups).

3.3.1 Social and ethical factors: problems of age

At least with respect to new and emerging technologies, the major issues facing older people concern *inclusion* and *exclusion*. And these issues are extensions of, or variations on, existing challenges that older people face in other domains (social or economic inclusion, for example).

Biometrics is potentially an extremely valuable technology for older people for a number of reasons. As discussed above, identification systems premised upon the (re)production of physical or mental tokens are subject to a number of weaknesses. We classified these under two heads, Group 1 vulnerabilities and Group 2 vulnerabilities. Especially with respect to Group 1 vulnerabilities (i.e. that tokens may be forgotten, lost, or broken), biometrics – being in a certain sense a 'token-less' identification system – enjoys a clear advantage. With biometrics, one need not memorise a complicated password; with biometrics, one need not fiddle around with keys or cards. Tasks such as these can be difficult for people suffering from, say, early-stage dementia (and bear in mind that memory lapses are common among older people even in the absence of degenerative diseases) or arthritis respectively.

European legislation recognises the 'rights of the elderly'. As part of the drive towards an information society, the European Commission has called for 'e-Inclusion' in order that 'every person who so wishes [may] fully participate in the information society, despite individual or social disadvantages' [39]. Article 25 of the Charter of Fundamental Rights of the European Union [40] states the Union's commitment to 'the rights of the elderly to lead a life of dignity and independence and to participate in social and cultural life'. The biometrics community obviously has a commitment to respect all relevant legislation, but it ought also to strive where possible to promote best practice. This, as mentioned above, corresponds to both ethical and pragmatic/economic imperatives.

To illustrate, let us begin by indicating some issues and questions that can be of particular significance to older people in relation to technology. These issues – which are not mutually exclusive and often overlap – are discussed further below.

- *Social, economic, cultural, political (etc.) inclusion and exclusion*. Does the technology in question serve to promote the inclusion of older people in the wider life and concerns of society, or to exclude and isolate them?
- *Inter-generational (in)justice*. Does the technology promote fairness among different demographic groups?
- *Ageism*. Does the technology discriminate against older people on the grounds of age?
- *Equality of access*. Is there equality of access to the technology? Is the technology open to and available for users from all backgrounds?
- *Freedom, autonomy, independence*. Does the technology encourage the freedom, autonomy, and independence of its older users? If not, does it at least not restrict the freedom, autonomy, and independence of its older users?
- *Transparency and openness*. Is the technology, its role, its potential benefits and dangers, and the nature of the way in which users use or interact with it sufficiently clear to older users?

- *Data protection.* Are all relevant data management principles respected? Is data collected for stated purposes and used for no other purposes? Is the amount of data captured proportional to the value secured by its use? Has the amount of data needed been minimised as much as possible? Is the data appropriately stored, with suitable safeguards and recourse for data subjects? Is it necessary to have data subjects' consent for the processing of their data? Are clear and transparent procedures in place for rectifying errors and ensuring accountability?
- *Privacy.* Privacy is a wider concept than data protection, encompassing human dignity and integrity. Does the technology respect users' privacy? Has the technology been developed with privacy in mind (as per, say, the principles of *Privacy by Design*)?

As we shall see, biometrics is well-suited to addressing these issues.

Consider the issue of *inclusion*. One manifestation of the problem of age for biometrics is that older people are more likely to suffer from failure to enrol biases due to lower quality features. As Drousou and Tzovaras [41] point out, 'people whose biometrics cannot be recorded well for the creation of the database reference [i.e. template] [...] are de facto excluded by the system' (p. 130); they are, write Mordini and Ashton [38], 'made conspicuous by their inability to be authenticated by biometric technologies'. If biometrics cannot accommodate older people, it will tend to restrict their access to the various goods with which biometrics are now associated (such as international travel, access to rights and welfare). Restriction of access in this way is undesirable in itself, but it may further promote *self-exclusion* – individuals opting out of certain activities in order to avoid the embarrassment, inconvenience, or stigma associated with trying and failing to (efficiently) use a biometric system – which can in turn encourage wider social, economic, cultural, or political exclusion, as well as impinging upon freedom and autonomy. Thus the problem is ostensibly one of exclusion, but incorporates also inter-generational injustice, equality of access, ageism, and freedom, autonomy, and independence.

The problem is perhaps additionally acute since the several factors by which one's propensity to suffer failure to enrol bias converge around older people. Deravi [42] writes:

> Failure to enrol on biometric systems or to consistently provide usable samples for biometric comparison may be due to a range of factors including physical or mental disability, age, and lack of familiarity or training in the use of particular biometric systems. (p. 179)

These factors preponderate among older people. It follows that even if the effects of age do not, in themselves, restrict an individual's ability to use a biometric system, there remains a reasonable chance that either a lack of familiarity with the technology or a physical or mental disability will.

Deravi notes that in order to overcome this difficulty, it will likely be fruitful to pursue multi-modal solutions. These, he contends, may 'reduce and perhaps even eliminate the size of the outlier population' (p. 180). But of course, as he goes on to indicate, it is then necessary to ensure that users' privacy is not threatened by this expansion into multi-modality.

Privacy is often thought of as pertaining particularly to other people's knowledge or ignorance of certain information about an individual. Popular accounts of privacy cast it in terms of ignorance (i.e. x has privacy relative to y just in case y is ignorant of certain facts about x), *control* (i.e. x has privacy relative to y just in case x is able to control whether y knows certain facts about x), or *restricted access* (i.e. x has privacy relative to y just in case societal conventions or laws restrict y's access to certain facts about x).[17] This emphasis on information may cause data protection and privacy to seem more closely connected than they are. But privacy, as indicated above, is a concept that outstrips data protection.

Of course the two concepts, privacy and data protection, *are* closely connected. Etymologically, 'privacy' implies a kind of *removal*, or *being set apart*, and that setting apart includes, of course, the setting aside of personal information. Yet privacy goes beyond that as well, as Mordini [46] explains:

> Privacy is primarily a moral concept. It involves claims about the moral status of the individual self, about its dignity and relation to others. Philosophers and legal theorists [...] tend to talk about privacy in terms of ideas like 'inviolate personality'. This moral core, it is argued, is the origin of social values such as autonomy, integrity, independence. Such values form a foundation for contemporary notions of human rights, citizenship and civic obligation in European public affairs. (p. 331)

Respecting privacy not only involves adhering to appropriate standards of data management then but also requires respect for dignity and integrity. In practical terms this entails designing biometric systems whose modes of functioning in no way whatsoever leave older people feeling humiliated, embarrassed, belittled or unduly challenged.[18] A system which, albeit unintentionally, tends to single out older people – e.g. because older people are disproportionately likely to experience some hold-up, due to, say, a failure to enrol or failure to capture error – is likely to cause embarrassment and even humiliation for some users. Moreover, given the demographic shifts to which we have adverted, the number of people liable to such embarrassment or humiliation is set to rise sharply.

With this broader notion of privacy in mind, we see that the various ethical issues are all inter-weaved. The concepts of autonomy, freedom, and independence are no exceptions. If possible – and it ought to be possible in the vast majority of instances – new technologies ought to positively promote the autonomy, freedom, and independence of older people. Biometrics could thrive in this respect.

Biometrics could make a very positive contribution to the lives of older people through its role in the development of *ambient intelligence* ('AmI') and *smart environments* ('SmE'). At root, these involve the shrinking back of ICTs (information and communication technologies) from dedicated terminals (like

[17] On these three views see References 43, 44 and 45 respectively.

[18] And of course this applies equally for everyone, not just older people.

computers), and their incorporation and embedding in the physical environment via techniques of (e.g.) ubiquitous, haptic, and human-centred computing, incorporating technologies such as RFID and biometrics. Individuals will interact with and receive various kinds of support from their – now electronic and networked – environment. As Cabrera Giráldez and Rodríguez Casal [47] suggest, AmI and SmE have great potential to aid older people by enabling:

> remote monitoring of activity and physical well being (including smart clothing); adaptive interfaces for people with physical disabilities; and a responsive and proactive environment (rather than simply a collection of devices) which enables easy communication with healthcare professionals, friends, family and the wider community. (p. 269)

'Smart homes' may support older people through the use of remote monitoring of various forms of behavioural biometrics, or in conjunction with bodily biometrics monitored by smart clothing [48].

As society ages – and especially in the context of the current economic crisis – retirement ages are being pushed ever higher. It may be that, in consequence, older people are forced to stay in employment for longer, making good shortfalls in pension and welfare provision through extended economic activity. Indeed, to come at it from another angle, developments in ICT will increasingly allow people to stay active for longer, meaning that the 'extended middle age' group [49] – those who reach retirement age perfectly able, from a physical and mental perspective, to continue in employment – are able to stay in employment for a longer period.

Biometrics can contribute to making workplaces more accommodating for older people, simplifying interfaces with technologies and so on. But of course there are risks, and these must be borne in mind. For instance, a biometric system that could be used – even if this is not its primary purpose – to detect medical information or an individual's age would be unsuitable for the work environment (since it is does not require a huge leap to imagine that such data could be used to discriminate against older employees[19]). Function creep is, to some extent, unavoidable, yet steps can be taken to minimise risks [51]. Such steps are needed, for age-discrimination is on the rise [52, 53].[20]

One guard against function creep is to ensure that technology providers are regulated in some way (e.g. by law or by codes of conduct) and that users of the technology are aware of what the technology is, what it is for, and how it works. For biometrics, this might entail ensuring that systems are developed with ethical input and consultation at all stages, that industry standards such as *Privacy by Design* are in place and that impact assessments are carried out prior to rollout of the technology. In addition, users of the technology must be aware of which biometric modalities are being measured, how they are being measured and why. This need not involve every

[19] Discrimination on grounds of age is outlawed in EU law through Directive 2000/78/EC. For discussion see Reference 50.

[20] According to Reference 53, the UK Tribunal Service's figures indicated a 79% increase in age discrimination claims between 2008/09 and 2010/11.

traveller passing through an airport passing a Biometrics 101 exam, but it ought to be decided what minimum standards of awareness and consent are required.

The principle underwriting the previous point is a core tenet of European data protection legislation, namely that 'personal data may be processed only if: the data subject has unambiguously given his consent' (barring certain defeating clauses, such as legal obligations and public interest).[21] However, this principle sits uneasily alongside AmI and SmE technologies. AmI and SmE technologies require that biometric systems become increasingly 'transparent' [54, 55]. A biometric system is 'transparent' in this sense if 'no explicit action is required from the user during authentication (in contrast to for instance presenting a finger during the crossing of a border)' ([55], p. 202). The tension is plain: the further biometric systems retreat from view, the more difficult it becomes to ensure that adequate consent for data acquisition and processing has been obtained.

Further difficulties in obtaining appropriate consent derive first from the fact that older people can be less aware of new technologies (including assistive technologies) [48] – which suggests that their understanding of the related data processing issues is correspondingly low – and, second, by the alleged fact of a difference between the attitudes of 'digital natives' and 'digital immigrants' to privacy and data protection. Digital natives (those having grown up with digital technologies), it is said, are less concerned with personal data protection (especially online) than digital immigrants (those having come to digital technologies as adults). If this trend is more than just anecdotal, it suggests that older people are more likely than younger generations to be uneasy about consenting to the processing of their data. The existence of a generational divide here is disputed [56, 57]. But, in fact, what is actually at stake in this dispute is the claim that *digital natives are not concerned about privacy*; hence even if there is no generational divide, it is likely still true that older people are reticent when it comes to sharing data. This could be an impediment to equality of access to (or at least of uptake of) biometric technologies.

To this point we have considered age and ageing as factors that challenge standard biometric applications by their consequences for biometric features or templates. Yet age itself is a biometric feature *per se*. Age is usually included among the so-called 'soft biometrics': those features not usually sufficient for unique personal identification, but which can be fruitfully used to support standard biometric features. Suppose one knows that a person who is to be identified is female, and that a given biometric system detects a feature which is equally distributed among men and women (e.g. fingerprint). If the biometric application were able to detect also the gender of individuals, you could exclude males from your search, thereby improving by around 50% the effectiveness of the whole system. Age can be used as a soft biometric in this way.

But what do biometric systems actually measure when they ascertain age? The word 'age' indicates three very different phenomena:

[21] See Directive 95/46/EC, Article 7.

- *Chronological age*, i.e. age measured by the length of time that someone has existed. Chronological age is a purely conventional variable which totally escapes biometric sensors and systems, but which is, however, implied by primary identifiers which usually include date of birth.
- *Biological age*, i.e. the measurable changes that take place in us all as we age. These changes cover a wide variety of measurements. Some are simple (e.g. changes in eyesight or hand-grip strength); others are more complex (e.g. lung capacity, or changes in hormone levels). These 'biomarkers' are the only reliable biological signs of ageing, but they cannot be easily captured and for now they are not used in biometric identification (except in one case discussed below).
- *Social age*, which is a way of grouping people based on particular roles, cultural experience, performances, social networks of belonging, and so forth. All societies possess tacit norms for determining social functions. These age rules are often described as 'social clocks', the culturally determined moments in time for specific social activities to be carried out. Schooling, marriage, employment, childbirth, grandparenthood (and so on) are some of the human activities ruled by social clocks.

Most biometric systems attribute age by using generic biological signs mixed with social clocks. For instance some systems detect hair colour, secondary skin texture properties, gait-speed, crow's-feet, warts, baldness – even clothing – which are all very imprecise signs of age with strong cultural connotations. Today the sole attempt to use a real biomarker in biometrics dates back to a few years ago, when an Israeli company patented a frequency ultrasound scanner which was able to detect not only fingerprints but also bone maturity by scanning finger cartilage. This supposedly allowed it to identify children (approximately under the age of 14).[22]

It is likely that future technology will be able to include more accurate sensors, which can measure hormones, lung capacity, or genetic biomarkers in an unobtrusive manner. Apart from obvious privacy considerations, there is a specific reason for concern here. If biometric systems were able to detect biological age, they could allow a comparison between it and chronological age. In practice one could evaluate the level of 'wear' of an individual's body and from this infer details of lifestyle, medical condition, and major diseases. This information could be misused in various ways.[23]

The factors addressed in this subsection suggest that biometrics faces several issues with respect to older people. Broadly speaking – and remembering that these terms have a wide scope and intersect at many points – these factors may be considered to fall into two main camps: *issues of inclusion and exclusion* and *privacy concerns*.

[22] Cf. http://cyber.law.harvard.edu/sites/cyber.law.harvard.edu/files/Verificage_ISTTFTAB_submission.pdf.
[23] Health insurance companies could be interested in this kind of information; one can also imagine employers determining the biological age of their employees in order to sack those who are older than their chronological age because they are likely to be less fit.

3.3.2 Social and ethical factors: the problem of ageing

Here we discuss a slightly subtle ethical and social factor concerning the terminology and characterisation (or conceptualisation, or 'discourse') in which the problem of ageing is embedded.

A common claim among some advocates is that biometrics are very strong identifiers because they authenticate individuals in terms of who or what they are (in some very fundamental sense of that verb).[24] 'Strong' biometrics are, in that case [58]:

> features that can be considered unique (very unlikely to be found equal in two individuals) and permanent (enduring in time). [. . .] This allows using these qualities as though they were almost tokens. In other words, their presence is considered almost univocally linked with a given subject. (p. 8)

'Weak' biometrics are features considered less than unique or permanent. Typically, features considered strong biometrics include *fingerprint, iris, hand vein*, and *retina*, while weaker biometrics typically include *body shape, odour, gait* (and a host of other behaviours), *voice*, and *electrophysiological signals* (e.g. heart rate, brain activity) [58].

Confidence in the permanence of the so-called strong biometrics is not what it was. As described above (section 3.2.2.1), biometrics such as fingerprint, face, and iris have all been shown vulnerable to problems of template ageing. Human ageing is problematic, therefore, because it produces a class of people – often older people, but more precisely anyone for whom a suitable length of time[25] has passed since the production of their enrolment template – for whom the strong biometrics are less reliable, and so for whom effective identification is more difficult.

The effectiveness of an identification system depends upon two key factors:

- The nature of the connection between the token and the individual associated with that token;
- The likelihood of the token being corrupted or compromised (e.g. stolen, lost, broken and faked).

The nature of the relationship between a token and the individual with whom it is associated may be fruitfully thought of in terms of Charles Sanders Peirce's theory of signs. Peirce is particularly useful here for two of the sets of distinctions he drew. First, Peirce distinguishes within a sign three elements:

- *The signifier*: i.e. that which, conventionally speaking, does the signifying;
- *The object*: i.e. what is signified;
- *The interpretant*: i.e. the understanding of the sign by an interpreter.

Unlike binary distinctions of signifier and object alone, this tripartite division ensures emphasis upon the role of the interpretant of a sign. Second, Peirce [59] famously divides signs into three kinds:

[24] Second generation biometrics ask not who or what, but *how* you are [58].

[25] What constitutes a *suitable* length of time will vary according to the biometric modality in question (as well as upon a host of other factors such as interoperability between sensors).

- *Icon*: a sign whose relation to its object (i.e. the thing it is a sign of) is interpreted in terms of resemblance or likeness in some respect (e.g. the relation of a portrait to its subject);
- *Index*: a sign whose relation to its object is interpreted in terms of brute fact or causal connection (e.g. the relation between the direction of a weather vane and the direction of the wind);
- *Symbol*: a sign whose relation to its object is interpreted via convention or fiat (e.g. the relation between '∨' and disjunction in elementary logic).

Traditional identification systems rely on tokens whose relation to their object is iconic, indexical or symbolic, but it is important to notice that the recognition of a token as a token of a particular individual is always a function of *both* the relation of sign to object *and* interpretant to sign and object.[26]

Typically the identifier will rely on both the nature of the relation of the token to the object – that is to say, on whether the token is an icon, index, or symbol – and the fact and manner of presentation of the token. A traditional paper passport or ID card, to take an example, is primarily iconic. Re-identification by such means relies upon the resemblance of the photograph to the person of whom it was taken. But indexical and symbolic elements enter into the evaluation also. Re-identification by photographic ID relies not only upon resemblance but also upon the causal (indexical) relation between the camera and subject, as well as the various institutionalised mechanisms by which the photograph comes to be an approved icon of the subject. These mechanisms are assumed reliable on the basis of the confidence that resides in the societal practice of taking and approving photographs for passports. In this way, the relation of the photographic ID to the subject of it is symbolic. Hence the traditional paper passport with photograph is primarily iconic, but is also partly indexical and symbolic. Other identificatory tokens will be indexical and symbolic, even if not all are iconic.

Reflection upon the Peircean theory of signs reveals that, as regards ageing, the fault or problem lies not with the individual or their biometric feature, but with the apparatus designed to identify the former by way of the latter. The 'strength' of a biometric is dependent upon the reliability of the link between it and its object (i.e. the subject from whom the feature derives). Biometrics rely, primarily, upon an indexical relation between a feature and its owner. One might, therefore, quite naturally conclude that:

(a) A feature is considered 'strong' if its indexical link to its owner is such that one may reliably infer the presence of the owner from the presence of the feature. And that being so . . .
(b) The problem with ageing is that it renders the inference in (a) less reliable.

It is necessary, however, to recall that in our discussion of Peirce's semiotics we made reference to *two* sets of distinctions. The 'natural conclusions' (a) and (b) use

[26] Peirce's account of the (constitutive) interrelations between the various elements distinguished in his system is complex. We abstract and adapt to need here. For a good introduction to Peirce's theory of signs see Reference 60.

only the distinction between icons, indices and symbols (biometrics being primarily indexical). This overlooks the distinction of the elements within a sign: *signifier, object*, and *interpretant*. Employing these, we may schematise the identification of a subject S in virtue of a biometric feature f by a biometric system B as follows (using t for times and subscripts to distinguish actors at different times):

- t_1 (*enrolment*): enrolment template x created by biometric system B from S_{t1}'s biometric feature f_{t1};
- t_2 (*presentation*): later template y created by B from S_{t2}'s biometric feature f_{t2};
- t_3 (*authentication*): judgement by B that f_{t2} indicates the presence of S_{t1} (on the basis of matching x and y).

Within this schema:

- Signifiers = the biometric features f_{t1} and f_{t2}
- Object = the subject S (i.e. S_{t2} *and* S_{t1}, since in this case, $S_{t2} = S_{t1}$)
- Interpretant = the biometric system B (including technology and any human overseer) responsible for sensing/reading f_{t2}, matching x and y, and judging therefrom the presence of S

Both f_{t1} and f_{t2} are signs of S, and the semantic relations between f_{t1} and S and between f_{t2} and S are primarily indexical, as the 'natural conclusions' (a) and (b) suggest. The natural conclusions omit to mention, however, the role of the interpretant B. B's judgement at t_3 can be considered to follow from roughly this argumentative pattern:

1. S_{t2} is the source of f_{t2} [from observation of S_{t2} at the scanner at t_2];
2. f_{t2} is a close-enough match with f_{t1} [from B's matching x and y];
3. The source of f_{t2} is the source of f_{t1} [from 2];
4. S_{t1} is the source of f_{t1} [from observation of S_{t1} at the scanner at t_1]; *So*...
5. S_{t2} is S_{t1}.

The strength of f_{t2} as a sign of S_{t1} is very much dependent upon the ability of the interpretant B to run through something like this argumentative pattern.

Notice that the strength of f_{t2} as an index of S_{t1} is unaffected by B's abilities to recognise it as such. The indexical relation is simply a matter of the factual or causal link between the two – and that link is entirely unaffected by the ageing process. The fact that a biometric template was created from f_{t2} by way of S_{t1}'s presenting her feature to B's scanner is a timeless historical fact.[27] What does alter is the character of S's feature. Hence if the time difference between t_1 and t_2 is many years, it may well be that lines 2 and 3 of the pattern – the matching of templates – are difficult or impossible for B to accurately judge. *That* is the real problem of template ageing, but it is plainly not a problem of the relation between f_{t2} and S_{t1}: that relation is unchanged. The problem stems entirely from B, from the interpretant.[28]

[27] It also being a timeless historical fact that (at t_1 and t_2) $S_{t1} = S_{t2}$.

[28] In all the above schemas we have omitted to include temporal subscripts for B. Such subscripts could have been used to highlight the changes that may occur as, say, sensors are changed, or more (or less) highly trained overseers are employed (etc.).

This excursus into the semiotics of biometrics shows that it is neither acceptable nor accurate to think of older people as in any sense 'problematic': as we have shown, their biometric features are as strongly linked, indexically, to their identities as are anybody else's. Similarly, the characterisation of older people as having poor quality biometric features – if 'poor quality' is accorded any kind of negative connotation – is equally unacceptable. The language of 'problem' and 'failure' is, at best, unhelpful. Concerning biometric systems' limitations when it comes to dealing with difference, van der Ploeg [61] writes:

> In biometric discourse the set of problems connected with this issue is referred to with a number of concepts, such as, for example, 'usability', 'accessibility', 'failure to enrol', 'exception handling' and 'template ageing'. In the discourses of social theory and politics, these matters invoke considerations in terms of distinctions made between *normal* and *abnormal* or *rule* and *exception*. [...] When a biometric system fails to cope with variations in human features falling outside a certain range, it thereby categorises and excludes, with more or less serious consequences to the people concerned. (pp. 288, 289)

It can be difficult to capture the kind of harm that discourses of 'failure to enrol' or 'exception handling' do. For one thing, to begin to speak of 'discourses' at all can give the impression that these are just academic problems, remote from 'real-life' experience. For another, the harm that this kind of language and conceptualisation can do is harm for which no individual or group of individuals could reasonably be blamed: biometricians, or the biometrics community or industry, are not *guilty* of anything – these conceptualisations are not intentional. What we are adverting to is the fact that what might very reasonably *appear* to be problems brought about by age and ageing – and so quite naturally associable with older people – are in fact wholly technical problems internal to biometric systems.

Failure to enrol is an illustrative example. Consider the statement (p):

(p) Older people pose a problem with respect to increased failure to enrol rates.

In a certain sense, (p) is true: if, to an initial class of 100 subjects, among whom a biometric system suffers an overall failure to enrol rate of r, one adds another 50 elderly subjects, one would expect r to rise. But insofar as such a case would tend to confirm (p), it would presuppose that the original value of r was more or less normal (older people posing a 'problem' because they push the failure to enrol rate *above the norm*). And that presupposition, even if it is reasonable – and we are not convinced it is especially reasonable given that our society is already quite an old one – is clearly grounded in contingent claims about the demographic makeup of the society we happen to be in. A 'normal' failure to enrol rate will depend on the kinds of subjects normally encountering the system. If we lived in a society of average age 80, 'normal' failure to enrol rates would be much higher. So although (p) is true, its truth is very much contingent. There are, *sub specie aeternitatis*, no normal failure to enrol rates. 'Normal' failure to enrol rates in our society ought to be representative of the kinds of people comprising it.[29] If that were the case, we would have something approaching an accurately representative 'normal' failure to enrol rate for our society as a whole. But in such a world, (p) would not be true.

For that reason, in our society, although we might grant that (p) is true *in a certain sense*, we must nevertheless concede that it is misleading. It is true that there exists a problem involving older people and increased failure to enrol rates. What is *not* true – and what is implied by (p) – is that older people are the cause or source of that problem. The source of the problem lies with the technology, which is – to the extent that (p) is in a certain sense true – out of touch with society.

If a technology is out of touch with society – if its deployment outpaces its refinement – there is a huge potential for function creep. For there is a strong likelihood of a 'policy vacuum', a lack of appropriately mature legislation governing the use of the technology. This is a major contributory factor in function creep [51]. Moreover, as technologies become more widely disseminated throughout society, they become embedded, normalised, and unremarkable. Under these conditions, a problem *q* concerning the technology will tend to provoke the first of the following two questions, while the second will be effectively silenced.

(Q1) How can we amend and improve this technology to overcome *q*?
(Q2) Knowing what we know now (and taking into account *q*), does this technology (still) constitute the optimal solution to the original problem for which it was introduced?

Suppressing (Q2) is dangerous because should its answer be negative, (Q1) turns out to be moot, and its consideration a waste of resources better spent on developing solutions to the original problem referred to in (Q2). However, when a technology is firmly embedded in the life of a society – when it is a commonplace of the daily routine – it is difficult to convince people by now accustomed to it that it should be abandoned.[30] The most natural approach to, for example, the problems of age and of ageing will *appear* to be captured in (Q1): i.e. amend and improve the technology. Alternative approaches are (relatively speaking) silenced. In this way, problems tend to be defined and solved within a conceptual framework shaped by the initial *technology*, rather than by the initial *need* for which the technology was introduced. The initial need may even be sidelined or forgotten, and the technology evolve to fulfil roles for which it was not intended. This is function creep of a potentially very dangerous form. It involves no active decision on the part of any identifiable individual or group to alter or subvert the purpose of the technology. The creep has the appearance of occurring naturally

[29] The difference in probability between the likelihood of a given 75-year-old encountering a failure to enrol, and the likelihood of a given 25-year-old encountering a failure to enrol, ought to be proportional to the difference in probability between the likelihood of a person selected at random being of age 75 and the likelihood of a person selected at random being of age 25.

[30] For instance, people having made an initial outlay for a product utilising the technology may be reluctant to write-off their investment.

alongside other more general developments in society. This can give it an air of inevitability.

We are not suggesting that, on the basis of the problem of ageing, biometrics should be abandoned. Indeed we will sketch an answer to (Q1) in section 3.4. But the discussion highlights two points. First that (Q2) is a valid question which should not be suppressed. Second, that it is very important that biometrics remains in touch with the demographic shape of society. Failure to do so can result in the exclusion of older people and in pernicious forms of function creep.

Responding to the problems of age and ageing will involve developing biometric systems which do not exclude older people in the ways that we have discussed in this subsection. That process might begin with some very simple measures. For instance, it would be worthwhile to move away from the practice of saying, more or less routinely:

- *'older people are problematic to biometrics because . . . '*;

and towards a practice of saying something along the following lines:

- *'this technology has trouble authenticating people who . . . '*.

This simple shift, from attributing a failing to a group of people to attributing it to the technology that fails to accommodate them, might seem mere wordplay (or, worse, a sort of political-correctness-gone-mad). It isn't. It gives a more accurate description of the problem, and it in no way stigmatises or alienates older people.

3.4 A sketch of an approach to 'age-blind' biometrics

In the final section of this chapter we conclude by sketching a rough outline of a hypothetical biometric system for border control, which, in our view, would be respectful of the ethical and social factors discussed above.[31,32]

We distinguished four approaches to the problem of equipping biometric systems to deal with the current demographic balance and dynamics:

1. Updating templates;
2. Simulation of the effects of ageing;
3. Use of age-invariant biometric features;
4. Multi-modality.

In sketching a hypothetical system we have steered clear of (1) on the grounds that – unless it can be carried out independently of the cooperation of the subject [27, 28] (which would anyway raise significant data protection concerns, particularly

[31] In what follows we assume that appropriate measures have been taken to ensure respect for privacy and high standards of data protection. There are no easy answers in this regard. The application and enforcement of existing regulations and norms could stand improvement (and still one might question their efficacy and adequacy). Nonetheless, given that such issues are very broad (in that they apply not only to biometrics, but to a host of other technologies, and in that they apply to everybody equally, not only older people), we do not discuss them here.
[32] Aspects of this section are discussed more briefly in Reference 64.

pertaining to consent and the right to rectification of inaccurate data) – it is liable to be inconvenient, and have steered clear of (3) on the grounds that, as far as we are aware, all biometrics without exception are mutable, and there is no guarantee that the least mutable will not be inconvenient in other respects. Our proposal combines (2) and (4). Primarily we propose a system based on multi-modality, but it is an adaptive multi-modality, informed by awareness of the effects of ageing.

We outlined in the previous section the danger of the problem of ageing coming to be seen not simply as a technical failing, but as implying a categorisation of older people as exceptional and problematic. At root, that problem stemmed from an erroneous conception of the 'normal' ('optimal') user of biometric systems. Against this standard, older people and other outliers appear as unusual and problematic. We therefore propose a system that dispenses entirely with the idea of a normal or optimal user. With respect to age factors, this requires treating no particular age group as any more or less remarkable than any other.

Careful selection and combination of modalities ought to provide an optimally inclusive system. If we know that the current state of the art in sensor technology for a certain modality is such that there is a greater incidence of some problem (e.g. failure to enrol) among older people, then the best solution would be to include modalities considered more reliable with older people. In our imagined system, crossing a border might involve negotiating a 'walkthrough' which includes sensors for fingerprint, iris, face and gait.[33] The subject would approach the walkthrough, and scan their biometric passport. The passport reader would ascertain the subject's age and the amount of time since the templates recorded on the passport were created. Based on the subject's age and on the latest expertise with respect to feature mutability and template ageing, the walkthrough would automatically calibrate itself in order to determine the appropriate matching scores of each modality, and their relative weightings.

This would, then, be a form of 'agent-based' biometrics, as Deravi [42] describes:

> With an agent-based biometric system, the dynamic selection of a variable set of biometric modalities can be accommodated to match the demands of a particular task, or the availability of particular sensors. For example, a multi-modal system should be able to deal with situations where a user may be unwilling or simply unable to provide a certain biometric sample, or where a preferred biometric modality cannot support a required degree of accuracy. The deployment of a multi-modal approach, where it is possible to choose from a menu of available modalities and modes of interaction, can therefore help to overcome barriers to access. (p. 184)

A decision would be taken by the time the passenger reaches the end of the walkthrough: all being well, they are cleared to cross the border. It could be that the score from the subject's fingerprint would not have been accepted by a normal mono-modal fingerprint system. But because the agent-based system knows the subject's age, and the 'age' of their enrolment template, it can reflect that, say,

[33] Or whichever modalities are deemed best by those in the know.

fingerprint is generally reliable only to a certain extent for people of *that* age with an enrolment template of *that* age. It is able to give the score from fingerprint matching less weight than, say, the score from the iris matching.[34]

We trust that such a system would constitute a robust and reliable form of identification. (We provide here only the briefest sketch. More and more reliable modalities could be added as experts see fit.) More importantly for present purposes, it would also be a system which avoided the main social and ethical issues that we have outlined in this chapter. By linking the calibration of the system to the age of the particular subject, no age or age group becomes exceptional. *There would be no 'normal' calibration of the system because calibration would be individually tailored.* In effect, the fact of ageing, and that change across time is normal, would be written into the design of the system.

This system would of course still face challenges. On a practical level, it could be that such a system is difficult to engineer or prohibitively expensive. Moreover, it is not obvious to us how our approach, of actively eschewing the concept of a 'normal' or 'standard' user, could be realised in use cases other than border control.[35] These are technical difficulties for others to address. Our aim has been to try to illustrate how a system could be inclusive of all age groups. Even if the system we have sketched is technically difficult – even if it is impossible – the principle motivating it is clear: there are no standard users.

Ethical and social challenges would remain also. For instance, as mentioned in footnote 31 above, we have made no mention of privacy or data protection measures. Equally as important, it would be important to judge, on a case-by-case basis, whether the implementation of such a system was a proportionate response to the problem for which it is introduced. (That is to say, serious consideration should be given to question (Q2) above.) The main potential problem with multi-modal biometric systems is that they gather more data from individuals than mono-modal systems. Proportionality is important. One of the main principles ruling data protection is the 'data minimisation' principle, as outlined in Reference 62:

> Data minimization means that first of all, the possibility to collect personal data about others should be minimized. Next within the remaining possibilities, collecting personal data should be minimized. Finally, the time how long collected personal data is stored should be minimized. Data minimization is the only generic strategy to enable anonymity, since all correct personal data help to identify [. . .]. Furthermore, data minimization is the only generic strategy to enable unlinkability, since all correct personal data provide some linkability [. . .].

[34] In principle, it would be possible also to include other information on the passport in order to allow for more precise calibration. For example, a subject having a medical condition, or having undergone treatment which affects certain biometric features, could have this data logged in their passport. Such measures would of course have to comply with the highest standards of data protection and informed consent.

[35] One possibility would be to employ one of the techniques discussed in Reference 1, namely to use a facial recognition system which is able to estimate the subject's age, and then to calibrate itself accordingly.

Biometric multi-modality should be justified not only on the basis of its effectiveness but also in terms of personal data minimisation. This does not prevent using multi-modality, but it demands that biometric templates must be protected by using pseudonymous identifiers or similar strategies, which imply that only a transformed version of the original biometric is stored [63].

These ethical and societal challenges are challenges for all biometric systems, and many other non-biometric systems besides. Our intention has been only to sketch a system which overcomes the ethical and social factors described in this chapter. This we have done by imagining a system which, by steadfastly refusing to consider any age or age group as 'normal', 'standard' or 'optimal', ensures that no age or age group is problematised or excluded.

Acknowledgements

This work has been partly funded by two grants from the European Commission within the scope of the FP7: TABULA RASA – Trusted Biometrics under Spoofing Attack (Grant Agreement no. 257289), and VALUE AGEING – Incorporating European Fundamental Values into ICT for Ageing: A Vital Political, Ethical, Technological, and Industrial Challenge (Grant Agreement no. 251686).

References

[1] Lanitis, A.: 'A survey of the effects of aging on biometric identity verification', *International Journal of Biometrics*, **2**(1), 2010, pp. 34–52.

[2] Carone, G. and Costello, D.: 'Can Europe afford to grow old?', *Finance and Development*, **43**(3), 2006, http://www.imf.org/external/pubs/ft/fandd/2006/09/carone.htm, accessed May 2012.

[3] UNPD: 'World population to 2300' (United Nations Department of Economic and Social Affairs, Population Division, 2004), http://www.un.org/esa/population/publications/longrange2/WorldPop2300final.pdf, accessed May 2012.

[4] Mordini, E. and Rebera, A.P.: 'Conceptualizing the biometric body', in Bus, J. *et al.* (eds), *Digital Enlightenment Yearbook 2012* (IOS Press, Amsterdam, 2012), pp. 265–273.

[5] Mordini, E. and Tzovaras, D. (eds): *Second Generation Biometrics: The Ethical, Legal and Social Context* (Springer, Dordrecht, 2012).

[6] Fakourfar, H. and Belongie, S.: 'Fingerprint recognition system performance in the Maritime Environment', *Workshop on Applications of Computer Vision (WACV)* (Snowbird, 2009).

[7] Maltoni, D., Maio, D., Jain, A.K. and Prabhakar, S.: *Handbook of Fingerprint Recognition* (Springer, 2009).

[8] Modi, S.K. and Elliot, S.J.: 'Impact of image quality on performance: Comparison of young and elderly fingerprints', *Proceedings of the 6th International Conference on Recent Advances in Soft Computing (RASC)*, 2006, pp. 449–454.

[9] Grant, H. and Domokos, J.: 'Dublin regulation leaves asylum seekers with their fingers burnt', *The Guardian* (October 7, 2011), http://www.guardian.co.uk/world/2011/oct/07/dublin-regulation-european-asylum-seekers, accessed July 2012.

[10] Harmon, K.: 'Can you lose your fingerprints?', *Scientific American* (May 29, 2009), http://www.scientificamerican.com/article.cfm?id=lose-your-finger prints, accessed July 2012.

[11] Bolle, R.M., Connell, J.H., Pankanti, S., Ratha, N.K. and Senior, A.W.: *Guide to Biometrics* (Springer, New York, 2004).

[12] Kent, J.: 'Malaysia car thieves steal finger', *BBC News* (March 31, 2005), http://news.bbc.co.uk/2/hi/asia-pacific/4396831.stm, accessed July 2012.

[13] Van der Putte, T. and Keuning, J.: 'Biometrical fingerprint recognition: Don't get your fingers burned', *Proceedings of the Fourth Working Conference on Smart Card Research and Advanced Applications* (Kluwer, Norwell, 2000), pp. 289–303.

[14] Pan, G., Wu, Z. and Sun, L.: 'Liveness detection for face recognition', in Delac, K., Grgic, M. and Bartlett, M.S. (eds), *Recent Advances in Face Recognition* (IN-TECH, Vienna, 2008).

[15] He, Z., Sun, Z., Tan, T. and Wei, Z.: 'Efficient iris spoof detection via boosted local binary patterns', in Tistarelli, M. and Nixon, M. (eds), *Advances in Biometrics, vol. 5558 of Lecture Notes in Computer Science*, pp. 1080–1090 (Springer, Berlin, 2009).

[16] Machado, A.F. and Queirozm M.: 'Voice conversion: A critical survey', *Proceedings of Sound and Music Computing (SMC)* (Barcelona, Spain, 2010).

[17] Qin, B., Pan, J-F., Cao, G-Z. and Du, G-G.: 'The anti-spoofing study of vein identification system', *Proceedings of the International Conference on Computational Intelligence and Security* (Beijing, China, 2009), pp. 357–360.

[18] Ashton, H., Rebera, A.P. and Mordini, E.: 'Environmental scanning report', Tabula Rasa Deliverable D7.1, 2011 (https://www.tabularasa-euproject.org/).

[19] Rebera, A.P. and Mordini, E.: 'Interviews report', Tabula Rasa Deliverable D7.2, 2012 (https://www.tabularasa-euproject.org/).

[20] Modi, S.K., Elliot, S.J., Whetstone, J. and Hakil, K.: 'Impact of age groups on fingerprint recognition performance', *IEEE Workshop on Automatic Identification Advanced Technologies* (Alghero, Italy, 2007), pp. 19–23.

[21] Albert, M., Sethuram, A. and Ricanek, K.: 'Implications of adult facial aging on biometrics', in Albert, M. (ed.), *Biometrics—Unique and Diverse Applications in Nature, Science, and Technology* (Intech, Rijeka, 2011), pp. 89–106.

[22] Fenker, S.P. and Bowyer, K.W.: 'Analysis of template aging in iris biometrics', *CVPR Biometrics Workshop* (2012), http://nd.edu/~kwb/Baker EFAI_2012.pdf, accessed May 2012.

[23] Baker, S., Bowyer, K.W. and Flynn, P.J.: 'Empirical evidence for correct iris match score degradation with increased time-lapse between gallery and probe matches', *Proceedings of the International Conference on Biometrics* (Alghero, Italy, 2009), pp. 1170–1179.
[24] Fenker, S.P. and Bowyer, K.W.: 'Experimental evidence of a template aging effect in iris biometrics', *IEEE Computer Society Workshop on Applications of Computer Vision* (Kona, Hawaii, 2011).
[25] Lanitis, A.: 'Facial biometric templates and aging: Problems and challenges for artificial intelligence', *AIAI-2009 Workshop Proceedings*, 475 (Thessaloniki, Greece, 2009), pp. 142–149.
[26] Lanitis, A.: 'Comparative evaluation of automatic age-progression methodologies', *EURASIP Journal on Advances in Signal Processing* (2008), pp. 1–10.
[27] Poh, N., Wong, R., Kittler, J. and Roli, F.: 'Challenges and research directions for adaptive biometric recognition systems', *Lecture Notes in Computer Science*, 5558/2009 (2009), pp. 753–764.
[28] Rattani, A., Freni, B., Marcialis, G.L. and Roli, F.: 'Template update methods in adaptive biometric systems: A critical review', *Lecture Notes in Computer Science*, 5558/2009 (2009), pp. 847–856.
[29] UNFPA: 'State of world population 2011: People and possibilities in a world of 7 billion' (United Nations Population Fund, 2011), http://foweb.unfpa.org/SWP2011/reports/EN-SWOP2011-FINAL.pdf, accessed May 2012.
[30] The Royal Society: *People and the Planet* (The Royal Society, London, 2012).
[31] Muzzi, M.: 'UNICEF good practices in integrating birth registration into health systems (2000–2009)', (UNICEF, 2010), http://www.unicef.org/protection/Birth_Registration_Working_Paper(2).pdf, accessed May 2012.
[32] Mordini, E. and Rebera, A.P.: 'No identification without representation: Constraints on the use of biometric identification systems', *Review of Policy Research*, **29**(1), (2012), pp. 5–20.
[33] Gelb, A. and Decker, C.: 'Cash at Your fingertips: Biometric technology for transfers in developing countries', *Review of Policy Research*, **29**(1), (2012), pp. 91–117.
[34] UNPD: 'World population ageing: 1950–2050' (United Nations Department of Economic and Social Affairs, Population Division, 2002), http://www.un.org/esa/population/publications/worldageing19502050/index.htm, accessed May 2012.
[35] European Commission: 'The demographic future of Europe – From challenge to opportunity' (Office for Official Publications of the European Communities, Luxembourg, 2006).
[36] The Economist: 'China's Achilles heel', *The Economist* (April 21, 2012), http://www.economist.com/node/21553056, accessed July 2012.
[37] Jain, A.K. and Kumar, A.: 'Biometric recognition: An overview', in Mordini, E. and Tzovaras, D. (eds), *Second Generation Biometrics: The Ethical, Legal and Social Context* (Springer, Dordrecht, 2012), pp. 49–79.
[38] Mordini, E. and Ashton, H.: 'The transparent body: Medical information, physical privacy and respect for body integrity', in Mordini, E. and

Tzovaras, D. (eds), *Second Generation Biometrics: The Ethical, Legal and Social Context* (Springer, Dordrecht, 2012), pp. 257–283.

[39] European Commission: 'To be part of the information society', Communication from the Commission to the European Parliament, the Council, the European Economic and Social Committee and the Committee of the Regions COM (2007) 694 Final, Brussels, 8 November 2007.

[40] European Parliament (2000). Charter of fundamental rights of the European Union: Solemn Proclamation (2000/C364/01), pp. 1-22, http://www.europarl.europa.eu/charter/pdf/text_en.pdf.

[41] Drousou, A. and Tzovaras, D.: 'Activity and event related biometrics', in Mordini, E. and Tzovaras, D. (eds), *Second Generation Biometrics: The Ethical, Legal and Social Context* (Springer, Dordrecht, 2012), pp. 129–148.

[42] Deravi, F.: 'Intelligent biometrics', in Mordini, E. and Tzovaras, D. (eds), *Second Generation Biometrics: The Ethical, Legal and Social Context* (Springer, Dordrecht, 2012), pp. 177–191.

[43] Matheson, D.: 'Unknowableness and informational privacy', *Journal of Philosophical Research*, 32, (2007), pp. 251–267.

[44] Westin, A.: *Privacy and Freedom* (Atheneum, New York, 1967).

[45] Gavison, R.: 'Privacy and the limits of law', *Yale Law Journal*, 77, pp. 475–493.

[46] Mordini, E.: 'Conclusions', in Mordini, E. and Tzovaras, D. (eds), *Second Generation Biometrics: The Ethical, Legal and Social Context* (Springer, Dordrecht, 2012), pp. 329–334.

[47] Cabrera Giráldez, M. and Rodríguez Casal, C.: 'The role of ambient intelligence in the social integration of the elderly', in Riva, G., Vatalaro, F., Davide, F. and Alcañiz, M. (eds), *Ambient Intelligence* (IOS Press, Amsterdam, 2005), pp. 267–282.

[48] Wadhwa, K. and Wright, D.: 'A survey of technology for the elderly', in Mordini, E. and de Hert, P. (eds), *Ageing and Invisibility* (IOS Press, 2010), pp. 143–172.

[49] Mordini, E.: 'Ethical recommendations', in Mordini, E. and de Hert, P. (eds), *Ageing and Invisibility* (IOS Press, Amsterdam, 2010), pp. 195–218.

[50] De Hert, P. and Mantovani, E.: 'The EU legal framework for the e-inclusion of older persons', in Mordini, E. and de Hert, P. (eds), *Ageing and Invisibility* (IOS Press, Amsterdam, 2010), pp. 83–120.

[51] Mordini, E. and Massari, S.: 'Body, biometrics and identity', *Bioethics*, 22 (2008), pp. 488–498.

[52] Allen, K.: 'Ageism is back as unemployed over-50s struggle to get back into work', *The Guardian* (April 15, 2012), http://www.guardian.co.uk/society/2012/apr/15/unemployed-older-workers-struggle-to-find-work, accessed July 2012.

[53] Woods, D.: 'Age discrimination claims increase by 79%, tribunal service statistics show', *HR Magazine* (July 1, 2011), http://www.hrmagazine.co.uk/hro/news/1019716/age-discrimination-claims-increase-79-tribunal-service-statistics, accessed July 2012.

[54] Tangelder, J. and Schouten, B.: 'Transparent face recognition in an unconstrained environment using a sparse representation from multiple still images', *Proceedings of the ACSI Conference* (Toronto, Ontario, 2006).

[55] Schouten, B.A.M., Salah, A.A. and van Kranenburg, R.: 'Behavioural biometrics and human identity', in Mordini, E. and Tzovaras, D. (eds), *Second Generation Biometrics: The Ethical, Legal and Social Context* (Springer, Dordrecht, 2012), pp. 195–214.

[56] Hoofnagle, C.J., King, J., Li, S. and Turow, J.: 'How different are young adults from older adults when it comes to information privacy attitudes and policies?', *Social Science Research Network* (2010), http://ssrn.com/abstract=1589864 or http://dx.doi.org/10.2139/ssrn.1589864, accessed July 2012.

[57] Clarke, R.: 'The privacy attitudes of the iGeneration', http://www.rogerclarke.com/DV/MillGen.html, accessed July 2012.

[58] Mordini, E., Tzovaras, D. and Ashton, H.: 'Introduction', in Mordini, E. and Tzovaras, D. (eds), *Second Generation Biometrics: The Ethical, Legal and Social Context* (Springer, Dordrecht, 2012), pp. 1–19.

[59] Peirce, C.S.: 'On a new list of categories', in Houser, N. and Klousel, C. (eds), *The Essential Peirce, Volume 1* (Indiana University Press, Bloomington, 1992), pp. 1–10.

[60] Atkin, A.: 'Peirce's theory of signs', in *Stanford Encyclopedia of Philosophy* (2010), http://plato.stanford.edu/entries/peirce-semiotics/, accessed June 2012.

[61] Van der Ploeg, I.: 'Security in the danger zone: Normative issues of next generation biometrics', in Mordini, E. and Tzovaras, D. (eds), *Second Generation Biometrics: The Ethical, Legal and Social Context* (Springer, Dordrecht, 2012), pp. 287–303.

[62] Pfitzmann, A., Hansen, M. and Tschofenig, H.: 'Terminology for talking about privacy by data minimization: Anonymity, unlinkability, undetectability, unobservability, pseudonymity, and identity management', Internet Engineering Task Force (2007), http://tools.ietf.org/id/draft-hansen-privacy-terminology-00.html, accessed July 2012.

[63] Yang, B., Busch, C., Gafurow, D. and Bours, P.: 'Research findings for standardisation', TURBINE deliverable D2.3.3, http://www.turbine-project.eu/dowloads/TURBINE-GUC-D2.3.3-Standardisation.R1.pdf, accessed July 2012.

[64] Rebera, A.P. and Guihen, B.: 'Biometrics for an ageing society: Societal and ethical factors in biometrics and ageing', in Brömme, A. and Busch, C. (eds), *BIOSIG 2012: Proc. 11th International Conference of the Biometrics Special Interest Group* (Gesellschaft für Informatik, Bonn, 2012), pp. 409–416.

Chapter 4
Using age to enhance performance in biometrics

Márjory Da Costa-Abreu[1] *and Michael Fairhurst*[2]

4.1 Introduction

Security systems which provide a high accuracy in identifying individuals, or in confirming their claimed identity, are becoming more important than ever in modern society. New techniques that can improve the performance of systems for the identification of individuals are very important to on-going efforts to protect identity, as opportunities for fraudulent claims and 'identity theft' increase [1]. A reliable identification process is therefore an essential part of a security system, and biometric information has been widely used for this purpose, often with considerable success, in criminal identification, workplace monitoring, access control, etc. [2].

Among the many issues of concern to a system designer, one of the most crucial in an environment that requires high security is that of error rate, and especially the balance between false positive and false negative rates [3]. This has implications for choice of which modality to use. It is possible to select from a considerable variety of modalities and there are many ways of using these, either individually or combined. In many applications it is also important to offer flexible user options in cases where either the user is not able to give a required sample or social or convenience factors preclude the use of a specific modality in practice. The most common solution in such circumstances is to create a more versatile biometric-based system which can allow the deployment of more than one modality (the multimodal approach), simultaneously offering the user the possibility of a degree of choice and decreasing the chances of the system delivering an incorrect identity prediction [4].

Although biometrics have been in practical use for some considerable time, the often complex analysis required to choose optimal modalities for an application in the face of conflicting demands of a multimodal system has been one of the major issues with which a system designer can be faced. The basic requirements of a unimodal system also become significantly magnified when considered in comparison with a multimodal context [5].

[1] DIMAp/UFRN, Brazil
[2] University of Kent, UK

The choice of each parameter of a practical biometric-based system will inevitably bear a close relation with a particular set of application requirements. However, there is no agreement on which classification/matching method works best with any specific modality or which modality represents better a specific problem solution opportunity [2].

Therefore, the availability of multiple information sources for biometric data processing can suggest various different strategies by means of which to achieve enhanced performance. These include, for example, selecting an optimal processing technique from among many options (it is possible, for example, that a multiclassifier system will produce better results than a single classifier), combining processors to create a multiple processor system to work on a unimodal source (which can combine the benefits of more than one source of information to optimise outcome) and, ultimately, combining multiple biometric modalities to overcome the shortcomings of any one individual modality. In each case, however, there are obvious questions to be asked about the processing engines implemented, and the performance of which they are inherently capable [6, 7].

4.2 Fusion of relevant information

The use of multiple categories of information in biometric-based applications is not new, notwithstanding the fact that this topic is perhaps relatively unrepresented in the literature compared with other structural approaches. The term 'multicategory' denotes, as the term suggests, the use of more than one category of information. Here, we use the term to imply using both biometric data and also some other information sources, usually that typically designated 'soft-biometrics'.

Essentially, any information which is not unique to an individual but, nevertheless, is specifically characteristic of the user can be used as a soft-biometric. However, in the same way as with conventional biometric information, soft-biometrics need to be measurable and categorisable. Hence, data about characteristics such as age [8, 9], gender [8, 9], handedness [10], iris colour [11], height [8, 9], percentage of body fat [12] and so on are typical examples of this type of information source.

Using such soft-biometric information can be a powerful tool for the designer of a biometric system. It is often easy (and usually relatively inexpensive) to acquire (it does not need another sensor for its capture), it is not invasive (the user often provides some of the most used soft-biometrics readily and often routinely), it is usually inclusive (everybody has this information) and, as will be demonstrated here, it is relatively simple to incorporate productively into the identification process.

Also it is important to emphasise that, while the primary objective of biometric processing is to establish individual identity, it is increasingly recognised that there is a close link between identification of individuals and situations commonly targeted by biometric systems where the prediction of important characteristics from that individual is also necessary [13–15].

Part of the analysis of data might well include the need to estimate a more general characteristic (such as age or gender, for example) of the 'owner' of the specific piece of information under consideration. Predicting an individual's characteristics has wider application, but the investigation reported in this chapter will be focused on the relationship between this predicted information and the prediction of the user's identity.

Thus, this chapter will discuss both the impact of the use of age information on overall performance, complexity and usability in the design of biometric-based systems as well as the predictive capabilities of biometric systems in relation to age.

Furthermore, because of the nature of the data in this specific category (population data characteristics) and as a consequence of a lower level analysis, it is possible to point to more specific strategies both for potentially improving error-rate performance achievable in a biometric system while also increasing the flexibility with which specific implementation configurations can be designed, and for considering a range of solutions of different levels of complexity.

4.3 A brief survey of the literature

The use of age information to improve identity verification has not been as widely investigated as most other options for improving accuracy and reliability in biometric processing. Nevertheless, it is possible to find some interesting relevant work in the literature.

The different approaches which have been considered are generally not exclusive to the use of age information, but are often generic methods applicable to the exploitation of any appropriate soft-biometric such as age, gender, handedness, height and so on. Furthermore, a considerable body of work is concerned not with using this information to enhance identification of individuals, but to determine this information in the first place. The research related with these methods can be divided into three different broad categories:

- Simple integration of the additional soft-biometric information with conventional 'hard' biometrics in the identification process:
  - iris and fingerprint fused with iris colour [11];
  - fingerprint fused with gender, ethnicity and height [8, 9, 16];
  - fingerprint fused with body weight and body fat [12].

  In studies in this area, the soft-biometric information is generally used in order to decrease the search space, limiting the identity search to users who belong to that specific category.
- Estimation of the soft-biometric information:
  - gender estimation from face [17, 18];
  - age estimation from face [19–25];
  - age and gender estimation from voice [26];
  - gender and handedness estimation from handwritten signature [27];
  - gender estimation from iris [28];
  - age estimation from voice [29, 30].

  In this approach, therefore, the conventional biometric data is used to predict the soft-biometric information of interest rather than the identity of the user. This supports some very important applications, for example, in forensic investigations.

- The implications for the biometric system of the age of the template:
  - fingerprint [31–34];
  - fingerprint, hand geometry and palmprint [35];
  - face [36–40];
  - handwritten signature [41].

  More specifically, studies in this area deal with the huge issue of what are the effects of the ageing of the biometric templates. Problems such as defining how often and what is the necessary time interval within which to update biometric templates are some of the important issues typically addressed by this type of work.

In References 8 and 9, the authors investigate characteristics such as gender, ethnicity, and height, which are extracted automatically to provide some information about the user in a fingerprint verification task. This information is incorporated into the decision-making process of the primary biometric system, using the probability related to each specific soft-biometric. The database adopted contains 160 users, each providing 4 samples of their left index finger, left middle finger, right index finger and right middle finger, where minutiae features are extracted for the biometric-based processing.

In Reference 16, a hybrid verification biometric system which uses face and fingerprint as the primary characteristics and gender, ethnicity, and height as the soft characteristics is described. The fingerprint database used contains 160 users, each providing 4 samples from their left index finger, left middle finger, right index finger and right middle finger, where minutiae features are again extracted. The face database contains images of 263 users, with 10 images per user where linear discriminant analysis (LDA) features were extracted (score vector of length 8). The experiments use an ethnicity classifier (Asian or non-Asian), a gender classifier and the height information already available. As in Reference 8, this information is incorporated into the decision-making process of the primary biometric system and uses the probability related to that specific soft-biometric characteristic. The results show that neither ethnicity nor gender information improves the performance of the systems.

In Reference 11, a framework for integrating the colour of a human iris within a multimodal biometric system is described, combining fingerprint and iris in a verification task. Steerable pyramid filters and multi-channel log-Gabor filters are used for extracting features of the fingerprints and iris respectively. Weighted averaging and a Parzen classifier are used for fusion of these features. The DSP-AAST Iris database and FVC fingerprint database are used.

In Reference 12, a fingerprint verification system which also makes use of body weight measurements is presented. Weighted sum of scores, support vector machines (SVM), multilayer perceptron (MLP) as well as the simple logical OR and AND operations are the methods employed in data fusion. A database containing fingerprints (specifically, extracted minutiae), body weight and fat percentage data for 62 individuals was collected for this study.

The literature reports many studies related to the prediction of gender and age using modalities such as the face [18, 38], voice [26] and iris [28]. Nevertheless, none of the work in the literature uses this predicted information in the identification process. However, all these approaches to processing which involve information about factors such as subject age show that just by using very simple information sources, an increase in performance can be achieved.

The age of an individual is clearly particularly important in some applications, for example, in scenarios involving entitlement determination, some security transactions, and in applications with legal implications. Thus, this type of information is often readily available without imposing further requirements on a system user, and therefore generates a very limited impact on the overall design of the system, the planning of the enrolment strategies, and so on. The relatively low uptake of this type of system to date is thus perhaps a little surprising.

4.4 Representation of age information

Data representation in classification scenarios and exactly how a system is configured to exploit available data are very important considerations, which can have a major impact on eventual performance. In the present context, there are three ways of integrating age information, as follows:

- the use of an absolute age parameter;
- the use of groups of users in different age bands;
- the use of fuzzy sets in different age bands.

The first option is to use directly the specific number which represents the (biological) age of the user. This is generally not a precise and absolute measurement, but rather an approximation, because age is a continuous variable, which clearly continuously changes at a rate dependent on the 'resolution' adopted. The second approach divides the population into age bands creating a broader categorical feature which indicates to which group each user belongs. The number of groups (which can be designated in many different ways: for example, child, adult, elderly, <25, 25–60, >60, and so on) is still very empirical, as there has not been an extensive and comprehensive study of the effects of age in biometrics and, therefore, we have a relatively incomplete knowledge of exactly how age affects each different modality or, more specifically, how optimally we can use in a biometric system the information we do have. For example, although using broad age groupings avoids the misleading approach of regarding each specific age as a separate and meaningful factor, it still relies on a 'sharp' division between categories, specifically at the group boundaries.

As physical age is not such a 'sharp' characteristic (i.e. age is a continuous variable) and the way people age is clearly relatively different in each age group and certainly across individuals, in order to retain this idea of the continuous nature of this information source, the third approach uses fuzzy groupings to represent age, as illustrated in Figure 4.1.

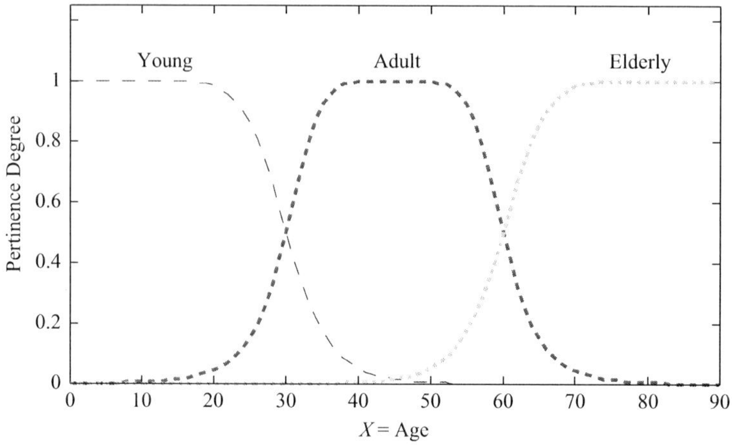

Figure 4.1 Fuzzyfication of subject age

Figure 4.1 shows the fuzzy representation of age as a continuous feature. In this approach, degrees of pertinence will be given to all the different age groups to each user. As an example, if the user is 25 years old, the values relating to each of 3 possible fuzzy sets might be Young = 0.2 (20%), Adult = 0.8 (80%) and Elderly = 0.0 (0%).

4.5 System design using age information

It appears, therefore, that age has the potential to be a very versatile source of information in a variety of scenarios. In general, there are two ways in which this information can be obtained:

- working with the information given by the system user;
- predicting the information from the biometric data extracted from the user.

In the first approach, the user gives the required information at the same time at which biometric data are collected. The problem with this approach is that the user might provide incorrect information, while in some situations the data may be examined in the absence of the user (an example of such a situation might be a forensic investigation, for instance), which can make additional information from the user impossible to obtain.

The second approach, where the system predicts the age from the biometric sample, is less directly dependent on the user. This is a very convenient opportunity to maximise the information extracted from the user without any additional overhead at the interaction level.

We will therefore consider the practicalities of the most common ways of using age information in order to enhance identity prediction.

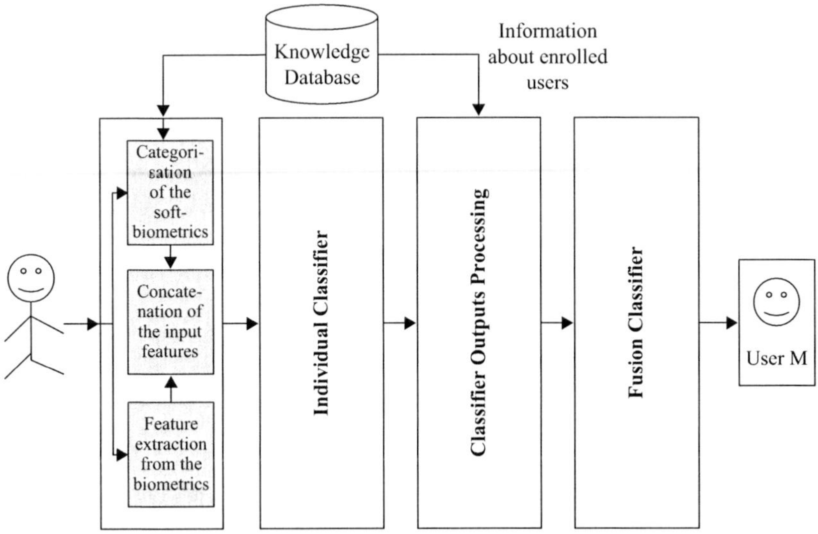

Figure 4.2 Age as input features

4.5.1 Age as an extra input feature

In this (perhaps most common) approach, the soft-biometric information (specifically, in the case of most interest here, age) functions, in essence, as an extra input feature. The soft-biometric information is used in the same way as any other biometric feature and is simply added to the relevant input vector. Figure 4.2 shows a schematic illustration of this process.

The modules of the system may be described as follows:

- Inside the first module shown in Figure 4.2 (corresponding to the *Input Data Processing*):
  - *Categorisation of the soft-biometrics*: This module, as its name suggests, performs the categorisation of age.
  - *Feature extraction from the biometrics*: This module is responsible for extracting the features from the real (hard) biometric sample(s) collected from the user through the appropriate sensor.
  - *Concatenation of input features*: This module will organise all the input features (hard biometrics and the age information) in a vector which will be passed to the next module of the system.
- In the *Individual Classifiers* module, the feature vector will be used as input to the classifiers and these classifiers will produce their outputs.
- The *Classifier Outputs Processing* module will organise all the outputs from the classifiers and will generate a vector which will be sent to the *Fusion Classifier* module.
- The final output of the system will be produced by the *Fusion Classifier* module (contains one classifier which will decide according to the base classifier decisions what is the best identity prediction for the input sample).

Thus, the individual classifiers will be trained with all the biometric input features as well as the categorised age information. The flow of information during the test phase in the system can be explained as follows:

1. The input sample F is extracted from the user in the *Feature extraction from the biometrics* in the *Input Data Processing* module, where

   $$F = [f_1, f_2, f_3, \ldots, f_n]$$

   where n is the number of features from the modality.
2. F is sent to *Concatenation of input features* module.
3. Also, the age information is collected and directed to the *Categorisation of the soft-biometrics* also in the *Input Data Processing* module.
4. The categorisation is performed and this information is sent to the *Concatenation of the input features* module as

   $$F_{age} = [c_1, \ldots, c_{cat}]$$

   where *cat* is the number of categories used in the system (depending on how many age band are used and how the age is represented).
5. F_{age} is also sent to *Concatenation of input features* module.
6. The *Concatenation of the input features* module simply organises the features which will be fed to the *Individual Classifiers* module. This vector will be formed by

   $$F_{withAge} = [f_1, f_2, f_3, \ldots, f_n, c_1, \ldots, c_{cat}]$$

   where n is the number of features from the modality and the remaining features are the ones corresponding to the age information.
7. *Individual Classifiers* module will contain only classifiers which will label the input sample as one of the enrolled users, producing the following confidence degrees:

   $$CD_c = [cd_{1,1}, cd_{1,2}, \ldots, cd_{1,y}, \ldots, cd_{c,1}, \ldots, c_{c,y}]$$

   where y is the number of enrolled users and c represents the number of the classifier.
8. All the confidence degree values are sent to *Classifier Outputs Processing* module which will then organise this information and send the relevant data to the fusion technique used in the *Fusion Classifier* module.
9. The *Fusion Classifier* module will then produce a final output of the system which will label the input sample as one of the enrolled users.

It is evident then that this is a very simple technique, which just includes any additional information (in this case, the age information) in order to improve the system performance. It does not have any restrictions as it can be used either in unimodal or in multimodal systems. Also, its complexity is very low when compared with techniques where the focus is on fusion techniques or classifier algorithms.

4.5.2 Age as a tool for feature selection

In this approach, the age information works as a feature selector, the selection being related to the demographic information which is saved in the *Knowledge Database* module of the system. This technique differs slightly from the one presented in section 4.5.1 in the training phase of the individual classifiers and one of the modules of the *Input Data Processing*. Figure 4.3 shows how the process is realised.

During the training phase, it is important to understand the relationship between the features and the age information, from which the system is able to choose the most suitable features for each user. This is one element of the information which will be stored in the *Knowledge Database* module and will be used in the *Select the feature using soft-biometrics* (in our case age) module. The feature analysis is carried out, after training the system with all the biometric features, as follows:

1. Categorise the age information which is going to be used.

   $$Age = [c_1, \ldots, c_{cat}]$$

   where *cat* is the number of age categories. For example, in our case, the vector will appear as follows:

   $$Age = [c\_young;\ c\_adult;\ c\_elderly]$$

2. From the validation set, group the samples into the defined categories.
3. Using the groups from the validation set, test the system and save the error rates for each classifier. The confidence vector for each classifier will be as follows:

   $$CD - validation = \left[cdVal_{c_1}, \ldots, cdVal_{c_{cat}}\right]$$

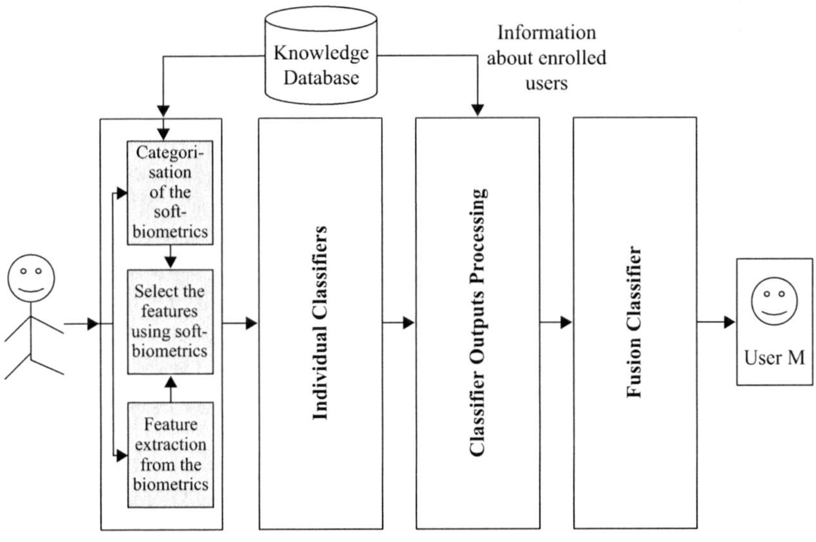

Figure 4.3 Age as feature selector

where *c* are the age categories and *cdVal* are the error rates of this classifier for the samples from each group category.

4. For each age group:
   (a) Select all the samples from users in this category of age information.
   (b) For each biometric feature, test all the validation set samples ignoring this one feature. The value of this evaluated feature will not be considered by the classifiers.
   (c) Save the error rates to each feature related to each age, which should be as follows:

$$CD - validation_f = \left[cdVal_{f,c_1}, \ldots, cdVal_{f,c_{cat}}\right]$$

   where *c* are the age categories and *cdVal* are the error rates of this classifier with respect to the samples from each group category making the feature *f* a null feature.

5. Once the system is tested with all different feature combinations and the error rates saved, for each feature *f* in each age category *c* the following process is executed:
   (a) Calculate the difference between the confidence when all the features are used and when feature *f* is null:

$$diff_f = cdVal_{f,c_{cat}} - cdVal_{c_{cat}}$$

   (b) If $diff_f < threshold$ (which is defined by the system designer), then this feature is not representative for this group (category) and it is not going to be used in the classification. Otherwise, the feature is representative and will be included in the classification.
   (c) Save this information in the *Knowledge Database* module.

The definition of the *threshold* will depend on the level of security associated with the identification process. Having a large value will mean that less representative features will be used in the process and, therefore, more information (even though not necessary) will be taken into consideration. Having a small value will mean that only more representative features will be taken into consideration, but less information will be available during the identification process.

The *diff* values will be compared with a *threshold* value (defined by the designer/operator of the system). If there is any gain in accuracy with respect to the feature-dependent error rates, then this feature is seen to improve performance. The information stored in the *Knowledge Database* module (Figure 4.3) will be the list of features which should be used when a user who belongs to a specific age category is tested.

4.5.3 Addition of age-prediction classifiers

By assuming that the users will provide additional information necessary to assist the identification process, there is a risk of being given incorrect information either by mistake or as an attempt deliberately to circumvent the use of the system. One way to try to overcome this problem is to use only the biometric data collected from the sensor(s) and use this to predict all the other information which is sought to support the identification process.

In this approach, the age information is predicted using the biometric data which is also used to predict identity. This means that the use of the age information will be in the *Individual Classifiers* module, where the *identity prediction classifiers* and the *age-prediction classifiers* will also be incorporated.

Age information, as we have noted, is not a unique identifier for an individual and does not itself carry enough information to identify a single user when used alone. Nevertheless, in the same way that there is potential for benefit in using different classifiers to predict identity from one sample, there is also possible benefit in predicting supplementary information from this same sample. The diversity of opinions generated by this different set of 'experts' (classifiers) can give a broader view of the (identity) searching problem, enhancing the final performance rate.

Figure 4.4 illustrates the system structure when using classifiers to predict age information.

The base classifiers (for identity prediction and age prediction respectively) each process the available input data to give their prediction. In pattern recognition terms, the identity prediction classifiers will classify any input samples as defining one of the enrolled users. On the other hand, the *age-prediction classifiers* will classify any input sample as one of the age options defined. In our case, the classifiers will try to classify the input sample into one of 3 possible age bands: <25, 25–60 or >60.

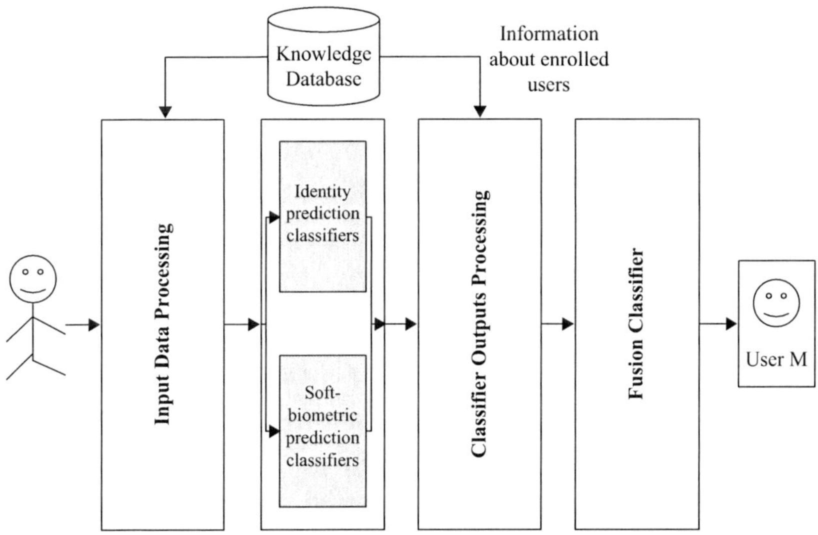

Figure 4.4 System when using age-prediction approach

The process in this system can be described as follows:

1. The input sample F is extracted from the user in the *Feature extraction from the biometrics* in the *Input Data Processing* module (shown in Figure 4.3), where

$$F = [f_1, f_2, f_3, \ldots, f_n]$$

where n is the number of features extracted.
2. F is sent to the *Individual Classifiers* module.
3. This feature F will feed both groups of classifiers in the *Identity prediction classifiers* and *age-prediction classifiers* modules.
4. Each classifier inside the *Identity prediction classifiers* module will produce a vector of confidence degrees as follows:

$$CD_c = [cd_{c,1}, cd_{c,2}, \ldots, cd_{c,y}]$$

where y is the number of enrolled users and c represents the number of the classifier.
5. Each classifier inside the *age-prediction classifiers* module will produce a vector of confidence degrees as follows:

$$CD_s = [cd_{s,1}, cd_{s,2}, \ldots, cd_{s,l}]$$

where l is the number of categories for the age information and s represents the number of the classifier.
6. Both these vectors CD_c and CD_s will be sent to the *Classifier Outputs Processing* module.

Although some investigations have been reported on how to predict age, this type of study has not been extensive. The other use of the age-prediction classifiers is in the *Input Data Processing* module as can be seen in Figure 4.5.

In this approach, the available data can be used either as a feature selector or simply as an extra feature which will be used by the individual classifiers.

4.5.4 Majority weighted vote-based fusion method

Majority Voting [42] is a non-linear fusion-based classifier combination method that takes into account only the top-ranked outputs of the component experts. The outputs of the classifiers are represented in a winner-takes-all form (for each classifier, the output of the winner is 1 and the remaining outputs are 0) and the weights for all the component experts are set equal to 1.

A schematic of this fusion-based configuration can be seen in Figure 4.6. After all the features are extracted and sent to the *Individual Classifiers* module, an adaptation of the general method can be described as follows:

1. Each classifier inside the *Identity prediction classifiers* module will produce a vector of confidence degrees as follows:

$$CD_c = [cd_{c,1}, cd_{c,2}, \ldots, cd_{c,y}]$$

where y is the number of enrolled users and c represents the number of the classifier.

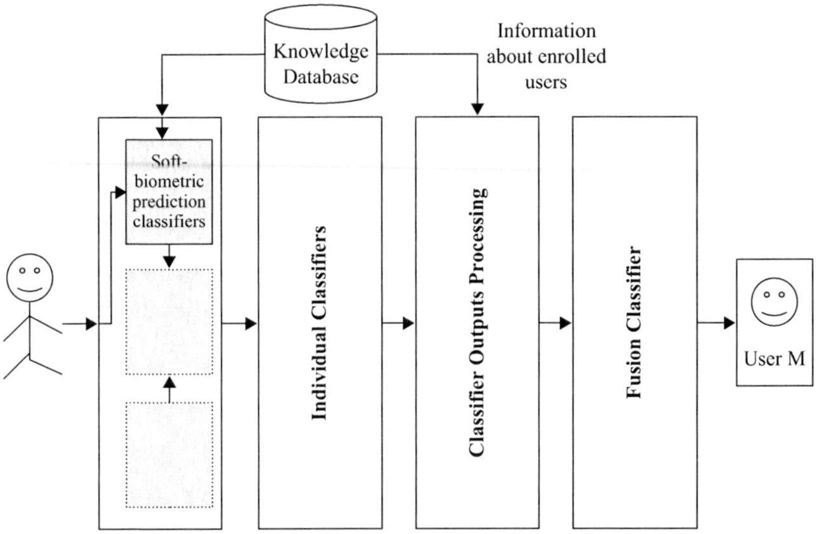

Figure 4.5 System when using age-prediction approach in the Input Data Processing module

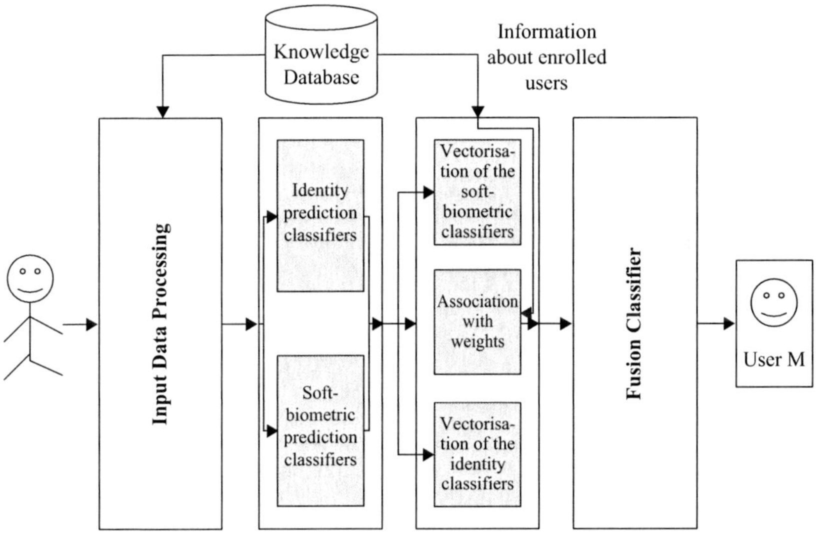

Figure 4.6 System when using age-prediction approach

2. Each classifier inside the *soft-biometric prediction classifiers* module will produce a vector of confidence degrees as follows:

$$CD_s = [cd_{s,1}, cd_{s,2}, \ldots, cd_{s,l}]$$

where l is the number of categories for the age information and s represents the number of the classifier.

3. In the *Classifier Outputs Processing* module, only the user and the age-prediction classifier with the highest confidences will be considered. From this, both *Vectorisation* modules will generate two vectors:

$$ID_{winners-identity} = [id_{1,winner}, id_{2,winner}, \ldots, id_{c,winner}]$$

where *id* is the label of the winner for each identity prediction classifier and *c* is the number of identity prediction classifiers:

$$SB_{winners-age} = [cat_{1,winner}, cat_{2,winner}, \ldots, cat_{s,winner}]$$

where *cat* is the value of the winner for each age-prediction classifier and *s* is the number of age-prediction classifiers.
4. The *Knowledge Database* module provides the respective characteristics of each winner user, which is as follows:

$$SB_{winner-age-real} = [catReal_{1,winner}, catReal_{2,winner}, \ldots, catReal_{c,winner}]$$

where *catReal* is the real age category for each identity prediction classifier winner and *c* is the number of identity prediction classifiers.
5. Each identity prediction classifier will 'vote' for its winner user and this winner will also receive the votes from the age-prediction classifiers where its winner is a characteristic from that user.
6. Because the classifiers vote based on different information (which differs from a traditional voting system, where all the classifiers vote based on the same information), it is necessary to assign weights to each group of classifiers.
7. The output of this method will be the winner identity label with the majority of votes.
8. When there is a tie, the identity prediction classifier with the greatest confidence provides the output of the system.

The use of age information in this context is very interesting and provides another example of how to use additional information which may be available in a particular scenario. The use of weights is explained in more detail in section 4.5.5.1.

4.5.5 Weighted sum-based fusion method

Sum-based fusion [43] is a linear fusion-based method that takes into account the confidence degree towards each class for each classifier. Thus, when an input pattern is presented to the base classifiers, the degrees of confidence for each class output are added to the other related outputs giving an overall score for that class. The winner class, and hence the identity label of the system, is the class with the highest score.

The same general system structure as for the majority voting approach is used, as shown in Figure 4.6. We also need to adapt this method to our current purpose, as follows:

1. In the same way as in the voting technique, each classifier inside the *Identity prediction classifiers* module will produce a vector of confidence degrees as follows:

$$CD_c = [cd_{c,1}, cd_{c,2}, \ldots, cd_{c,y}]$$

where y is the number of enrolled users and c represents the number of the classifier.

2. Each classifier inside the *soft-biometric prediction classifiers* module will produce a vector of confidence degrees as follows:

$$CD_s = [cd_{s,1}, cd_{s,2}, \ldots, cd_{s,l}]$$

where l is the number of categories for the age information and s represents the number of the classifier.

3. Differently from the voting system, in the *Classifier Outputs Processing* module, all the users and the age categories will be considered. Hence, both *Vectorisation* modules will generate two vectors:

$$CD_{identity} = \left[\sum_{i=1}^{c} cd_{c,1}, \sum_{i=1}^{c} cd_{c,2}, \ldots, \sum_{i=1}^{c} cd_{c,u}\right]$$

where cd is the confidence of the each user calculated by the classifier c and u is the number of enrolled users:

$$CD_{age} = \left[\sum_{i=1}^{s} cd_{s,1}, \sum_{i=1}^{s} cd_{s,2}, \ldots, \sum_{i=1}^{s} cd_{s,l}\right]$$

where cd is the confidence of the winner for that classifier l is the number of categories and s is the number of age-prediction classifiers.

4. The *Knowledge Database* module provides the respective characteristics of each enrolled user as follows:

$$SB_{age-real} = \left[catReal_1, catReal_2, \ldots, catReal_u\right]$$

where *catReal* is the real age group for each identity prediction classifier winner and u is the number of enrolled users.

5. The confidences related to the age information will be added to the users who have that characteristic.
6. Because the confidence values added are based on different information (which differs from a traditional sum-based implementation, where all the classifiers' confidence values correspond to the same information), it is necessary to assign weights to each group of classifiers.
7. The output using this method will be the label which has the highest confidence degree.
8. When there is a tie, the identity prediction classifier with the greatest confidence provides the output of the system.

The use of age information in this context is also very interesting but can be seen to take into account even more detail than occurs in the previously presented method.

Here, not only is the opinion of each classifier observed but the confidence with which this classifier makes its decision is also taken into consideration.

4.5.5.1 Definition of the weights

The weights can be chosen in different ways: taking into account important information about the quality of the database, the population characteristics can be related to a specific task, and so on.

The rationale for using these classifiers for different purposes (identity prediction and age prediction) giving them different degrees of importance (weights) is that, in many cases, the least reliable classifier can nevertheless still give valuable information about specific samples that the most reliable classifier might not easily be able to provide in its own right.

The general approach assumes that each group of classifiers may have a degree of confidence in, or relative importance with respect to, a specific identification task. Analysing these relative considerations has to take into account the characteristics of the database (information about the target population, for example) and the effects of including age information in the classification is very important.

4.5.6 Multiagent approach using age information

This method was originally based only on identity classifier agents [44]. However, in this adaptation, the aim is to incorporate age-prediction classifiers to help the agents make their decision and thus refine the decision-making process [45]. The whole system can be seen in Figure 4.7.

The *Classifier Module* in this model (that can be seen in Figure 4.8) contains not only identity prediction classifiers but also supplementary age-prediction classifiers as well. Some other modifications in the overall process are as follows:

- The sensitivity analysis is performed not only with respect to all features of the input patterns in the identity prediction classifier for the agents but also in the age-prediction classifier. Step 1.(a) will be performed with respect to all classifiers.
- In the negotiation process, the agents will try to show the other agents that their results are not good ones, using both information about the predicted identity and information about the age information related to the predicted identity.
- In step 2.(b), the same ranking of the dissimilarity of the features between the test samples and the training mean is carried out for the age-prediction classifiers.
- In step 2.(c), however, there will be an extra step, after step 2.(c)(ii) which is as follows:
  - Check the age information related to the predicted identity.

If there is evidence in the age-prediction classifiers that the agent is wrong, using the same idea of feature sensitivity, this information is used in a punishment protocol. In order to take into account the age information, the new punishment value is calculated according to (4.1):

$$Pun_i = \frac{D_i{}^* S_i}{R_i{}^* C_a} + \sum_{j=1}^{n} \frac{D_{j,i}{}^* S_{j,i}}{Rage_i{}^* C_{mb}} \tag{4.1}$$

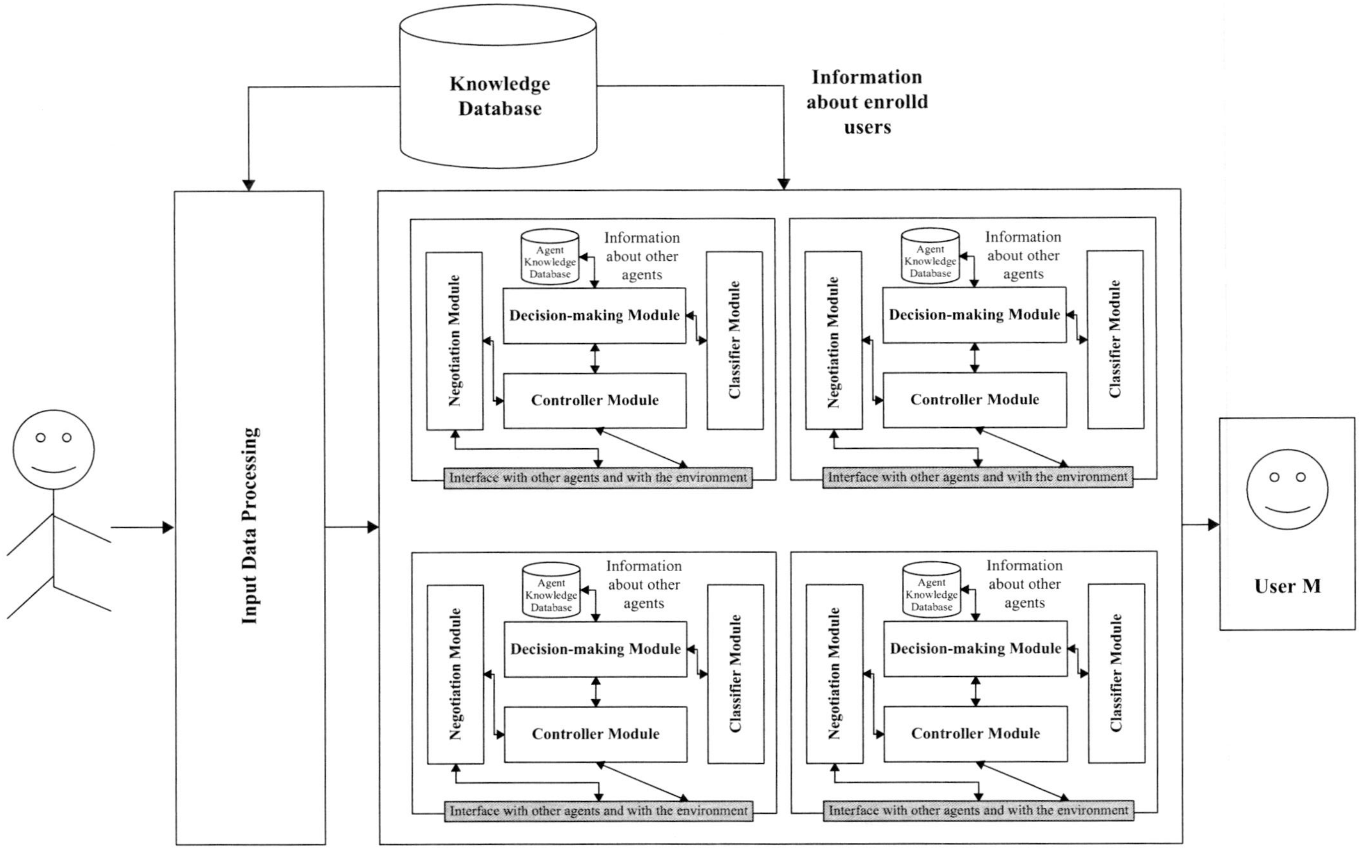

Figure 4.7 Architecture of the agent

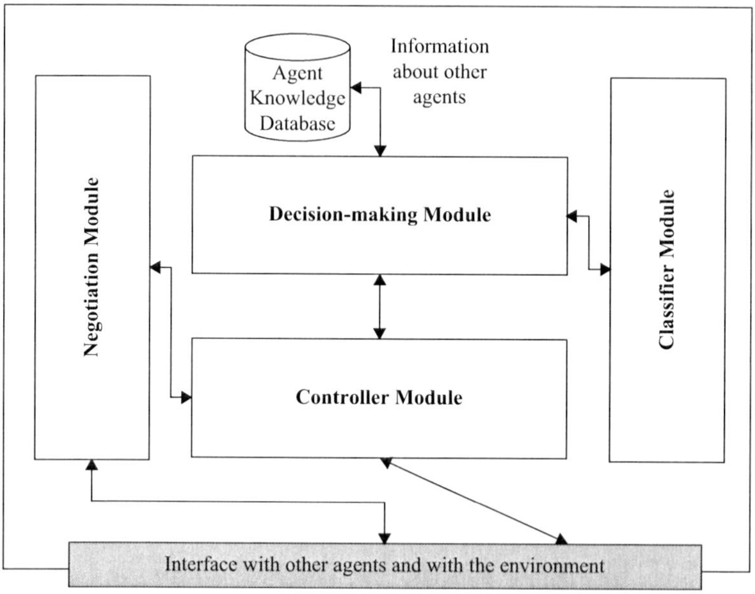

Figure 4.8 Multiagent-based system workflow

where:

- D_i is the difference between the current i feature of the test pattern and its training mean for the identity classifier;
- $D_{j,i}$ is the difference between the current i feature for the current winner class j of the test pattern and its training mean for the age-prediction classifier;
- S_i is the sensitivity of the classifier to the corresponding i feature of the chosen class for the identity classifier;
- $S_{j,i}$ is the sensitivity of the classifier to the corresponding i feature for the current winner class j of the chosen class for the age-prediction classifier;
- R_i is the ranking of the i feature in its difference from the training mean for the identity classifier;
- $Rage_i$ is the ranking of the i feature in its difference from the training mean for the age-prediction classifier;
- C_a and C_{mb} constants define the intensity of the punishments;
- n is the number of age-prediction classifiers that do not agree with the attacked agent.

Once again, the constants C_a and C_{mb} will depend on the level of acceptance when an agent disagrees with another agent. A low value, such as a value which will cause a small punishment in the confidence degree of the agent, for example, will produce a slow and less flexible negotiation process, while a high value, such as a value which will cause a large punishment in the confidence degree of the agent, for example, will produce a more rapid negotiation process.

The sensitivity analysis and the training mean, along with further environmental information related to the age-prediction classifiers, are transformed into rules and constitute the overall domain knowledge base of the classifier agent.

The key to this approach is a strategy which is the opposite of that which has been dominant in multisource processing hitherto. Here, instead of incorporating additional age information directly into a conventional decision-making framework, we have instead adopted a procedure which uses predictions from the acquired biometric data to support a broader-based 'network' of soft decisions which are then subject to scrutiny and negotiation in order to reach an informed overall decision. This is an approach which appears to be well suited to the nature of the generalised biometric-based identification problem, and one which can be readily modified to incorporate the addition of novel sources of information not currently exploited.

4.6 Experimental results and discussion

As an illustrative example of the potential offered by the approaches discussed in this chapter, some results will be presented which show their application across different biometric modalities. Specifically, we will consider the handwritten signature, fingerprint and hand-geometry data. All the samples are drawn from the database compiled as part of the European BioSecure project, of which more details can be found in Reference 46. Our particular dataset contains samples from the same 79 subjects in each of the 3 modalities noted. The following should be noted:

- The fingerprint dataset contains samples from sensors based on both thermal and optical fingerprint capture methods. In each case samples corresponding to right thumb, index and middle fingers and left thumb, index and middle fingers are collected (two samples from each of these fingers). In our working database we extracted features based on the minutiae of the fingerprint samples as follows: X-pixel Coordinate, Y-pixel Coordinate, Minutia Type, Direction, Ridge Curvature and Ridge Density. The minutiae were extracted using the VeriFinger [47] software. As each fingerprint image generates a different number of detectable minutiae, while the classifiers adopted need a common number of entries, it is necessary to normalise the number of minutiae. Here we use a standard algorithm for core detection [48] and identify the 15 minutiae closest to the core to use as input to the classifier, this being chosen as a sound strategy on the basis of extensive experimental testing. We use both optical and thermal samples of all 6 fingerprints.
- The handwritten signature dataset contains 25 samples for each subject, where 15 are samples of the subject's true signature and 10 are attempts to imitate another user's signature. In this investigation we have used only the 15 genuine signatures of each subject, but collected in 2 sessions (30 samples per user in total). The data were collected using an A4-sized Wacom graphics tablet with a density of 500 lines per inch. Twenty-one representative biometric features were extracted from each signature sample, chosen to be representative of those known to be commonly adopted in signature processing applications.

- The hand geometry database contains 4 hand-image samples for each subject, where 2 samples are from the right hand and the other 2 from the left. A digital camera was used for image capture, in each case with the hand in a wide open position [49]. From each sample, 19 geometric distance measures were extracted as follows: 5 distances from the tip of the finger to its border with the palm of the hand (5 features). From each finger, except the thumb, we use the medial-lateral distance of the distal end of the middle phalanx, the medial-lateral distance of the proximal end of the proximal phalanx and the medial-lateral distance of the distal end of the proximal phalanx (3 distances $\times$ 4 fingers = 12 features), and from the thumb, we use only the medial-lateral distance of the distal end of the proximal phalanx (2 features).

Figure 4.9 presents the identification error rates obtained using the different structures described earlier, when age, as a soft-biometric, is combined with hard biometric information in the three illustrative modalities. It is readily apparent that the inclusion of the age information can offer benefits in terms of improving the recognition accuracy (more so for some modalities than others). Also, the results show that when age is used to select the input features, the performance is better than when it is used directly as an extra input feature. This happens because unnecessary features are thereby removed from the process and only important features used.

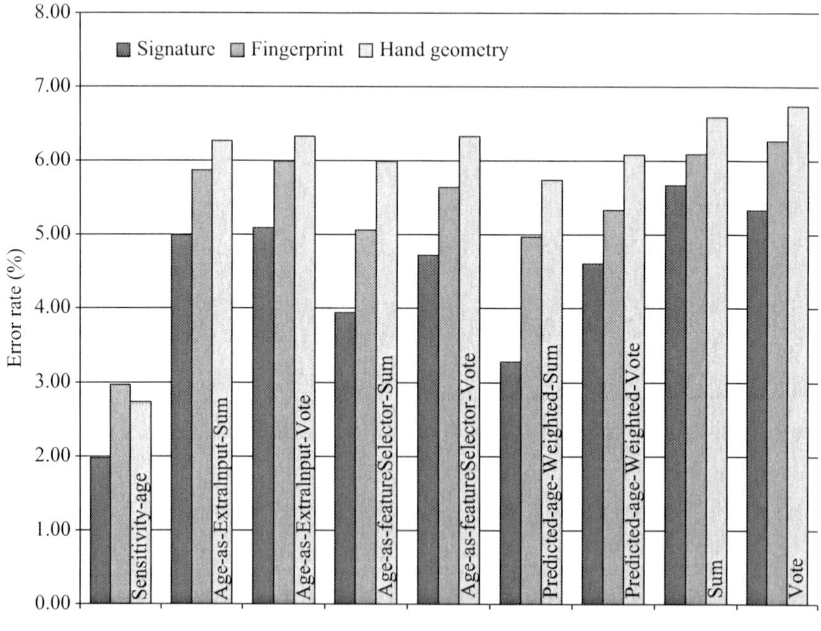

Figure 4.9 Experimental results for identity prediction from biometrics

The different age groupings tend to more effectively localise some particular characteristics, most importantly characteristics which arise because of the effect of ageing on changes in biometric data, which help the classifiers in the class separation.

The signature database revealed the best results in these experiments, perhaps because behavioural biometrics naturally inherit more individual characteristics (psychological, for example) than physiological modalities. This type of information can create relatively unstable sample characteristics but, on the other hand, may often provide more detailed information about the user, and which may be more sensitive to factors such as age.

Figure 4.10 presents the results of the age-prediction classifiers. The difference in performance increases when comparing the fusion techniques with and without predicted age. This is an encouraging reason for using age information, since this data can overcome the limitations of information available from a single modality.

Another important issue addressed in this chapter is the impact of using a more sophisticated representation of the age information, in improving the identification of individuals from their signature, and we will analyse the error rates for the individual age groups adopted (<25, 25–60 and >60) specifically, for the signature database. A Mamdani rule aggregation method is used to represent the fuzzy age groups [50] using a simple trapezoidal function, while a simple division in age band and the absolute age are also used. The results can be seen in Figure 4.11. In a

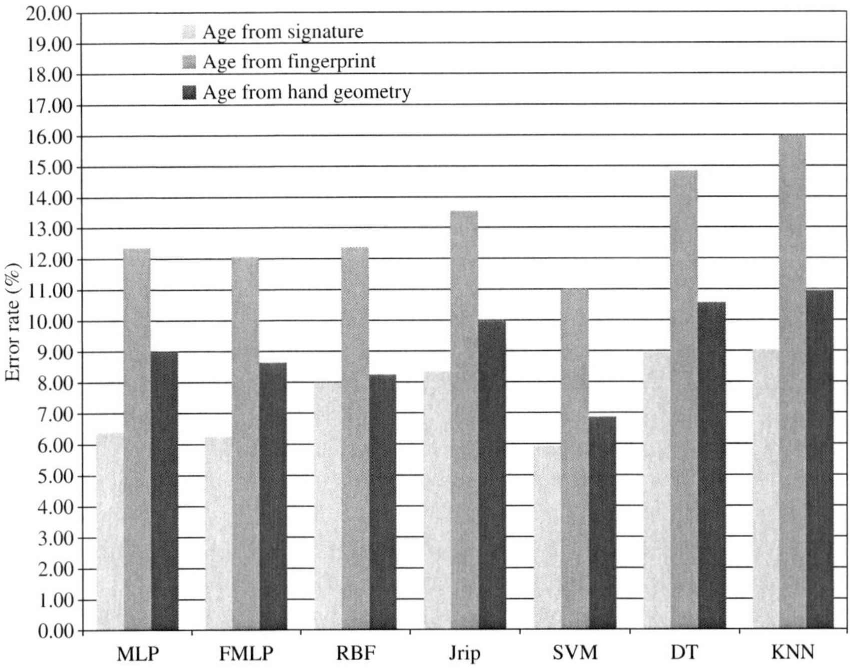

Figure 4.10 Experimental results for age prediction from biometrics

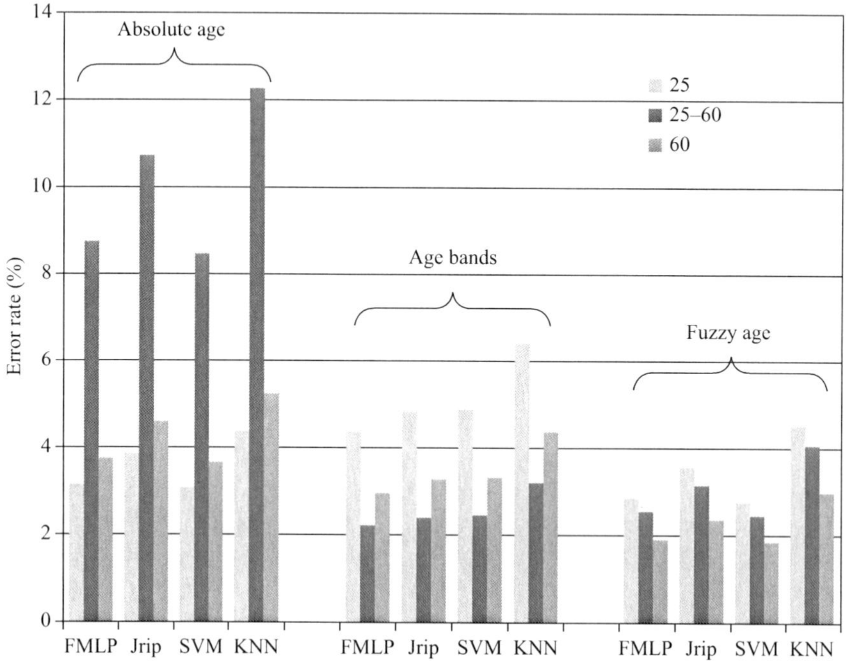

Figure 4.11 Age relative error percentages for the BioSecure database for the handwritten signature database

comparison with work previously reported in the literature these results raise some interesting considerations as follows:

- Clearly, the use of the *absolute age* representation produces the worst results. This is hardly surprising, and reflects the fact that a single number representation of age is not sufficiently subtle, and does not provide an appropriate timescale across which to capture changes in human characteristics related to the complexities of the ageing process.
- When the individual error rates are obtained by representing age with a broader set of *age bands*, it can be seen that the least affected group is the <25 group, followed by the >60 group, while the 25–60 group generates the lowest error rates. One effect here, however, is that the test population is unevenly distributed and the errors are broadly proportional to the number of subjects in each group (34% for the <25 band, 59% for the 25–60 band and 7% for the >60 band).
- The results using the *age bands* representation are very different from the results using the *absolute age* representation. The error rate is seen to decrease to approximately half, which in itself shows that using a more continuous representation of age can improve the accuracy of the system.
- Another very interesting change is the individual error distribution across the age population. The highest error rate is generated by the <25 group followed

by the >60 group and then by the 25–60 group, even though the majority of the population is in the middle group.

- Finally, the error rates obtained using the *fuzzy age* representation are the lowest of the three cases. Also, it is important to notice that the difference among the individual error rates associated with the different groupings is less than 0.5%, which indicates that the technique works relatively uniformly across the diverse population.
- The highest individual error rate with respect to the *fuzzy age* representation occurs in the <25 group. It might be suggested that this is because in this age group the signature is relatively unstable and is still forming, and that this is highlighted by the use of a more accurate representation of the subject age, but there is clearly further investigation to carry out here.

It is easy to underestimate the potential benefits of integrating the age information into biometric-based identification processing. However, although this kind of information is not unique to a specific individual, we have shown here that its incorporation within the process, for example to enhance different fusion algorithms, can be a very promising approach.

4.7 Conclusions

Natural ageing processes can have a detrimental effect on biometric processing, since ageing introduces variability of key biometric features over time. However, age and ageing can also be a positive issue for biometrics research. On the one hand, knowledge of a subject's age is a further piece of identity 'evidence' which can potentially be exploited and, on the other hand, there are many applications where it can be very useful to predict the age of a subject where this is not known, and biometric data can be starting point for attempting this. This chapter has explored each of these views of age information, drawing them together in an integrated perspective.

We have explored some basic approaches which allow us to enhance identity prediction performance by incorporating age information, and we have discussed a number of important issues concerning the relationships between biometric solutions in system implementations which involve biometric data and age, and where many possible configurations can be considered.

More specifically, a number of different techniques by means of which to include age information to improve identity prediction accuracy have been introduced. The results presented demonstrate the real potential value of this type of strategy, and show how additional information which is often available in practical situations is already explicitly collected or which can even be predicted explicitly from biometric data can be exploited in a way which can enhance the identification process.

Also, when resource limitations are a main factor in a system design, it may be particularly beneficial to be able to use cheap, easy-to-collect and well accepted information to enhance unimodal system performance, rather than moving immediately to a full multimodal configuration.

References

[1] Kittler J, Maltoni D. 'Introduction to the special issue on biometrics: Progress and directions'. *IEEE Transactions on Pattern Analysis and Machine Intelligence*. 2007;**29**(4):513–516.

[2] Fairhurst MC. 'Document identity, authentication and ownership: The future of biometrics verification'. *The 7th International Conference on Document Analysis and Recognition. ICDAR 2003*. Washington, DC: IEEE Computer Society; 2003. pp. 1108–1116.

[3] Dass SC, Zhu Y, Jain AK. 'Validating a biometric authentication system: Sample size requirements'. *IEEE Transactions on Pattern Analysis and Machine Intelligence*. 2006;**28**(12):1902–1319.

[4] Snelick R, Indovina M, Yen J, Mink A. 'Multimodal biometrics: Issues in design and testing'. *The 5th International Conference on Multimodal Interfaces. ICMI 2003*. New York, NY: ACM; 2003. pp. 68–72.

[5] Abreu MCC, Fairhurst MC. 'An empirical comparison of individual machine learning techniques in signature and fingerprint classification'. *The First COST 2101 Workshop on Biometrics and Identity Management. Lecture Notes in Computer Science. Roskilde*, Demark; 2008. pp. 134–143.

[6] Jain AK, Ross A, Prabhakar S. 'An introduction to biometric recognition'. *IEEE Transactions on Circuits and Systems for Video Technology*. 2004;**14** (1):4–20.

[7] Jain AK, Bolle R, Pankanti S. *Biometrics: Personal Identification in Networked Society*. Secaucus, NJ: Springer-Verlag; 2005.

[8] Jain AK, Dass SC, Nandakumar K. 'Can soft biometric traits assist user recognition?' *The 1st International Conference on Biometric Authentication. Lecture Notes in Computer Science.* Hong Kong; 2004. pp. 561–572.

[9] Jain J, Dass S, Karthik N. 'Soft biometric traits for personal recognition systems. *The 1st International Conference on Biometric Authentication. Lecture Notes in Computer Science.* Hong Kong; 2004. pp. 731–738.

[10] Abreu MCC, Fairhurst MC. 'Analysing the impact of non-biometric information on multiclassifier processing for signature recognition applications'. *The 2nd IEEE International Conference on Biometrics: Theory, Applications and Systems. BTAS 2008*. Washington, DC; 2008. pp. 1–6.

[11] Zewail R, Elsafi A, Saeb M, Hamdy N. Soft and hard biometrics fusion for improved identity verification. *The 47th Midwest Symposium on Circuits and Systems. Vol. 1 of MWSCAS 2004*; 2004. pp. 225–228.

[12] Ailisto H, Vildjiounaite E, Lindholm M, Mäkelä S, Peltola J. 'Soft biometrics-combining body weight and fat measurements with fingerprint biometrics'. *Pattern Recognition Letter*. 2006;**27**(5):325–334.

[13] Jain AK, Griess F, Connell S. 'On-line signature verification'. *Pattern Recognition Letter*. 2002;**35**(1):2963–2972.

[14] Pervouchine V, Leedham G. 'Extraction and analysis of forensic document examiner features used for writer identification'. *Pattern Recognition Letter*. 2007;**40**(3):1004–1013.

[15] Dyer AG, Found B, Rogers D. 'Visual attention and expertise for forensic signature analysis'. *Journal of Forensic Science*. 2006 (November);**51**(6): 1397–1404.

[16] Jain AK, Nandakumar K, Lu X, Park U. 'Integrating faces, fingerprints, and soft biometric traits for user recognition'. *ECCV Workshop BioAW*; 2004. pp. 259–269.

[17] Mäkinen E, Raisamo R. 'An experimental comparison of gender classification methods'. *Pattern Recognition Letter*. 2008;**29**(10):1544–1556.

[18] Matta F, Saeed U, Mallauran C, Dugelay J. 'Facial gender recognition using multiple sources of visual information'. *IEEE 10th Workshop on Multimedia Signal Processing*. Cairns, Queensland, Australia; 2008 (October);**1**:785–790.

[19] Takimoto H, Kuwano T, Mitsukura Y, Fukai H, Fukumi M. 'Appearance-age feature extraction from facial image based on age perception'. *2007 Annual Conference SICE. SICE 2007*. Kagawa, Japan; 2007. pp. 2813–2818.

[20] Geng X, Zhou ZH, Smith-Miles K. 'Automatic age estimation based on facial aging patterns'. *IEEE Transactions on Pattern Analysis and Machine Intelligence*. 2007 (December);**29**(12):2234–2240.

[21] Lanitis A, Draganova C, Christodoulou C. 'Comparing different classifiers for automatic age estimation'. *IEEE Transactions on Systems, Man, and Cybernetics: Part B*. 2004 (February);**34**(1):621–628.

[22] Sun Y, Zhang JM, Wang LM, Zhan YZ, Song SL. 'A novel method of recognizing ageing face based on EHMM'. *International Conference on Machine Learning and Cybernetics. Vol. 8 of ICMLC 2005*. Guangzhou, China; 2005. pp. 4599–4604.

[23] Guo G, Fu Y, Dyer CR, Huang TS. 'A probabilistic fusion approach to human age prediction'. *IEEE Computer Society Conference on Computer Vision and Pattern Recognition Workshops. CVPRW 2008*. Anchorage, AK; 2008. pp. 1–6.

[24] Guo G, Fu Y, Dyer CR, Huang TS. 'Image-based human age estimation by manifold learning and locally adjusted robust regression'. *IEEE Transactions on Image Processing*. 2008 (July);**17**(7):1178–1188.

[25] Guo G, Fu Y, Huang TS, Dyer CR. 'Locally adjusted robust regression for human age estimation'. *IEEE Workshop on Applications of Computer Vision. WACV 2008*. Copper Mountain, CO; 2008. pp. 1–6.

[26] Metze F, Ajmera J, Englert R, Bub U, Burkhardt F, Stegmann J, *et al.* 'Comparison of four approaches to age and gender recognition for telephone applications'. *IEEE International Conference on Acoustics, Speech and Signal Processing. Vol. 4 of ICASSP 2007*. Honolulu, Hawaii, USA; 2007. pp. 1089–1092.

[27] Liwicki M, Schlapbach A, Loretan P, Bunke H. 'Automatic detection of gender and handedness from on-line handwriting'. *The 13th Conference of the Graphonomics Society. CGS 2007*. Melbourne, Australia; 2007. pp. 179–183.

[28] Thomas V, Chawla NV, Bowyer KW, Flynn PJ. 'Learning to predict gender from iris images'. *The 1st IEEE International Conference on Biometrics: Theory, Applications, and Systems. BTAS 2007*; 2007. pp. 1–5.

[29] Nishimoto M, Azuma Y, Miyamoto Y, Fujisawa TX, Nagata N. 'Subjective age estimation using speech sounds: Comparison with facial images'. *IEEE International Conference on Systems, Man, and Cybernetics. SMC 2008*. Singapore; 2008. pp. 1900–1904.

[30] Ajmera J. 'Effect of age and gender on LP smoothed spectral envelope'. *IEEE Odyssey 2006: The Speaker and Language Recognition Workshop. Odyssey 2006*. San Juan; 2006. pp. 1–4.

[31] Modi SK, Elliott SJ, Whetsone J, Kim H. 'Impact of age groups on fingerprint recognition performance'. *IEEE Workshop on Automatic Identification Advanced Technologies. WAIAT 2007*. Alghero, Italy; 2007. pp. 19–23.

[32] Theofanos M, Micheals R, Scholtz J, Morse E, May P. 'Does habituation affect fingerprint quality?' *CHI'06 Extended Abstracts on Human Factors in Computing Systems. CHI 2006*. New York, NY: ACM; 2006. pp. 1427–1432.

[33] Uludag U, Ross A, Jain AK. 'Biometric template selection and update: A case study in fingerprints'. *Pattern Recognition*. 2004 (July);**37**(7):1533–1542.

[34] Fierrez-Aguilar J, Muaoz-Serrano LM, Alonso-Fernandez F, Ortega-Garcia J. 'On the effects of image quality degradation on minutiae- and ridge-based automatic fingerprint recognition'. *The 39th Annual 2005 International Carnahan Conference on Security Technology. CCST 2005*. Buffalo, NY; 2005. pp. 79–82.

[35] Uhl A, Wild P. 'Comparing verification performance of kids and adults for fingerprint, palmprint, hand-geometry and digitprint biometrics'. *The 3rd IEEE International Conference on Biometrics: Theory, Applications and Systems. BTAS 2009*. Piscataway, NJ: IEEE Press; 2009. pp. 397–402.

[36] Miyoshi T, Hyodo M. 'Aging effects on face images by varying vertical feature placement and transforming face shape'. *IEEE International Conference on Systems, Man and Cybernetics. Vol. 2 of SMC 2006*. Taipei, Taiwan; 2006. pp. 1548–1553.

[37] Liang Y, Li C, Yue H, Luo Y. 'Age simulation in young face images'. *The 1st International Conference on Bioinformatics and Biomedical Engineering. ICBBE 2007*. Wuhan, China; 2007. pp. 494–497.

[38] Patterson E, Sethuram A, Albert M, Ricanek K, King M. 'Aspects of age variation in facial morphology affecting biometrics'. *The 1st IEEE International Conference on Biometrics: Theory, Applications, and Systems. BTAS 2007*. Washington, DC; 2007. pp. 1–6.

[39] Ramanathan N, Chellappa R. 'Face verification across age progression'. *IEEE Computer Society Conference on Computer Vision and Pattern Recognition. Vol. 2 of CVPR 2005*. San Diego, CA, USA; 2005. pp. 462–469.

[40] Biswas S, Aggarwal G, Ramanathan N, Chellappa R. 'A non-generative approach for face recognition across aging'. *The 2nd IEEE International Conference on Biometrics: Theory, Applications and Systems. BTAS 2008*. Washington DC, USA; 2008. pp. 1–6.

[41] Guest RM. 'Age dependency in handwritten dynamic signature verification systems'. *Pattern Recognition Letter*. 2006;**27**(10): 1098–1104.

[42] Kuncheva LI. *Combining Pattern Classifiers: Methods and Algorithms*. Wiley-Interscience. Indianapolis, IN; 2004.

[43] Kittler J, Alkoot FM. 'Sum versus vote fusion in multiple classifier systems'. *IEEE Transactions on Pattern Analysis and Machine Intelligence*. 2003 (January);**25**(1): 110–115.

[44] Abreu MCC, Fairhurst MC. 'Analysing the benefits of a novel multiagent approach in a multimodal biometrics identification task'. *IEEE Systems Journal*. 2009 (December);**3**(4): 410–417.

[45] Da Costa-Abreu MC, Fairhurst MC. 'Combining multiagent negotiation and an interacting verification process to enhance biometric-based identification'. *The COST 2101 European Conference on Biometrics and ID Management. BioID 2011*. Berlin, Heidelberg: Springer-Verlag; 2011. pp. 95–105.

[46] Ortega-Garcia J, Fierrez J, Alonso-Fernandez F, Galbally J, Freire MR, Gonzalez-Rodriguez J, *et al.* 'The multiscenario multienvironment BioSecure multimodal database (BMDB)'. *IEEE Transactions on Pattern Analysis and Machine Intelligence*. 2010;**32**: 1097–1111.

[47] Neurotechnologija. 'Verifinger fingerprint identification SDK'; 2008. Available from: http://www.neurotechnologija.com/verifinger.html

[48] Khan NY, Javed MY, Khattak N, Chang UMY. 'Optimization of core point detection in fingerprints'. *The 9th Biennial Conference of the Australian Pattern Recognition Society on Digital Image Computing Techniques and Applications. DICTA 2007*. Washington, DC: IEEE Computer Society; 2007. pp. 260–266.

[49] Amayeh G, Bebis G, Erol A, Nicolescu M. 'Peg-free hand shape verification using high order Zernike moments. *The 2006 Conference on Computer Vision and Pattern Recognition Workshop. CVPRW 2006*; 2006. p. 40.

[50] Damghani KK, Sadi-Nezhad S, Aryanezhad MB. 'A modular decision support system for optimum investment selection in presence of uncertainty: Combination of fuzzy mathematical programming and fuzzy rule based system'. *Expert Systems with Applications: An International Journal*. 2011 (January);**38**: 824–834.

Section 2

Modality-related approaches to management of age factors

Chapter 5

Human face ageing: a perspective analysis from anthropometry and biometrics

Karl Ricanek Jr.[1], Gayathri Mahalingam[1], A. Midori Albert[1] and Richard W. Vorder Bruegge[2]

5.1 Introducing the face

The human face is a marvel in its simplicity: two eyes, one nose, and one mouth; however, this confluence of tissue, bone, fat, and muscle lends itself to complex communication. Our brains are wired to seek out faces, albeit in the most mundane of scenes. Carl Sagan hypothesized that humans are "hard-wired" from birth to identify the human face as an evolutionary entrenched survival technique. He believed that the human ability to interpret a face in random, or otherwise mundane, images is an evolutionary by-product as survival depended on knowing the face of friend versus foe. The work by Hadjikhani *et al.* (February 2009) [1] on the early activation in the ventral fusiform cortex using a magnetoencephalography supports Sagan's hypothesis by demonstrating that objects that resembled a face (no matter how slight) had very similar early activation signals relative to those found for real faces. This work was further supported by Voss *et al.* (2010) [2] who used functional magnetic resonance imagery (fMRI) to suggest that face perception evoked by face-like objects is a relatively early process. This means that humans can see faces in the most mundane images as illustrated in Figure 5.1.

The power of the face to convey ancestry and emotion is best summarized by a quote by U.S. poet Ralph Waldo Emerson, "A man finds room in the few square inches of the face for the traits of all his ancestors; for the expression of all his history, and his wants." This chapter examines the diachronism of the face as a function of human ageing from birth to senescence to illuminate the challenges in developing systems to model this highly fluid process. It further investigates the challenges faced by researchers of automatic face recognition systems to solve the problem of template ageing. As you read through this chapter the authors leave you with a quote from Milan Kundera, "The serial number of a human specimen is the face, that accidental and unrepeatable combination of features.

[1] University of North Carolina, Wilmington, USA
[2] Federal Bureau of Investigation, Science and Technology Branch, USA

Figure 5.1 A fake face from simple geometric objects. This image is courtesy of wikipedia, Fakeface.svg

It reflects neither character nor soul, nor what we call the self. The face is only the serial number of a specimen."

5.2 Physical manifestations of facial ageing

Approaches to accurate and reliable computerized face modeling require an understanding of the various physical manifestations of facial ageing. Three areas of face ageing are discussed in this section, beginning with a summary of the background work in comprehending growth and development of the human craniofacial complex, followed by adult age-related facial changes, and culminating with considerations of face ageing anomalies, namely asymmetry, extreme photoageing, and deviations of similarity in identical twins.

5.2.1 Growth and development

Much of the research conducted on facial growth and development from infancy through adolescence and early adulthood centers on applications to oromaxillary and facial reconstructive or plastic surgeries. Thus, much of the published literature is characterized by techniques to correct abnormalities stemming from genetic, congenital, or traumatic etiologies with an emphasis on the timing of treatment during growth (e.g., References 3 and 4). For example, Reference 5 provides a historical overview of concepts of craniofacial growth with a theoretical perspective toward addressing the surgical correction of anomalies. There are also studies addressing various treatments for abnormalities classified into different categories, such as Class I and Class II growth patterns in the dentoalveolar region of the face [6].

Regarding growth patterns, Chvatal *et al.* [7] worked on creating and testing multilevel models for predicting longitudinal craniofacial growth. Some studies provide population-specific data regarding facial growth, which includes recognition of any sex differences [8]. An example of this type of work is a study by Norquay *et al.* [9] who focused on Western Australian children and adolescents. Specifically, Norquay *et al.* [9] assessed three-dimensional facial profiles and analyzed size and shape differences during growth and development. Averaged facial growth trajectories showed mainly vertical growth, with differences between females and males. At around age 15 years, sexual dimorphism was evident in the growth of the nose and jaws. Other sexually dimorphic features were manifest by around age 18 years. Between the ages of 10 and 15 years, females displayed accelerated growth, with males displaying similar growth at a later age and over a longer period of time. Results from this study on Western Australian individuals were compared with similar data published on Canadian and Italian individuals, with significant differences detected. That population differences exist is a testament to the need to consider the relevance of population-specific data in face modeling.

In general, some of the craniofacial developmental changes occurring in infancy through adolescence and young adulthood are similar for all individuals – females and males – and across different populations. Taylor [10] summarizes some of the most common developmental changes in the face, emphasizing proportional changes. For example, children's faces grow longer and downward. While much of the face occupies the lower portion of the head in infancy and young childhood, older children and adolescents show marked elongation of the face, largely due to growth in the maxilla and mandible to accommodate the dentition. Figure 5.2 shows fetal skulls at early stages of development of child growth exhibiting the characteristics of prenatal development.

Taylor [10] describes some fundamental facial changes occurring from infancy through young adulthood: The round head of the infant with its large cranium, flattened bridge of the nose, epicanthic folds of the eyelids, and large eyes begins to change at around age 3–4 years. In the pre-school age child, the eyes have lengthened horizontally, the maxilla and mandible have grown larger to house the erupted primary dentition, and within the alveolar bone, the developing permanent dentition. Further, the inter-orbital distance has widened, the chin has taken on some definition, and the nasal bridge has begun to form, though the nose is still small. Figure 5.3 shows the skull of a 4-month-old infant and Figure 5.4 shows the skull of a 14–22 month old child.

In young childhood, roughly ages 5–8 years, according to Taylor [10], the bridge of the nose has moved higher up, the face has lengthened, nose and chin length have increased, the nasal tip is seen, and the maxilla and mandible have continued to grow larger as the permanent dentition has erupted. In the juvenile stages of growth, around 11 to 12 years of age, the bridge of the nose continues to move higher up on the face, the chin acquires a more definitive shape, the nasal tip

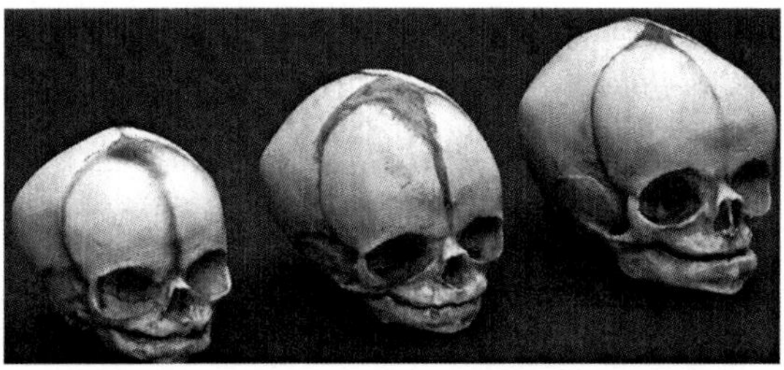

Figure 5.2 Fetal skulls exhibiting the characteristics of prenatal development

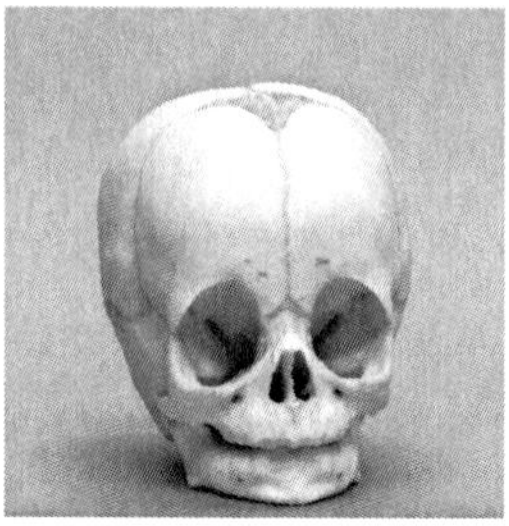

Figure 5.3 4-month-old infant skull

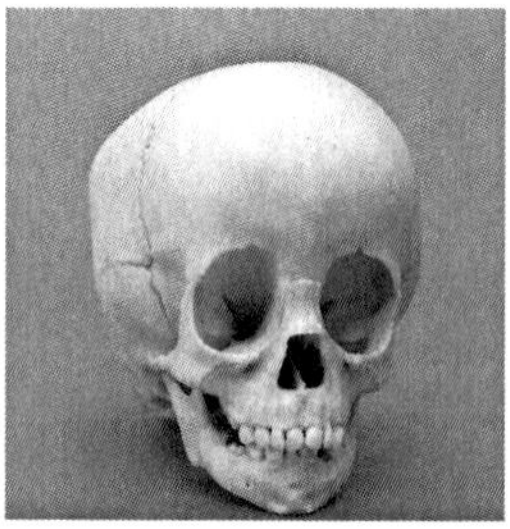

Figure 5.4 14- to 22-month-old child skull

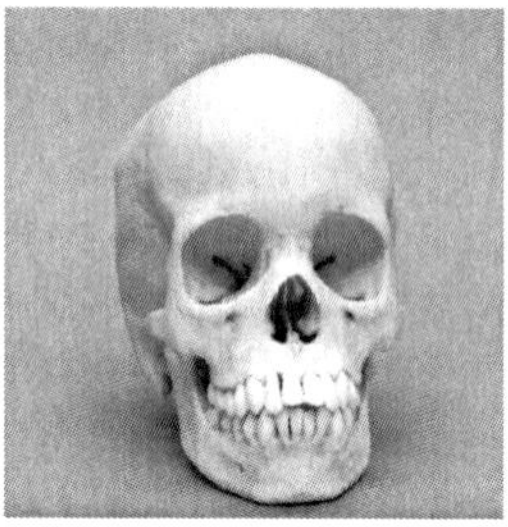

Figure 5.5 13-year-old skull

grows, and the teeth appear rather large relative to the size of the face. By the adolescent years, the teeth take on a more proportionate look in terms of their size relative to the face, and the cheekbones begin to show definition. At the young adult stage of development, the bridge of the nose is fully mature, and the bony features of the face are well formed. Figures 5.5 and 5.6 show the images of skulls of a 13-year-old and adults.

Taylor [10] provides a solid, distilled list of changes seen in growing children. As mentioned earlier, much of the other published literature on facial changes during growth and development in infancy, childhood, and adolescence deals with issues of malocclusion of the jaw, sleep apnea and its effects on sinus development, or other pathologies where the focus is geared toward the timing of surgical

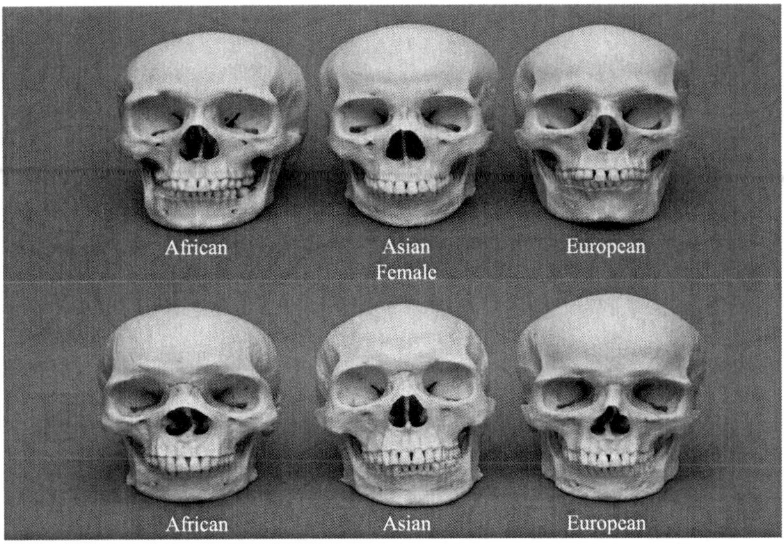

Figure 5.6 Adult skulls

intervention so as to promote normal growth and alleviate obstructions to skeletal maturation [4]. Thus, there is a paucity of journal publications on normal growth, both skeletally and of the soft tissues, of the head and face. Of value to face modeling techniques would be face-ageing studies specifically aimed at searching for clear patterns in craniofacial morphological change in normal, healthy sub-adults, and looking for sex and or population (i.e., gender or race) differences.

5.2.2 Age-related craniofacial morphological changes: young adult to senescence

The appearance of an aged face in adults is an amalgamation of biologic and environmental factors that have left their mark over time. For a detailed, comprehensive understanding of diachronic changes in the adult face, see Albert *et al.* [11], who discuss major findings from the literature from the earliest publications on the topic up through 2007. Summarized here are key findings on adult face ageing from 2008 to 2012.

Many recent publications continue to discuss face ageing as a twofold process being a result of intrinsic (biological, genetic) and extrinsic (environmental) factors, which lead to the irreversible degradation of tissues, mainly the skin [12–14]. Although bony remodeling occurs throughout life, many of the shifts in shape are minute [11] with the exception of tooth loss, perhaps; most of the age-related changes that lead to the appearance of an aged face are noticeable in the soft tissues. Thus, the focus here will be on soft-tissue face ageing.

Friedland and Maffi [15] give an excellent chronology of major adult face and neck age changes. During the 1930s, the upper eyelid skin becomes redundant and crow's feet, wrinkles form lateral to the canthi (outer corners of the eyes).

During the 1940s, nasolabial folds become more prominent and transverse forehead furrows and vertical glabellar frown lines (lines between the eyes, above the nose) develop. In the 1950s, rhytides (wrinkles) form in the neck, the jawline becomes less distinctive, jowls develop, and the tip of the nose droops. In the 1960s, 1970s, and 1980s, there is atrophying of the cutaneous and subcutaneous tissues, which contributes to the formation of increased wrinkles and sagging of the skin. These changes are general patterns and serve as a guideline; however, it should be noted that individuals of the same chronological age do not necessarily show age-related features to the same degree [16]. Although individuals vary in terms of the specific ways their faces exhibit ageing, there are some patterns observed in ageing concerning the skin as well as volume and texture changes, which are highlighted here.

5.2.2.1 Ageing skin

Skin ageing that is intrinsic (i.e., not due to ultraviolet radiation) is characterized by changes at the histologic level of the many layers of the skin. There is a flattening at the interface of the epidermis and dermis as well as decreasing numbers of different skin cell types and blood vessels. Components of the skin atrophy are altered in terms of their organizational structure. Loss of elasticity, collagen, and fat results in sagging, creases, and wrinkling of the skin. Extrinsic factors such as ultraviolet radiation lead to skin wrinkling as well, along with hyper-pigmentation. These ageing skin changes are found in both females and males; however, it is interesting to note that there are sex differences. Female skin is thinner, less elastic, and less vascular than male skin [17]. Race differences in skin ageing are mainly due to pigmentation issues. More darkly pigmented skin is better buffered against the ultraviolet rays, thus photo damage is not as great compared to lightly pigmented skin types. A study by Wolff *et al.* [18] compared early menopausal women of European American descent with those of African American descent. Skin wrinkling was more pronounced in the European American women, as was loss of skin tightness the longer the time passed since menopause. Interestingly, for both groups, greater body weight was correlated with less skin wrinkling.

5.2.2.2 Texture and volume changes in ageing skin

Texture changes refer to a roughened appearance of the skin due to wrinkling caused by dryness [16]. As mentioned above, skin changes include organizational disturbances, thus dermal ridges and hills form, which alters the texture (topography, in a sense) of the skin. Findings by Lai [19], who studied texture versus shape in identifying aged faces, found that texture played a more significant role over shape in a face appearing aged. Aside from texture changes, volume loss also leads to the appearance of an aged face.

Areas of the face where ligaments attach muscle to bone are prone to fat atrophy, resulting in volume loss. When these areas lose volume, the ligaments are less supported, and sagging occurs. The main area affected that contributes to the appearance of an aged face is the cheek region. Volume loss is also common in the brow area (between the eyebrows above the nose), periorbital region, temples, lower jaw, chin, and perioral region [13]. Lips, in particular, thin as age advances, and more so in women compared to men according to Friedland and Maffi [15]; Ibher *et al.* [20]; Sforza *et al.* [21]; and De Menezes *et al.* [22].

The above features of face ageing pertain to changes that occur relatively naturally through the passage of time. While environmental factors undoubtedly affect face ageing, particularly excessive exposure to solar radiation, the extent to which extrinsic factors cause premature or anomalous ageing is not well known. Research that also considers anomalies in face ageing, therefore, would be beneficial to the face modeling approaches that need to manage age factors. Some anomalous conditions that raise awareness about the complexity of face ageing are described below.

5.2.2.3 Face ageing anomalies

Asymmetry, extreme photo-ageing, and deviations in facial appearance in adult identical twins are three anomalous conditions that are relevant to the pursuit of better understanding face ageing. Each of these is discussed briefly below.

It is generally understood that complete facial structural symmetry is non-existent. As Sandor *et al.* [23] note, the causes of asymmetry are many and include, but are not limited to, the following: deformational plagiocephaly (a head that is flattened or misshapen), synostotic plagiocephaly (a misshapen head due to premature suture closure of the cranial bones), clefting conditions, hemifacial hyperplasia, hemifacial atrophy, hemifacial microsomia, costochondral graft overgrowth, condylar and hemimandibular hyperplasia, idiopathic condylar hypoplasia, traumatic condylar hypoplasia, torticollis (wry neck), hamartomas (benign, abnormal growths), vascular malformations, and neoplasms. Asymmetries are typically classified into three broad categories: congenital, developmental, and acquired. Further subclassifications of congenital asymmetries are those due to malformations, deformations, or disruptions in growth [23]. Sandor *et al.* [23] explain that malformations, deformations, and disruptions may be interrelated and are not necessarily mutually exclusive; they all may result in facial asymmetries. Asymmetries may serve as an individualizing feature when assessing faces across the lifespan.

Regarding asymmetry in facial ageing, perhaps one of the best illustrated cases, is that of William McElliott unilateral dermatoheliosis [24], popularized in a published photograph through the *New England Journal of Medicine*. Sixty-nine year old McElliott's asymmetric photo damage was sustained during his work as a truck driver on the same route for 28 years. Chronic and excessive exposure to the sun on the left side of his face has resulted in a noticeable asymmetric photo-damaged aged-face appearance, with the left side of this individual face looking significantly older than the right side of his face. Deep wrinkling, skin sagging, and texture changes characterize the left side of McElliott's face to a far greater extent than that of the right side of his face. Case studies such as this one demonstrate the impact environmental factors have on face ageing. Further corroborating the effects of environmental influences on face ageing are the cases of adult identical twins whose facial dissimilarities seem to correlate with lifestyle dissimilarities. Figure 5.7 shows an example of an asymmetric dermatoheliosis effect caused by constant exposure of one side of the face to ultraviolet rays from the sun.

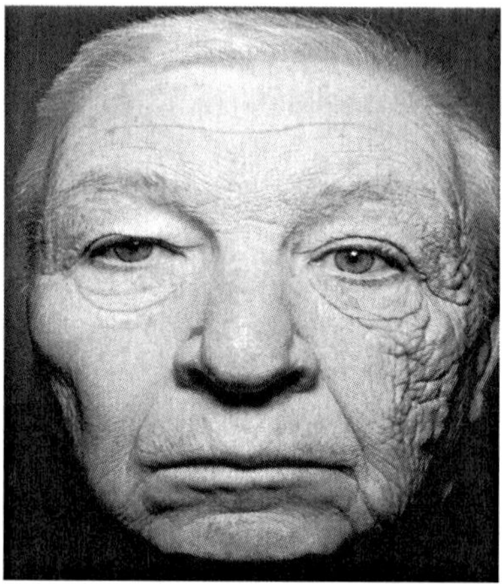

Figure 5.7 Asymmetric dermatohelios wrinkling [24]

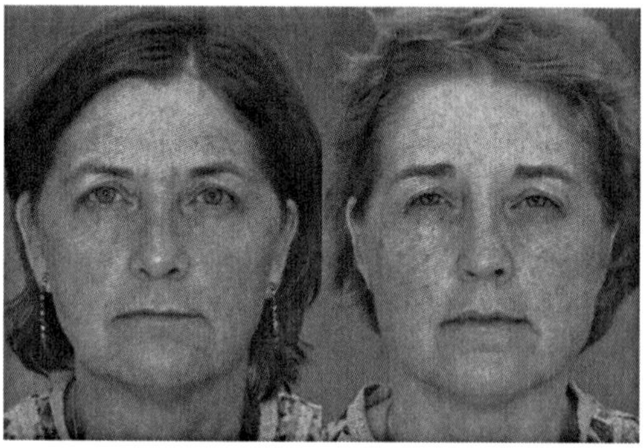

Figure 5.8 Effects of ageing in identical twins

In an online presentation of facial images, where some disparities in the facial appearance of adult identical twins are seen, Parker-Pope [25] explains that factors such as greater sun exposure, smoking, depression, and weighing less may contribute to one twin looking older than the other twin. While these observations are largely subjective and anecdotal, studies with greater rigor may certainly arise out of these findings. New directions for exploring the effects of the environment on ageing, which include physical, emotional, cultural, and socioeconomic factors, etc., could be researched in future studies. Information from studies of this nature would enhance and expand our perspective beyond our current understanding of the biologic processes of face ageing and the environmentally induced photo-ageing. Both general patterns and individualized projections of face ageing could be ascertained, leading to improved reliability and accuracy in automated face recognition and facial age progression and regression technologies. Figure 5.8 shows the effects of ageing and environmental factors on the face of identical twins.

The above anomalies in face ageing, due to environment or lifestyle factors, should be considered along with the general patterns of face ageing due to degradation across the adult lifespan, and understood as distinct from the changes in facial appearance due to growth and development from infancy through adolescence and young adulthood. Awareness of the features of the ageing face from this holistic perspective integrating anatomical, biological, environmental, and observational avenues is imperative to improving computer-automated face recognition and facial age progression/regression technologies. Endeavoring to stay abreast of the latest findings that reveal patterns and variability within the complexity of face ageing ensures that computerized modeling of face ageing utilizes the best possible approaches.

5.3 Computerized modeling of physical manifestations of facial ageing: a survey

5.3.1 Introduction

The changes in appearance of a face with respect to age progression is contributed to by both genetic and environmental factors such as exposure to ultraviolet light, food habits, climate (arid vs. wet or cold vs. hot), smoking, and exercise. The changes in appearance due to these factors are mainly attributed to the shape and texture variations of the face. However, the rate at which these variations are manifested vary for different groups based on ancestry (biogeographic origin), gender, and finally age. Significant facial shape changes are observed during the formative years and major texture variations are observed during the middle to late stages of adulthood. The major shape variations during the formative years are due to the growth of the cranium causing its elongation [26]. This in turn influences the position of the nasomaxillary segment, which affects the muscles of the lower face causing an overall shape change. Additional details on the growth and development are found in section 5.2.1.

Texture changes are manifested as drooping (ptosis) facial features, forehead wrinkles, nasolabial folds, broken jawline, crow's feet, upper eyelid skin wrinkles, elongation of lips, etc. in adult faces. Such ageing traits are attributed to factors such as the effects of gravity, elasticity of the skin, facial fat loss/gain, and elongation of the facial muscles.

A recent study by Shaw *et al.* [27] conducted on the mandible (jaw bone) of individuals from three age categories, young (20–36 years), middle (41–64 years), and old (65 years and older), using three-dimensional CT scan revealed that the angle of the jaw increases markedly with age. This work was the first major effort to describe bone remodeling of the mandible across adulthood. Shaw *et al.* found that bone remodeling, regardless of tooth loss, dramatically modified the mandible resulting in significant appearance changes for senior adults. They noted that the jaw length decreased significantly in comparison between the young and middle age groups, whereas a decline in the jaw height was observed from the middle to old age group. The decline in jaw height results in "sagging of the skin" due to less support provided for the soft tissues of the lower face region. These findings

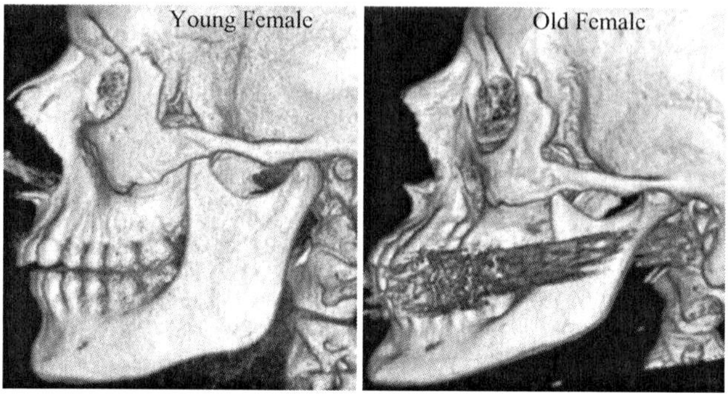

Figure 5.9 Differences in mandibles between young female and old female. This illustration originally appeared in Reference 27

suggest that the representation of facial ageing effects must include the temporal variations in both the shape and texture changes of the face. Figure 5.9 shows the differences in the mandible of young and old females.

Researchers in the field of computer vision have primarily investigated facial ageing from the perspective of an identification paradigm, which includes both a forensic identification perspective and an automatic identification perspective. Forensic identification includes the development of synthesized face likenesses for the purpose of human-centered person recognition. While age progression for automatic face recognition attempts to improve automatic face recognition against the very challenging problem of ageing or template ageing.[1] Face (signal) ageing is modeled by the shape and texture changes that occur as the human face gets older.

The proposed approaches follow a general framework in which the facial ageing effects are modeled by characterizing the temporal variations in shape and texture as a function of time. These approaches can be broadly classified as model-based and template-based approaches. The model-based approaches characterize the variations in shape and texture of the face by identifying a set of fiducial facial features from a set of longitudinal face images of an individual or as cross-section of the population at specified age intervals. These fiducial facial features should represent well the age-related facial changes, and hence, are identified either using anthropometric landmarks or using facial feature point extraction techniques. The generated models are then used in the simulation of an older or younger face version or age transformation, as in template ageing, of the given face. Illustrations of model-based techniques for synthetic ageing or de-ageing can be found in References 28–30.

The template feature-based approaches follow a similar procedure, but characterize the facial ageing at the template level, not the signal level. A set of feature descriptors is computed from the facial features and is stored as templates for the

[1] Template ageing is different from the general problem of ageing in that "ageing" refers to ageing that occurs at the signal, in the case of the face the signal is the face image, while template ageing is a reference to the difference between extracted signal features formed as a template of a particular subject that is separated by time.

subject, which is then used for matching. The template-based approaches suffer from the problem called template ageing that refers to degradation of recognition performance that occurs with increased time between acquisition of an enrollment image and acquisition of the testing image. In his recent work, Kevin Bowyer *et al.* [31] suggests that template ageing effects are most prominent for face and fingerprint biometrics.

5.3.2 Ageing datasets

Several private and public face datasets have been collected over the years to aid studies related to face recognition. However, very few datasets are publicly available for studies related to facial ageing. This section provides a concise description of the publicly available ageing datasets utilized for the study of facial ageing.

5.3.2.1 FG-NET ageing database

The FG-NET (Face and Gesture Recognition Research Network) ageing database [32] is a publicly available ageing database that has been extensively used for evaluation by researchers. The database is composed of 1,002 images of 82 subjects (with an average of 8 images per subject) in the age range 0–69 years. The database predominantly includes images in the age range 0–18 years. The images are manually annotated with 68-landmark features and include information on image size, age, gender, presence of facial hair and eyeglasses, horizontal pose, and vertical pose. Many of the images are scanned from legacy photographs and as such these images exhibit common noise errors associated with old photographs. Figure 5.10 shows sample images from the FG-NET database.

Figure 5.10 Sample images from FG-NET ageing database

5.3.2.2 FERET database

The FERET database contains images of faces with variation in illumination, pose, age, facial expressions, etc. The dataset includes a few hundred longitudinal images. These temporally separated images are categorized into gallery and probe sets with the gallery set composed of 1,196 images. The probe set is further divided into 2 subsets with the first subset made up of 722 images of subjects whose gallery match was taken 0–1,031 days prior. The second probe subset includes 234 images of subjects whose gallery match was taken 540–1,031 days prior. Figure 5.11 shows sample images from the FERET database.

5.3.2.3 MORPH database

The MORPH database comprises age-separated face images of adults. The database includes 2 albums, Album 1 and Album 2. Album 1 consists of 1,690 scanned images of 515 subjects in the age range 15–68 years. The time span between the images of a subject range from a few months to a few years. All images in Album 1 are scanned photographs acquired before 1990. The Album 2 (non-commercial version) consists of more than 20,000 images of nearly 4,000 individuals. The images in Album 2 are high-resolution digital images with variations in pose, illumination, facial expressions, facial hair, eyeglasses, etc. In addition, the

Figure 5.11 Sample images from FERET database

database provides meta-information such as age, gender, and race. Tables 5.1 and 5.2 provide the statistics on MORPH Album 1 and Album 2 datasets. Figure 5.12 shows sample images from the MORPH database. MORPH can be downloaded from www.faceaginggroup.com.

Table 5.1 MORPH Album 1 statistics

MORPH Album 1												
Gender	**Race**			**Age group**					**# Samples**			
	African American	**Caucasian**	**Other**	**<18**	**18–29**	**30–39**	**40–49**	**50+**	**#1**	**#2**	**#3**	**#4+**
Male	1037	365	3	142	803	345	93	22	519	289	67	9
Female	216	69	0	15	182	70	18	0	105	58	13	2

Table 5.2 MORPH Album 2 statistics

MORPH Album 2									
Gender	**Race**			**Age group**			**# Samples**		
	African American	**Caucasian**	**Other**	**40–49**	**50+**	**#1**	**#2**	**#3**	**#4+**
Male	9165	1965	455	1804	9781	355	2431	3645	5154
Female	1393	662	33	229	1859	86	396	679	927

Figure 5.12 Sample images from MORPH ageing database

5.3.3 Survey

Several model-based approaches have been proposed to describe the changes in facial appearance as age progresses. These approaches follow a generative approach in which the shape and/or texture models are generated using extracted facial features. Burt and Perrett [33] utilized age-related facial cues to model facial ageing. These facial cues are identified using a set of fiducial features (facial feature points) from every face image and are utilized in the characterization of facial shape. The shape and texture are then averaged to form *composite faces* for each age group. The composite faces have been shown to be perceived as younger than their age group. The age transformation of a given gallery-probe image is performed by incorporating the differences between the composite faces from the age group of the gallery and probe images onto the probe image. The study of composite faces was extended in Reference 34 in which wavelet-based methods were used to retain edge information in the composite faces leading to better age transformation.

Generative modeling of the adult-face ageing process has been underway for more than a decade at the Face Aging Group laboratory where researchers have enhanced the work pioneered by Tim Cootes and Andreas Lanitis [35] to develop photorealistic age-progressed and age-regressed images. The body of this work is based upon modeling the texture and shape of the face via active appearance modeling (AAM). In the work by Patterson *et al.* [36] of the Face Aging Group, Patterson was able to demonstrate that generative models could outperform trained forensic sketch artist renditions of an aged subject. This work was based on human observers' ratings of the automatically generated synthetic aged images versus the forensic artist rendered drawings. The forensic artist required several photographs of family members to deduce a future look. Human observers overwhelmingly favored synthetic images over the artist drawings. In follow-up work, Sethuram *et al.* [37] of the Face Aging Group developed a ethno-gender hierarchical approach to AAM modeling. The objective of the work was to establish a framework for performing direct comparisons of signal-based ageing techniques and to demonstrate the efficacy of employing a hierarchical approach to generative modeling. The notion of using a hierarchical approach is to segment the model space, which in this case is the space of all possible faces, to a more homogenous face space based upon the same gender and ancestry, and hence, a more precise but constrained representation of the face. The premise is that it is easier to model a segmented space, e.g., to model the faces of Japanese females alone. The second objective of Sethuram *et al.*'s (2010) work was the establishment of a metric that could be used for direct comparison of synthetic ageing techniques. Until this work, researchers used either their impressions of goodness to likeness or a small cohort of human observers. Sethuram *et al.* suggested that automatic face recognition engines could be used as an unbiased arbitrator of likeness by comparing the modeled image to the original image. In her work she used a commercial matcher to perform the measure of likeness using the match scores from the verification test.

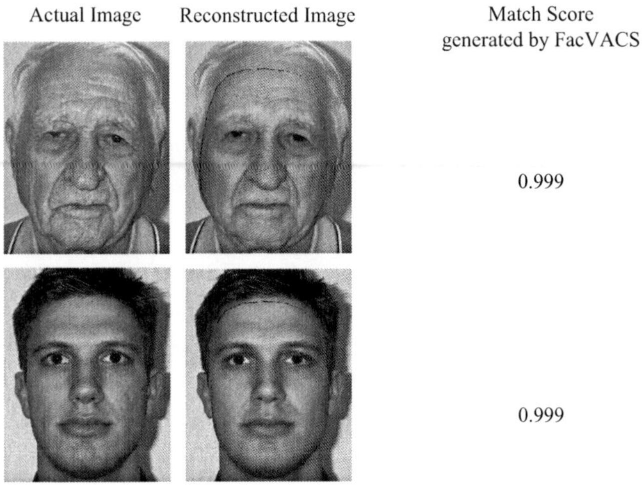

Figure 5.13 Comparison of hierachical ethno-gender model to original image using verification match score (max = 1.0) [37]

Figure 5.13 provides an example of the ethno-gender hierarchical approach and its normalized match score.

Ramanathan and Chellappa [29] proposed a craniofacial model that characterizes the facial shape variations observed in young aged faces (0–18 years). The proposed shape model adopts the "revised" cardioidal strain transformation model, which is fused with anthropometric measurements of the age-related facial features. The cardioidal strain model represents the face growth as a result of applying cardioidal strain to the skull, which stretches the face downward and outward (Figure 5.15). The model characterizes the amount of facial growth by the use of facial growth parameters for different age-related facial features. The model utilized 52 facial proportions across different ages to formulate the facial growth parameters as a non-linear optimization problem. Figure 5.15 illustrates the age transformation results by applying the "revised" cardioidal strain transformation model on images of two subjects. The model assumes that the degrees of freedom associated with facial feature deformations are directly related to the physical properties and geometric orientations of the underlying facial muscles. Further, the model assumes that the weight-loss or weight-gain are implicitly accounted for in the shape model. However, the model does not account for any facial deformations due to changes in fat padding and the dermis, which play a role in causing visible differences when there is weight loss or gain.

Similarly, the authors proposed a shape and texture variation model [38] that characterizes the facial ageing effects associated with adult ageing. The shape and texture variation models incorporate a physical model to account for the changing elastic properties of the facial muscles. The texture variations are modeled using a muscle model that considers three types of facial muscles: linear, sheet, and

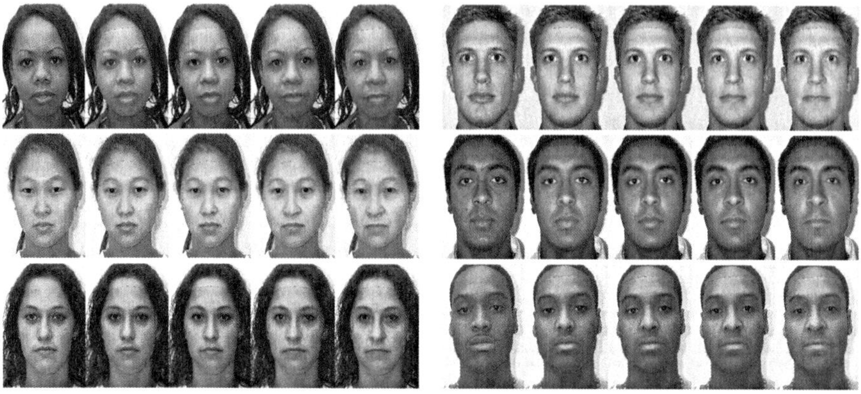

Figure 5.14 Example of Face Aging Group lab ageing models applied to various ethnicities and gender groups

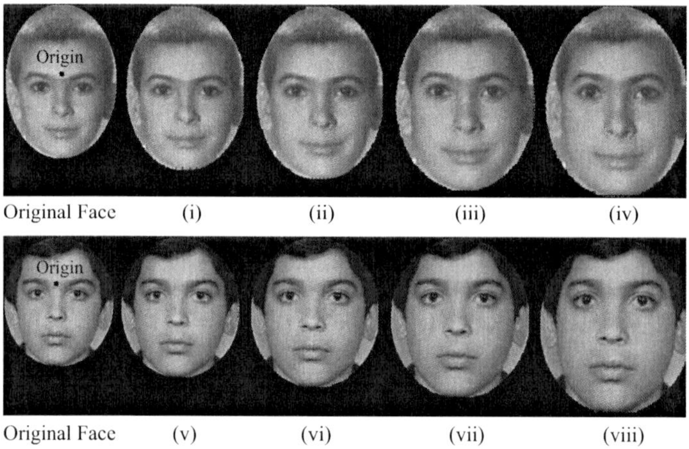

Figure 5.15 Age transformation results obtained by applying the "revised" cardioidal strain transformation on images of two subjects. This illustration originally appeared in Reference 29

sphincter based on their muscle functionality and the effects on them due to gravity. The subtle variations in skin wrinkle change are studied using the pixel-based differences in gradient magnitudes and orientations. Figure 5.16 provides an illustration of the application of the shape and texture model on real-world images of subjects.

Park *et al.* [39, 40] proposed an automatic ageing simulation technique that uses a three-dimensional ageing model. Mapping the set of two-dimensional face images

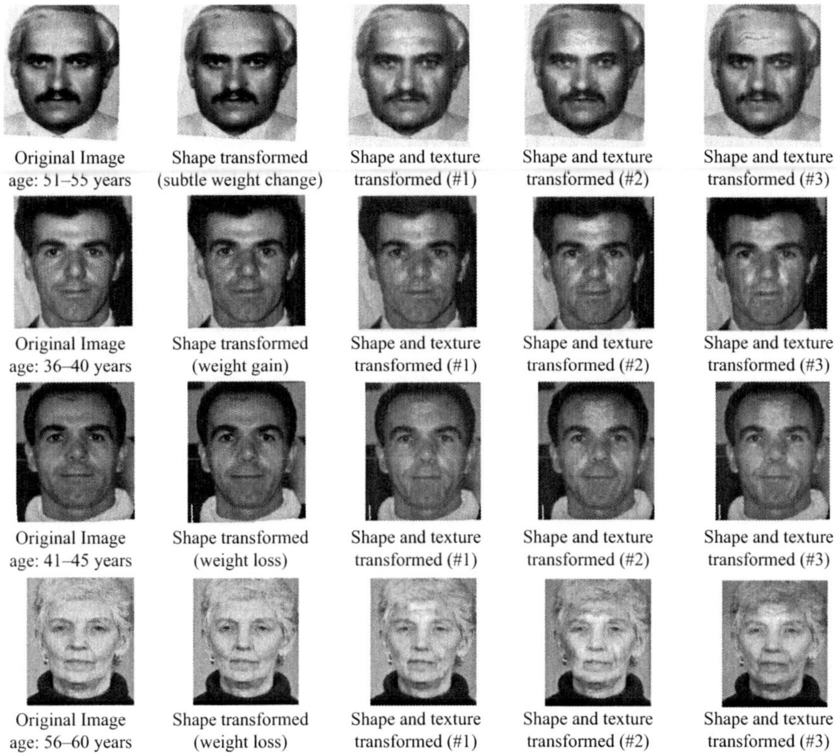

Figure 5.16 Results of application of shape and texture model proposed in Reference 38 on real-world images. The second column illustrates the shape transformation results and the third, fourth, and fifth columns illustrate the textural variations induced on the shape-transformed image. This illustration originally appeared in Reference 38

to the three-dimensional ageing model learns the variations in shape and texture. The ageing simulation technique modifies both the probe and the gallery images by means of the three-dimensional shape and texture model to compensate for the age variations. The three-dimensional model utilizes the shape vectors and the texture patterns extracted at 68 facial feature points from the two-dimensional face images. The ageing simulation is performed by fitting the generic three-dimensional model to the feature set of the face projected to two-dimensional and performing an ageing transformation, which computes the desired shape and texture to be the weighted sum of the shapes and the textures, respectively. Figure 5.17 illustrates the age-simulation results on a sample image of a subject from the FG-NET ageing database. Similar to Reference 40, Singh *et al.* [41] proposed an age transformation algorithm that minimizes the variations in facial features due to ageing by registering the gallery and probe face images in the polar coordinates domain. The accuracy in age

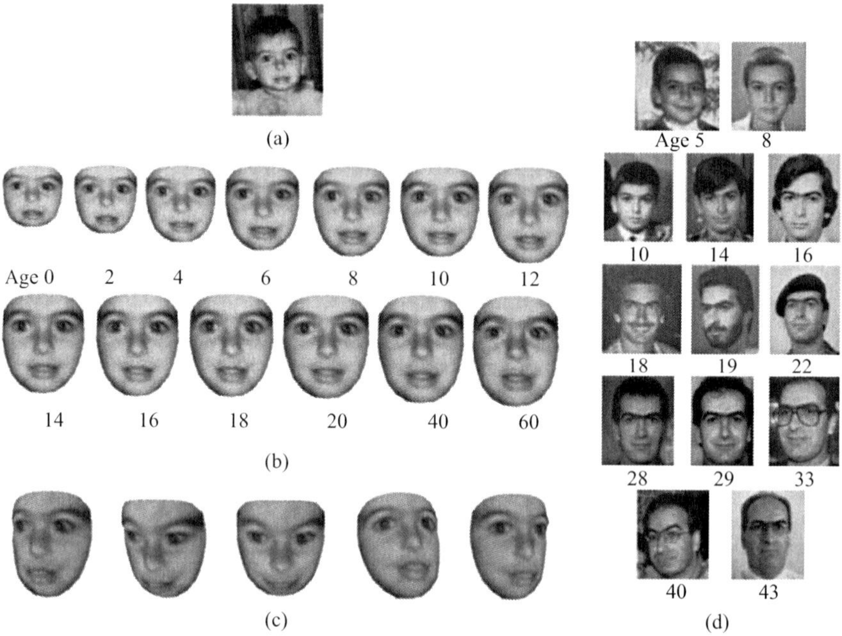

Figure 5.17 An example illustration of ageing simulation on an image from the FG-NET database. (a) The original image. (b) Age transformed images from ages 0 to 60 years. (c) Face images rotated at 5 different poses on an age simulated image at age 20 years. (d) Ground truth images of the subject at their respective ages. This illustration originally appeared in References 39 and 40

transformation of a face image using the model relies on the availability of images of a subject or the entire population for all possible age groups that reflects the major facial changes due to ageing effects.

The power of three-dimensional modeling lends itself to resolving pose issues. However, the aged faces in Figure 5.17 show very little aged texture and arguably does not match the real images of the subject in Figure 5.17(d). It is noteworthy that this work makes the attempt to age-progress individuals from childhood through middle-aged adulthood. The use of this technique saw an improvement in the rank-one matching for the FG-NET and MORPH datasets when used with a world-class commercial matcher [42]. The rank-one accuracy improved from 26.46% to 37.4% on FG-NET and from 57.8% to 66.4% on MORPH subset.

Suo *et al.* [30, 43] proposed a multi-resolution dynamic face ageing model in which the face images are represented by a multi-layer And-Or graph. The graph represents the age-related changes in hairstyle, ageing effects and deformations of facial components, and the changes in skin texture due to wrinkles in specific face zones. The face ageing is modeled as a dynamic Markov process on the generated graph. The model performed the ageing simulation of a probe image by constructing

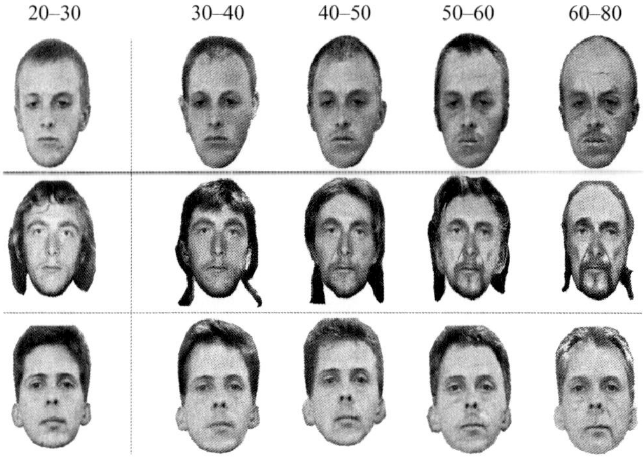

Figure 5.18 Example ageing simulation results on MORPH database. First column shows the original image and the second to fifth columns represent synthetically aged images using the original image. This illustration originally appeared in References 30 and 43

an And-Or graph for the probe image and by sampling the graph structures over various age groups. A new face image is generated using the sampled graphs. Figure 5.18 shows an example of ageing simulation results on the MORPH database. A database consisting of more than 50,000 images across age was used for the generation of the And-Or graph, which resulted in better recognition accuracies when compared with accuracies on smaller datasets. In addition, the authors showed the significance of using hairstyles and hair color in age perception. This work demonstrates the best visual effects of ageing of any of the studied techniques. The incorporation of hair ageing and hairstyle is powerful addition for human observers. However, the authors did not provide a comparison of their ageing technique relative to the task of recognition, either forensic or automatic. The generation of the aged update for the specific face region – there are six regions not including the hair – consists of one of several average wrinkle maps. Hence, the visually pleasing wrinkle maps may cause significant errors for face recognition matchers.

A graph-based shape and texture model was proposed in Reference 44 for modeling the changes due to facial ageing. Unique facial fiducial points were extracted using a modified local feature analysis and were used in the generation of image graphs to represent the facial shape changes. Local binary pattern descriptors are extracted around each feature point to represent the facial texture at these unique locations and were used to generate the texture model. A two-stage matching process was involved, where in the first stage, a Maximum a Posteriori (MAP) solution is derived using the probabilistic texture model in order to prune the search space. The shape model is used in the second stage where a graph matching procedure is involved to match the graphs from the pruned search space and that of the probe image. Their results in reference [43] on the FG-NET database were shown to be better than [29] and [39], especially for children.

5.3.3.1 Face verification across ageing

Several machine-learning approaches have been proposed in the past for the task of verifying age-separated images using their templates. These approaches can be broadly classified as "generative" and "non-generative" approaches. Generative approaches aim at reducing the age separation between the two images by age transforming one of the images using a computational ageing model. On the other hand, non-generative approaches generate an age-invariant template from the face images of a subject and use the same for verification purposes. The approaches proposed by Ling *et al.* [45], Ramanathan and Chellappa [46], Biswas *et al.* [47], and Mahalingam and Kambhamettu [48] are non-generative approaches. Ling *et al.* [45] proposed an operator for face representation and an SVM-based classification framework. The operator is derived based on image gradient orientations extracted from different image scales. The authors showed the robustness of the operator in representing age-invariant patterns and provided an insight into the effect of age in verifying face images of subjects from different age categories.

Ramanathan and Chellappa [46] proposed a Bayesian classification framework in which the face images are represented using their pixel intensities. The features are extracted from Point-Five face, which is generated by vertically reflecting the better illuminated half of the face assuming facial symmetry. Face verification is performed in two stages where, in the first stage, the given pair of images is identified as either intra-personal (belonging to the same person) or extra-personal (belonging to different people). The second stage involves classifying the intra-personal image pair based on the age gap between the images.

Biswas *et al.* [47] proposed a classification framework that utilized the coherency in facial feature drifts across age as a measure to perform face verification across ages. The authors proposed that the facial features drift as age progresses. However, while these drifts follow a coherent pattern for the same individual, they differ between individuals. The drift is measured by identifying features derived from the inner face regions, which are insensitive to pose variations when compared with fiducial features extracted from the outer contour of the face. Mahalingam and Kambhamettu [48] proposed a classification framework based on Adaboost that uses local binary pattern (LBP) features derived from different image scales. The authors showed the robustness of the hierarchical LBP operator in representing age-invariant features of the face across all age categories.

5.4 Conclusion

The management of age factors for the face is a very complex topic as it relates not only to the automatic algorithms for face matching but also to the human system.

Although humans have a remarkable capacity to recognize known persons across those persons' lifespans, this capacity does not carry over to unknown faces. It has become imperative to find a solution to ageing, and template ageing more specifically, for those market segments that deal with security, e.g., border security and criminal and terrorists apprehension. Reports from the U.S. National Institute of Standards and Technology (NIST) face recognition challenges indicate that commercial face recognition system performance suffers from the problem of template ageing as indicated in the Multiple Biometrics Evaluation 2010 Still Image. The full report is available at www.nist.gov/customcf/get_pdf.cfm?pub_id=905968. Details of MBE 2010 are found at www.nist.gov/itl/iad/ig/mbe.cfm.

This chapter has provided an overview of the mechanisms of ageing and an update on the current state of the art for synthetic ageing and for the template ageing problem. Although there have been some clear advancements in all areas of face ageing, a solution is still elusive. Yet, while the early years of this research area were starved of data, there are now massive datasets with longitudinal images available to the research community, with new datasets available online each year. As a result, young and seasoned researchers alike will be motivated to forge ahead to develop comprehensive solutions to the set of challenges inherent in face ageing.

References

[1] Hadjikhani, N., Kveraga, N., Naik, P., Ahlfors, S.: 'Early (m170) activation of face-specific cortex by face-like objects'. *Neuroreport* **20**(4) (2009) 403–407

[2] Voss, J., Federmeier, K., Paller, K.: 'The potato chip really does look like elvis! neural hallmarks of conceptual processing associated with finding novel shapes subjectively meaningful'. *Cerebral Cortex* **22**(10) (2010) 2354–2364

[3] Al-Thomali, Y., El-Bialy, T.H.: 'Cephalometric craniofacial features of growing patients with chronic renal failure'. *Archives of Oral Biology* **57** (2012) 257–263

[4] Reid, R.R.: 'Facial skeletal growth and timing of surgical intervention'. *Clinics in Plastic Surgery* **34** (2007) 357–367

[5] Carlson, D.S.: 'Theories of craniofacial growth in the postgenomic era'. *Seminars in Orthodontics* **11** (2005) 172–183

[6] Riesmeijer, A.M., Prahl-Andersen, B., Mascarenhas, A.K., Joo, B.H., Vig, K.W.: 'A comparison of craniofacial class i and class ii growth patterns'. *American Journal of Orthodontics and Dentofacial Orthopedics* **125** (2004) 463–471

[7] Chvatal, B.A., Behrents, R.G., Ceen, R.F., Buschang, P.H. 'Development and testing of multilevel models for longitudinal craniofacial growth prediction'. *American Journal of Orthodontics and Dentofacial Orthopedics* **128** (2005) 45–56

[8] Jacob, H.B., Buschang, P.H. 'Vertical craniofacial growth changes in french-canadians between 10 and 15 years of age'. *American Journal of Orthodontics and Dentofacial Orthopedics* **139** (2011) 797–805

[9] M. Norquay, M. Walters, K. Rooney, and D. Gillett. 'Patterns of facial growth in western australian children and adolescents'. *Journal of Comparative Human Biology* **61** (2010) 214–215

[10] Karen T. Taylor. 'Age progression: Growth'. In: Taylor, K.T., editor. *Forensic Art and Illustration*. Boca Raton (2001) pp. 235–249

[11] Albert, A.M., Ricanek, K., Patterson, E.: 'A review of the literature on the aging adult skull and face: Implications for forensic science research and applications'. *Forensic Science International* **172** (2007) 1–9

[12] Fillit, H. M., Rockwood, K., Woodhouse, K. 'Aging and the skin'. *Brocklehursts Textbook of Geriatric Medicine and Gerontology*, 7th edn. Elsevier, USA (2010) pp. 133–136

[13] N. S. Sadick, C. Karcher, and L. Palmisano. 'Cosmetic dermatology of the aging face'. *Clinics in Dermatol* **27** (2009) S3–S12

[14] Flynn, C. and McCormack, B.A. 'Simulating the wrinkling and aging of skin with a multi-layer finite element model'. *Journal of Biomechanics* **43** (2010) 442–448

[15] J. A. Fearon and T. R. Maffi. Evaluation of the Aging Face: Plastic Surgery Secrets Plus, 2nd edn. (2010) pp. 477–483

[16] Miyamoto, K., Inoue, Y., Hsueh, K., Liang, Z., Yan, X., Yoshii, T., Furue, M. 'Characterization of comprehensive appearances of skin aging: An 11-year longitudinal study on facial skin aging in japanese females at akita'. *Journal of Dermatological Science* **64** (2011) 229–236

[17] PL. Leong. 'Aging changes in the male face'. *Facial Plas Surg Clin of North Am* **6** (2008) 277–279

[18] Wolff, E., Pal, L., Altun, T., Madankumar, R., Freeman, R., Amin, H., Harman, M., Santoro, N., Taylor, H.S. 'Skin wrinkles and rigidity in early postmenopausal women vary by race/ethnicity: Baseline characteristics of the skin ancillary study of the keeps trail'. *Fert and Steril* **95** (2011) 658–662

[19] Lai, M., Oruç, I., Barton, J.J. 'The role of skin texture and facial shape in representations of age and indentity'. *Cortex* (2011) (In Press)

[20] Iblher, N., Kloepper, J., Penna, V., Bartholomae, J.P., Stark, G.B. 'Changes in the aging upper lipa photomorphometric and MRI-based study (on a quest to find the right rejuvenation approach)'. *Journal of Plastic, Reconstructice & Aesthetic Surgery* **61** (2008) 1170–1176

[21] Sforza, C., Grandi, G., Binelli, M., Dolci, C., De Menezes, M., Ferrario, V.F. 'Age- and sex-related changes in three dimensional lip morphology'. *Forensic Sci Intl* **200** (2010) 182.e1–182.e7

[22] De Menezes, M., Rosati, R., Baga, I., Mapelli, A., Sforza, C. 'Three-dimensional analysis of labial morphology: Effect of sex and age'. *International Journl of Oral and Maxillofacial Surgery* **40** (2011) 856–861

[23] Sándor, GK., McGuire, T.P., Ylikontiola, L.P., Serlo, W.S., Pirttiniemi, P.M. 'Management of facial asymmetry'. *Oral and Maxillofacial Surgery of Clinics of North America* **19** (2007) 395–422

[24] Jennifer, R.S. Gordon, M.D., and Joaquin, C. Brieva, M.D. 'Unilateral dermatoheliosis'. *N Engl J Med* **366** (2012) 425

[25] Parker-Pope, T.: 'Twins and the wrinkles of aging'. *The New York Times Health/Science,* February 2009

[26] Trenouth, M.J.: 'Shape changes during human fetal craniofacial growth'. *Journal of Anatomy* **139** (1984) 639–651

[27] Shaw, R.J., Katzel, E., Koltz, P., Kahn, D., Girotto, J., Langstein, H.: 'Aging of the mandible and its aesthetic implications: A three dimensional CT study'. *Plastic and Reconstructive Surgery* **1** (2010) 332–342

[28] Patterson, E., Ricanek, K., Albert, M., Boone, E.: 'Automatic representation of adult aging in facial images'. *Proceedings of the 6th IASTED International Conference on Visualization, Imaging, and Image Processing* (2006) pp. 171–176

[29] Ramanathan, N., Chellappa, R.: 'Modeling age progression in young faces'. *Computer Vision and Pattern Recognition* (2006) 387–394

[30] Suo, J., Zhu, S., Shan, S., Chen, X.: 'A compositional and dynamic model for face aging'. *TPAMI* **9** (2009) 385–401

[31] Fenker, S.P., Bowyer, K.W.: 'Analysis of template aging in iris biometrics'. *CVPR Biometrics Workshop* (2012)

[32] 'Face, gesture recognition working group: FG-NET aging database'. (2000)

[33] Burt, M., Perrett, D.I.: 'Perception of age in adult caucasian male faces: computer graphic manipulation of shape and colour information'. *Journal of Royal Society* **259** (1995) 137–143

[34] Tiddeman, M., Burt, B., Perrett, D.: 'Prototyping and transforming facial texture for perception research'. *Computer Graphics and Applications* **21**(5) (2001) 42–50

[35] Lanitis, A., Taylor, C.J., Cootes, T.F.: 'Toward automatic simulation of aging effects on face images'. *TPAMI* **24** (2002) 442–455

[36] Patterson, E., Sethuram, A., Albert, M., Ricanek, K.: 'Comparison of synthetic face aging to age progression by forensic sketch artist'. *Proceedings of the 6th IASTED International Conference on Visualization, Imaging, and Image Processing*, Palma De Mallorca, Spain (2007)

[37] Sethuram, A., Ricanek, K., Patterson, E.: 'A hierarchical approach to facial aging'. *Proceedings of IEEE Computer Society Workshop on AMFG in association with the 23rd IEEE Conference on Computer Vision and Pattern Recognition*, San Francisco, CA (2010) pp. 100–107

[38] Ramanathan, N., Chellappa, R.: 'Modeling shape and textural variations in aging adult faces'. *IEEE Conference on Automatic Face and Gesture Recognition*, Amsterdam, Netherlands (2008)

[39] Park, U., Tong, Y., Jain, A.K.: 'Face recognition with temporal invariance: A 3d aging model'. *Proceedings of the International Conference on Automatic Face and Gesture Recognition* (2008) pp. 1–7

[40] Park, U., Tong, Y., Jain, A.K.: 'Age-invariant face recognition'. *IEEE Transactions on Pattern Analysis and Machine Intelligence* **32**(5) (2010) 947–954

[41] Singh, R., Vatsa, M., Noore, A., Singh, S.K.: *Age Transformation for Improving Face Recognition Performance.* Springer-Verlag, USA (2007) pp. 576–583

[42] GmbH, C.S.: Facevacs software developer kit

[43] Suo, J., Min, F., Zhu, S., Shan, S., Chen, X.: 'A multi-resolution dynamic model for face aging simulation'. *Computer Vision and Pattern Recognition* (2007) 17–22

[44] Mahalingam, G., Kambhamettu, C.: 'Age invariant face recognition using graph matching'. *Proceedings of the IEEE International Conference on Biometrics: Theory, Applications, and Systems*, Washington, D.C. (2010) pp. 1–7

[45] Ling, H., Soatto, S., Ramanathan, N., Jacobs, D.: 'Face verification across age progression using discriminative methods'. *IEEE Transactions on Information Forensics and Security* **5** (2010) 82–91

[46] Ramanathan, N., Chellappa, R.: 'Face verification across age progression'. *IEEE Transactons on Image Processing* **15** (2006) 3349–3362

[47] Biswas, S., Aggarwal, G., Chellappa, R.: 'A non-generative approach for face recognition across aging'. *IEEE International Conference on Biometrics: Theory, Applications, and Systems*, Washington, D.C. (2008)

[48] Mahalingam, G., Kambhamettu, C.: 'Face verification of age separated images under the influence of internal and external factors'. *Image and Vision Computing* **30** (2012) 1052–1061

Chapter 6

Ageing in biometrics: a case study in online signature

Javier Galbally, Marcos Martinez-Diaz and Julian Fierrez

6.1 Introduction

Due to the fact that biometrics, as an automatic means of human recognition, constitutes a relatively novel field of research (Jain *et al.*, 2006), to date most efforts undertaken by the different parties involved in the development of this technology (researchers, industry, evaluators, etc.) have been mainly focused on the improvement of its performance (i.e. finding novel methods to obtain lower error rates) (Wayman *et al.*, 2005; Cappelli *et al.*, 2006). As a consequence, other important aspects closely related to this type of system such as the performance degradation effect known as *ageing* have been left partially uncovered (Jain *et al.*, 2008).

Although there always exists a certain variability among biometric samples of any given user (even when they have been acquired successively) (Doddington *et al.*, 1998), the term ageing is generally used to refer to the gradual decrease in a system performance caused by the changes suffered by the user's trait in the long term (which cannot be avoided as these are inherent to human nature) (Lanitis, 2010). These changes provoked by age entail that after a sufficiently long period of time, the initial enrolment template of a certain subject substantially differs from his current biometric samples, producing this way lower similarity scores and increasing the error rates of the system. Thus, ageing may be considered as a special type of large intra-user variability (i.e. variability within samples of the same user) caused by the inherent transformations of the human body over time.

The amount of time required for the stored template to become obsolete varies for each biometric trait depending, among other external aspects, on its own degree of permanence (Jain *et al.*, 2008). Thus, biometric traits such as fingerprints or the iris are more permanent over time, while others, such as the face (specially at early stages of life) or the signature, are much more sensitive to relatively short time variations. Furthermore, not every biometric system will be affected in the same way by ageing, as the features extracted from one given biometric trait are not all equally robust to the passing of time. In this context, the effect of ageing should be

Autonomous University of Madrid, Spain

analysed in a wide range of recognition systems working on a certain individual biometric modality in order to determine the approximate period of time in which the performance of that given trait will be consistent, before its discriminant capabilities start to drop.

In addition, once the consistent-performance time interval for a given recognition system has been set, an analysis of the best approach to overcome the effect of ageing should also be carried out. Among the different palliative methods that have been proposed in the literature, the ones that have received more attention from researchers and industry are the automatic template update strategies (Uludag *et al.*, 2003; Rattani *et al.*, 2009). These schemes use some type of target function (e.g. quality measure, similarity score) to automatically select from the most recent biometric samples given by the user to access the system, those which are most suited to be used to recompute (update) the subject's enrolment template.

In this context, for the definitive introduction of biometric technology in the security market, it is of utmost importance to take into account the problem of ageing in practical biometric applications, and to implement strategies that compensate for the gradual drift of their performance so that their valid life period (in which they offer a competitive performance) is increased.

However, in spite of their importance, studies regarding ageing and template update are difficult to carry out due to the lack of long-term biometric data. It is not easy to find databases where a statistically significant group of people have been captured over a sufficiently long period of time (Rawls and Ricanek, 2009). Furthermore, the acquisition process of such a database should be carried out under almost identical conditions (in terms of acquisition devices, level of control, supervision, etc.) so that the differences in the system performance can be attributed to the passing of time and not to the variability produced by other external factors.

It has not been until recently that different European and national efforts have led to the acquisition of compatible (regarding certain traits) multimodal databases with a large number of common users which have been captured in different sessions over a several year time span. Some examples include the Biosec (Fierrez *et al.*, 2007b), BiosecurID (Fierrez *et al.*, 2009) and Biosecure (Ortega-Garcia *et al.*, 2009) projects. For the current work, the signature modality of this common subset of users has been used to generate a new long-term dynamic signature dataset which has permitted us to analyse the effect of ageing on three competitive on-line signature verification systems working on totally different features and matchers. In addition to studying the sensitivity of the signature trait over time, several template update strategies have also been explored in order to assess their efficiency as a way to maintain the consistency of the system performance in the long term.

The present chapter constitutes pioneering work in the study of ageing in dynamic signature. We believe that it can be of great utility not only for researchers but also for developers and vendors in order to produce more secure and trusted applications based on the signature trait, to better understand its strengths, and to be able to foresee the weaknesses of this biometric modality. Furthermore, this type of study can also help to develop the ongoing biometric standards and to better define the requirements that real applications should comply with (ISO/IEC 19794-7, 2005; ISO/IEC 19794-11, 2005; ANSI-INCITS 395-2005, 2005).

The chapter is structured as follows. After the introduction, a selection of the most important related works may be found in section 6.2. The on-line Signature Long Term DB used in the experiments is presented in section 6.3. The experimental protocol followed is described in section 6.4, while results are given in section 6.5. Conclusions are finally drawn in section 6.6.

6.2 Related works

In the literature there exist different works where ageing has been studied from a medical point of view (Morgan *et al.*, 1994; Mueller, 1997; Coleman and Grover, 2006), or even its forensic implications (Albert *et al.*, 2007), however, not many studies can be found where ageing is analysed from a pure biometrical perspective (an exhaustive survey of these works was recently published by Lanitis (2010)). Furthermore, almost all of these biometric works on ageing are related to the face modality, but, to the best of our knowledge, none of them has been focused on the study of the signature trait.

Among these face-related contributions, there are works dealing with different aspects of ageing, for instance, its effect on the performance of face verification systems (Ling *et al.*, 2007), methodologies for the synthetic simulation of age (Lanitis *et al.*, 2002; Lanitis 2008), approaches for the compensation and modeling of the ageing effect (Suo *et al.*, 2007), automatic estimation methods of age (Geng *et al.*, 2007), or descriptions of long-term facial databases (FG-NET, 2007; Rawls and Ricanek, 2009). All this interest in the study of the effect of time on face recognition led in 2004 to the creation of a research group specialized in the analysis of the different factors related to face ageing (Face Aging Group, 2004).

Outside the face trait, Modi *et al.* have studied the correlation between the quality of fingerprint samples and the age of the users that produced them, and its impact on the final performance of fingerprint recognition systems (Modi and Elliott 2006; Modi *et al.*, 2007). In the same direction as the fingerprint works by Modi *et al.*, two studies have analysed the degree of the signing/drawing skill of people belonging to different age groups, and their ability to repeat certain valuable recognition features (Ketcham *et al.*, 2003; Guest, 2006). Although these works study an interesting issue related to ageing, they are not equivalent to the analysis carried out in the present work, as they do not track individuals over a significant period of their life, but they are focused on establishing a relationship between a certain group of people (e.g. the elderly, youngsters) and a given characteristic (fingerprint quality or signing skill) of their biometric samples (e.g. the elderly-bad quality-poor skill, youngsters-good quality-high skill).

Although it cannot be strictly considered as ageing, several works have analysed the short-term variability of signatures using samples captured in the same session (intra-session variability, within minutes), or in different sessions (inter-session variability, within weeks) of a regular acquisition campaign (Guest, 2006;

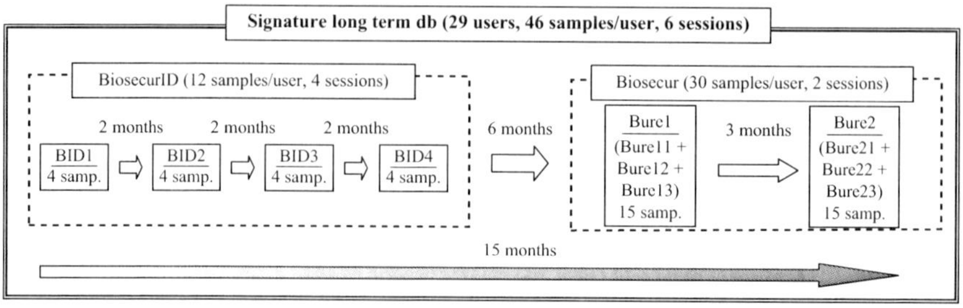

Figure 6.1 General time diagram of the different acquisition sessions that form the Signature Long Term DB

Galbally *et al.*, 2009; Scheidat *et al.*, 2012). In these cases, the differences in the systems performance can be attributed more to the inherent variability of the biometric samples (inter- and intra-user short-term variability) than to a real process of ageing, as the time interval between samples is in general too short (Doddington *et al.*, 1998; Poh *et al.*, 2006).

Regarding strategies that try to minimize the effect of ageing, among other possibilities such as using age invariant features (Guest, 2006), or compensating age changes (Park *et al.*, 2010), most efforts have been focused on the study of template update techniques (Rattani *et al.*, 2009) (i.e. using the most representative recent test samples of a user to update his enrolment template). In this field, different fully unsupervised or semiautomatic possibilities have been proposed in the fingerprint trait (Uludag *et al.*, 2003; Lumini and Nanni, 2006), face-based systems (Rattani *et al.*, 2008), or even multimodal biometric applications (Roli *et al.*, 2007).

6.3 The on-line signature long term database

The dataset used in the experimental section of this work comprises the on-line signature data of the 29 common users to the BiosecurID and the Biosecure databases. These 2 signature subsets, which were acquired in a 15 month time interval according to the general diagram shown in Figure 6.1, present some unique features that make them specially suited for the ageing evaluation performed in the present work.

- *The BiosecurID signature subset* (Fierrez *et al.*, 2009). It comprises 16 original signatures and 12 skilled forgeries per user, captured in 4 separate acquisition sessions (named here BID1, BID2, BID3 and BID4). The sessions were captured leaving a 2 month interval between them in a controlled and supervised office-like scenario. Users were asked to sign on a piece of paper inside a grid that marked the valid signing space, using an inking pen. The paper was placed on the Wacom Intuos 3 pen tablet that captured the time signals of each signature at a 100 Hz sampling rate (trajectory functions x and y with an accuracy of ± 0.25 mm, and pressure function p with a precision of 1,024 pressure levels). All the dynamic information is stored in separate text files following

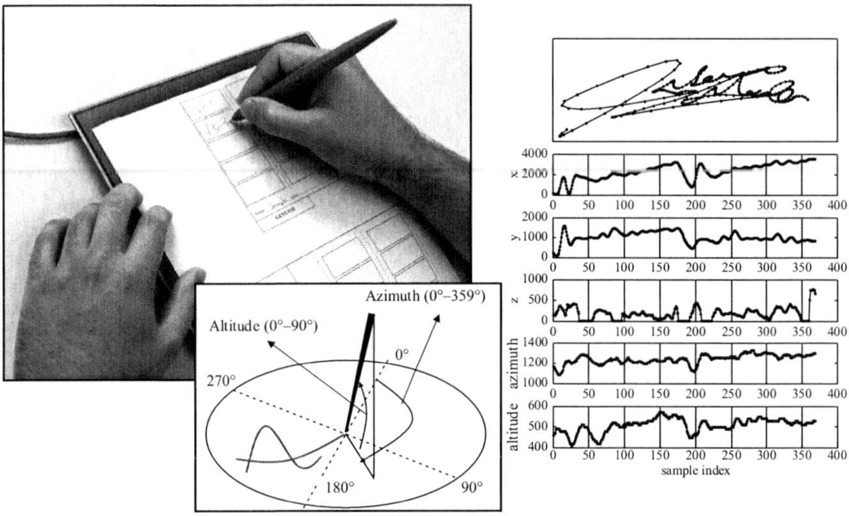

Figure 6.2 Example of a signature acquisition for the Signature Long Term DB

the format used in the first Signature Verification Competition, SVC (Yeung *et al.*, 2004). See Figure 6.2 for an acquisition example.

- *The Biosecure signature subset* (Ortega-Garcia *et al.*, 2009). This dataset was captured 6 months after the BiosecurID acquisition campaign finished (the time sequence of the 2 databases is shown in Figure 6.3). It comprises 30 original signatures per user, and 20 skilled forgeries, distributed in 2 acquisition sessions separated by 3 months (named here Bure1 and Bure2). The 15 original samples corresponding to each session were captured in 3 groups of 5 consecutive signatures with an interval of around 15 minutes between groups (named here Bure11-12-13 and Bure21-22-23 respectively). The signature dataset was designed to be fully compatible with the BiosecurID one. The acquisition scenario and protocol are almost identical: as in the BiosecurID case, users had to sign using an inking pen on a piece of paper with a restricted space placed over the Wacom Intuos 3 pen tablet. The dynamic information stored is the same as in BiosecurID and following also the SVC format.

For the final dataset used in the present work only the original signatures were considered; this way it comprises 1,334 signatures coming from the 29 common users of the 2 databases with 46 samples per user (16 from BiosecurID, and the remaining 30 from Biosecure) which are distributed in 6 sessions (BID1-2-3-4 and Bure1-2) according to the general time diagram shown in Figure 6.1.

The Signature Long Term DB is the first signature dataset where the signature of a given user can be tracked over a 15 month time span, as there are 6 almost uniformly distributed acquisition sessions in this interval. Therefore, the database is specially suited to assess if that period of time (15 months) is sufficient to detect a decrease in the verification performance of signature-based biometric systems.

Figure 6.3 Typical samples that can be found in the Signature Long Term DB. Each signature corresponds to each of the acquisition sessions of five different users

Furthermore, as all the samples of the same subject have been acquired under almost identical conditions we may discard external factors as the cause of a possible degradation in the recognition rates.

It should also be noted that all the users included in the database are adults. This means that their signature is a well-learned sequence of movements which may be considered as permanent and that has already gone through the transitional learning period which usually happens in youth. The effect of ageing during the time span in which the signature has not yet been fully fixed should be much greater and would be the subject of future work.

Some typical examples of the signatures that can be found in the different sessions comprising the Signature Long Term DB are shown in Figure 6.3.

6.4 Experimental protocol

The experimental framework has been designed with a twofold objective, on the one hand to determine to what extent the signature trait is affected by ageing and on the other hand to evaluate the efficiency of different template update approaches to thwart this effect. To achieve these goals, the protocol includes two sets of experiments in which the genuine scores (i.e. those affected by ageing) are computed on the Signature Long Term DB simulating different scenarios:

- *Ageing experiments: Fixed template and varying test.* In this case the user models enrolled to the system are always computed using the same samples (i.e. those belonging to the first session of the Signature Long Term DB, BID1), while the test signatures are taken from the following sessions (BID2-3-4 and Bure1-2).
- *Template update experiments: Varying template and fixed test.* In this case the test samples are always taken from one of the last sessions in the Signature Long Term DB (Bure13), while the enrolled models are updated with signatures coming from different previous sessions (BID2-3-4 and Bure11-12).

As mentioned in section 6.1, not all the systems working on a given trait are necessarily affected in the same way by ageing. In order to account for possible differences, we have carried out both sets of experiments on three different competitive on-line signature verification systems which use totally diverse feature sets (feature- and function-based) and matchers (distance measure, Hidden Markov Models, and Dynamic Time Warping). A brief description of each of the three systems is given next, while their DET curves evaluated on the BiosecurID DB (as an indication of their recognition capabilities) are shown in Figure 6.4.

- *System A: Feature-based + Mahalanobis distance.* This system models the signature as a holistic multidimensional vector composed of the best performing 40-feature subset extracted in Galbally *et al.* (2008) from the total set of 100 global features described in Fierrez-Aguilar *et al.* (2005). In the present study, we used this 40-feature representation of the signatures normalizing each of them to the range [0,1] using the tanh estimators described in Jain *et al.* (2005). Finally, the similarity scores are computed using the Mahalanobis distance between the input vector and a statistical model of the client estimated using a fixed number of training signatures.
- *System B: Function-based+HMM.* This function-based verification system applies a regional approach using a statistical model built using Hidden Markov Models (HMMs) (Rabiner, 1989) to a set of 10 time sequences selected

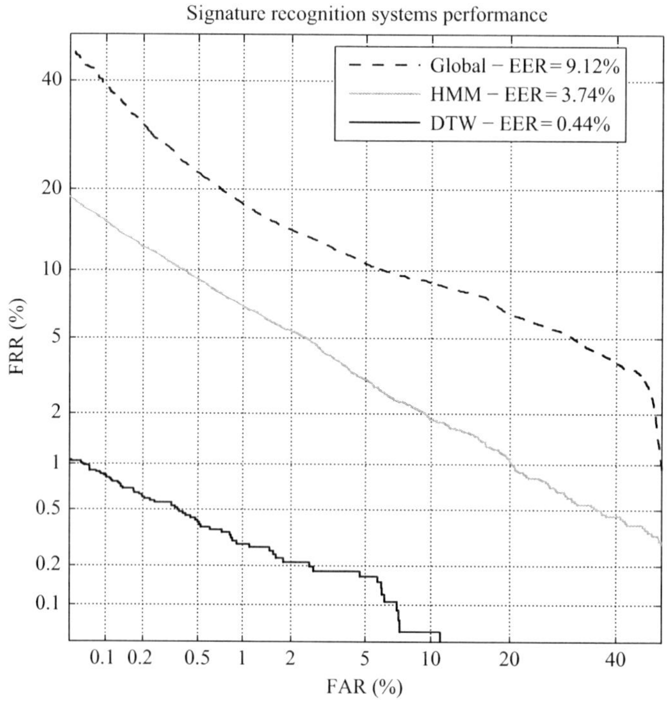

Figure 6.4 Performance evaluated on the BiosecurID DB of the three signature recognition systems used in the experiments, considering four signatures for enrolment

applying the SFFS feature selection algorithm to the total set of 34 functions defined in Martinez-Diaz *et al.* (2009a). This subset of 10 signals are derived from the coordinate (x and y) and pressure (p) functions, while no pen inclination signals are used as their utility for automatic signature recognition is at least unclear (Fierrez *et al.*, 2007a). After some preprocessing (position and rotation alignment) and the generation of the 10 functions, similarities are computed using 12 left-to-right HMM states and mixtures of 4 Gaussians per state. This system participated in the Signature Verification Competition 2004 with very good results (Yeung *et al.*, 2004), and the general configuration is detailed in (Martinez-Diaz *et al.*, 2008).

- *System C: Function-based + DTW.* In this function-based local approach a subset of 9 time functions (selected using SFFS from the total 34 feature set as in the case of system B) are directly matched using the elastic technique known as Dynamic Time Warping (DTW) (Kholmatov and Yanikoglu 2005). DTW is an application of Dynamic Programming to the problem of matching time sequences of different lengths, thus, the goal of DTW is to find an elastic match among samples of a pair of sequences that minimize a given distance measure. In this particular implementation, which is thoroughly described in

Martinez-Diaz *et al.* (2009b), we use the Euclidean distance as the measure to be optimized and only three correspondences among samples of the compared sequences are allowed, using symmetrical weighting factors. Although the DTW algorithm has been replaced by more powerful ones such as HMMs or SVMs for speech applications, it remains a highly effective tool for signature verification as it is best suited for small amounts of training data, which is a common case in signature verification (Houmani *et al.*, 2012).

6.5 Results

The results obtained for the three systems considered and for the two experimental scenarios (ageing and template update) are presented in this section.

6.5.1 Ageing experiments

As mentioned in section 6.4, these experiments are aimed at estimating the impact of ageing on signature recognition systems. For this purpose, the enrolled models of the 29 users present in the Signature Long Term DB are trained using the 4 signatures corresponding to the first session (BID1). Then, the genuine scores are computed matching these models against the signatures of the following 5 sessions: BID2-3-4 and Bure1-2.

This way, for each user we compute five different sets of scores: BID1 vs. BID2, BID1 vs. BID3, BID1 vs. BID4, BID1 vs. Bure1, and BID1 vs. Bure2 (see Table 6.1). For each of these sets, one final score is computed as the average of all the similarity measures in that set, so that, in the end, we obtain one final group of five scores, each of them representing the average of the genuine scores for each of the five experiments (A, B, C, D and E).

The results of these experiments for the three systems considered in the work are shown in Figure 6.5, where we can observe the evolution of the scores when the testing signatures move away (in terms of time) from the model. Several observations can be extracted from these results:

- All the three curves consistently decrease as the testing signatures are more distant from the model in terms of time, which means that the user's

Table 6.1 Enrolment and test signatures used to compute the genuine scores in the ageing experiments

	Ageing Experiments	
	Enrolment	**Test**
Exp. A	BID1 (4 sig.)	BID2 (4 sig.)
Exp. B	BID1 (4 sig.)	BID3 (4 sig.)
Exp. C	BID1 (4 sig.)	BID4 (4 sig.)
Exp. D	BID1 (4 sig.)	Bure1 (15 sig.)
Exp. E	BID1 (4 sig.)	Bure2 (15 sig.)

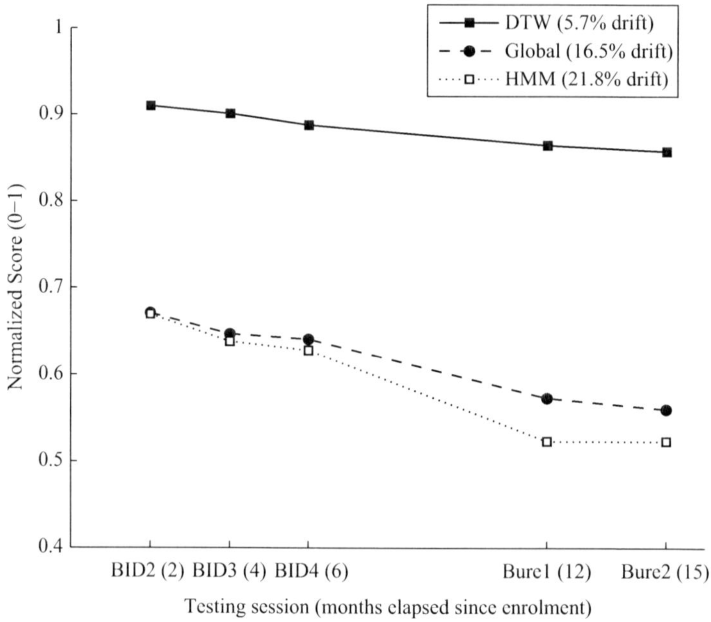

Figure 6.5 Evolution of the genuine scores through time for the three systems considered in the experiments

intra-variability increases with time or, in other words, that all the three recognition approaches are affected by ageing. Therefore, the results point out that this effect is not specific to a certain signature recognition technology, but that, as expected, it is inherent to the signature trait itself.

- Not all the systems are affected in the same way by the passing of time, that is, not all the curves in Figure 6.5 present the same decreasing slope. In particular, the system based on DTW presents a decrease in the scores between the first and the last testing set of signatures of 5.7%, compared to a 16.5% of the one based on global features and a 21.8% for the HMM. Thus, we may conclude that the signature recognition technology based on DTW is not only more accurate (see Figure 6.4) but also more robust to ageing.
- Another important observation to be made from the results shown in Figure 6.5 is that the effect of ageing on the signature trait is not negligible. There is a significant drift in the scores (from the first to the last signature test set) in a relatively short period of time (only 15 months).

6.5.2. *Template update experiments*

The results presented in section 6.5.1 confirm the necessity to develop strategies that can help to minimize the effect of ageing, specially on those behavioural or learned traits, such as the signature, which are more sensitive to the passing of time. Here, we analyse the efficiency of different template update approaches varying the enrolment

Table 6.2 Enrolment and test signatures used to compute the genuine scores in the template update experiments

	Template Update Experiments	
	Enrolment	**Test**
Exp. F (baseline)	BID1 (4 sig.)	Burel3
Exp. G (complete update)	Burel1 (4 sig.)	Burel3
Exp. H (mixed update)	BID1 (4 sig.) + Burel1 (4 sig.)	Burel3
Exp. I (complete update)	Burel1 (4 sig.) + Burel2 (4 sig.)	Burel3

signatures used to compute the users' models and testing always with the same set of samples, as shown in Table 6.2. In particular, the scenarios considered are as follows:

- *Baseline result (exp. F)*. This represents the ageing scenario to be corrected using template update strategies. There is a 12 month difference between the enrolled model (BID1) and the test set (Burel3).
- *Complete update (exp. G)*. The first template update approach considered is to discard the old enrolment samples (BID1) and replace them by new samples acquired very close in time to the test set (Burel1).
- *Mixed update (exp. H)*. In this case we do not discard the old samples but we update the enrolled model with newly acquired samples (BID1 + Burel1). In this case there will be more available data to train the model than in the previous two cases (experiments F and G).
- *Complete update (exp. I)*. Here, we take the same amount of training data as in experiment H, but all from recent acquisitions (Burel1 + Burel2).

The results of the previously described setups for the three considered systems are shown in Figure 6.6. As in the case of the ageing experiments, the final score shown for each of the four considered scenarios is the average of the scores obtained for all the users in the database. Although all the template update strategies studied improve the performance with respect to the baseline experiment, two different behaviours may be observed in Figure 6.6 depending on the signature recognition system considered:

- *HMM system*. HMM-based systems heavily depend on the amount of training data available (Fierrez *et al.*, 2007a). As a consequence, it is better to perform a mixed update (i.e. do not discard the old samples, exp. H) so that the model is trained with as many signatures as possible (eight signatures, in this particular case), instead of using only the most recent samples (i.e. exp. G, where only four signatures are used for enrolment).
- *Global features and DTW systems*. On the other hand, the systems based on DTW and global features do not rely as much on the amount of enrolment data, but on the quality of these data (Martinez-Diaz *et al.*, 2009b). Therefore, the performance achieved using four recently acquired samples (exp. G) is almost the same as that obtained using eight of those signatures (exp. I). This means

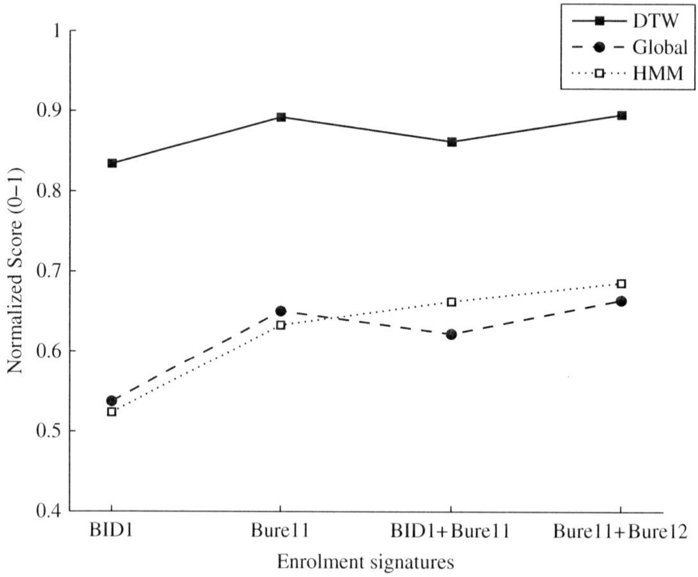

Figure 6.6 Average genuine score for the four different template update strategies tested and for the three systems considered in the experiments

that, as can be seen in Figure 6.3, in these cases it is preferable to perform a complete update with the most recent samples (i.e. exp. G) than to keep the old samples (i.e. exp. H) even if this means training the enrolled model with a smaller number of signatures.

As could be expected, in all cases the best possible template update strategy is to use for enrolment all the most recent samples available (i.e. exp. I). However, this may represent a somewhat unrealistic scenario, as we are assuming that we have access to as many as eight signatures captured in a time period very close to the test set. The amount of new collected data will rarely comply with this condition.

6.6 Conclusions

We have conducted the first systematic study on the degradation of on-line signature with time and how this ageing effect may be compensated. For this purpose, we have introduced the new Signature Long Term DB which contains dynamic signature samples of 29 users captured under very similar conditions over a 15 month time span. The experiments, carried out using 3 totally different state-of-the-art systems representing the most usual technologies in on-line signature recognition, have proven that the effect of ageing is present in this trait even for time lapses of several months. The recognition technology based on DTW has demonstrated that it is not only the most accurate (Houmani *et al.*, 2012) but also the most robust to the passing of time. Regarding the template update approaches analysed to minimize the effects of ageing it has been found that the most effective strategies may vary depending on the system being used, although, in general, using as much recently collected data as possible clearly improves the results.

Research studies such as the one presented here try to shed some light into the difficult problem of biometric ageing. Performing systematic studies of biometric systems sensitivity to the passing of time is essential before efficient approaches that minimize the impact of the detected effects can be developed, so that the user acceptability of this thriving technology is improved.

Acknowledgments

This work has been partially supported by projects Contexts (S2009/TIC-1485) from CAM, Bio-Challenge (TEC2009-11186) and Bio-Shield (TEC2012-34881) from Spanish MECD, *Cátedra UAM-Telefónica* and by the Spanish *Dirección General de la Guardia Civil.*

References

Albert, A. M., Ricanek, K., Patterson, E., 2007. 'A review of the literature on the aging adult skull and face: Implications for forensic science research and applications'. *Forensic Science International* **172**, 1–9.

ANSI-INCITS 395-2005, 2005. ANSI-INCITS 395-2005: 'Information technology – biometric data interchange formats: Signature/sign data'.

Cappelli, R., Maio, D., Maltoni, D., Wayman, J. L., Jain, A. K., 2006. 'Performance evaluation of fingerprint verification systems'. *IEEE Transactions on Pattern Analysis and Machine Intelligence* **28**(1), 3–18.

Coleman, S. R., Grover, R., 2006. 'The anatomy of the aging face: volume loss and changes in 3-dimensional topography'. *Anesthetic Surgery Journal* **26**, 4–9.

Doddington, G., Liggett, W., Martin, A., Pryzbocki, M., Reynolds, D., 1998. 'Sheep, goats, lambs, and wolves: A statistical analysis of speaker performance in the NIST 1998 Speaker Recognition Evaluation'. *Proceedings of the International Conference on Spoken Language Processing (ICSLP).*

Face Aging Group, 2004. http://www.faceaginggroup.com/

FG-NET, 2007. 'The Face and Gesture Recognition Research Network (FG-Net) Aging Database'. http://www.fgnet.rsunit.com/

Fierrez, J., Galbally, J., Ortega-Garcia, J., Freire, M. R., Alonso-Fernandez, F., Ramos, D., Toledano, D. T., . . . Gracia-Roche, J. J., 2009. 'BiosecurID: A multimodal biometric database'. *Pattern Analysis and Applications* **13**, 235–246.

Fierrez, J., Ortega-Garcia, J., Ramos, D., Gonzalez-Rodriguez, J., 2007a. 'HMM-based on-line signature verification: Feature extraction and signature modeling'. *Pattern Recognition Letters* **8**, 2325–2334.

Fierrez, J., Ortega-Garcia, J., Torre-Toledano, D., Gonzalez-Rodriguez, J., 2007b. BioSec baseline corpus: A multimodal biometric database. *Pattern Recognition* **40**, 1389–1392.

Fierrez-Aguilar, J., Nanni, L., Lopez-Pealba, J., Ortega-Garcia, J., Maltoni, D., 2005. 'An on-line signature verification system based on fusion of local and global information'. *Proceedings of the IAPR Audio- and Video-Based Biometric Person Authentication (AVBPA)*. Springer LNCS-3546, Springer-Verlag, Berlin, Germany; pp. 523–532.

Galbally, J., Fierrez, J., Martinez-Diaz, M., Ortega-Garcia, J., 2009. 'Improving the enrollment in dynamic signature verification with synthetic samples'. *Proceedins of the IAPR International Conference on Document Analysis and Recognition (ICDAR)*. Barcelona, Spain; pp. 1295–1299.

Galbally, J., Fierrez, J., Ortega-Garcia, J., 2008. Performance and robustness: A trade-off in dynamic signature verification. *Proceedings of the IEEE International Conference on Acoustics, Speech and Signal Processing (ICASSP)*. pp. 1697–1700.

Geng, X., Zhou, Z. H., Smith-Miles, K., 2007. 'Automatic age estimation based on facial aging pattern'. *IEEE Transactions on Pattern Analysis and Machine Intelligence* **29**, 2234–2240.

Guest, R., 2006. 'Age dependency in handwritten dynamic signature verification systems'. *Pattern Recognition Letters* **27**, 1098–1104.

Houmani, N., Mayoue, A., Garcia-Salicetti, S., Dorizzi, B., Khalil, M. I., Moustafa, M. N., Abbas, H.,... Vivaracho-Pascual, C., 2012. 'BioSecure signature evaluation campaign (BSEC'2009): Evaluating online signature algorithms depending on the quality of signatures'. *Pattern Recognition* **45**, 993–1003.

ISO/IEC 19794-11, 2005. 'ISO/IEC 19794-11:2005, information technology – biometric data interchange formats – part 11: Signature/sign processed dynamic data'.

ISO/IEC 19794-7, 2005. 'ISO/IEC 19794-7:2005, information technology – biometric data interchange formats – part 7: Signature/sign time series data'.

Jain, A. K., Nandakumar, K., Ross, A., 2005. 'Score normalization in multimodal biometric systems'. *Pattern Recognition* **38**, 2270–2285.

Jain, A. K., Ross, A., Flynn, P. (Eds.), 2008. *Handbook of Biometrics*. Springer: Springer, New York, NY.

Jain, A. K., Ross, A., Pankanti, S., 2006. 'Biometrics: A tool for information security'. *IEEE Transactions on Information Forensics and Security* **1**(2), 125–143.

Ketcham, C. J., Dounskaia, N. V., Stelmach, G. E., 2003. 'Control of multijoint drawing movements: a comparison of young and elderly adults'. *Proceedings of the Conference of the International Graphonomics Society (IGS)*. Scottsdale, AZ; pp. 42–44.

Kholmatov, A., Yanikoglu, B., 2005. 'Identity authentication using improved online signature verification method'. *Pattern Recognition Letters* **26**, 2400–2408.

Lanitis, A., 2008. 'Evaluating the performance of face-aging algorithms. *Proceedings of the IEEE International Conference on Automatic Face and Gesture Recognition (ICAFGR)*. Amsterdam, Netherlands.

Lanitis, A., 2010. 'A survey of the effects of aging on biometric identity verification'. *International Journal of Biometrics* **2**, 34–52.

Lanitis, A., Taylor, C. J., Cootes, T. F., 2002. 'Toward automatic simulation of aging effects on face images'. *IEEE Transactions on Pattern Analysis and Machine Intelligence* **24**, 442–455.

Ling, H., Soatto, S., Ramanathan, N., Jacobs, D. W., 2007. 'A study of face recognition as people age'. *Proceedings of the IEEE International Conference on Computer Vision (ICCV)*. Rio de Janeiro, Brazil; pp. 1–8.

Lumini, A., Nanni, L., 2006. 'A clustering method for automatic biometric template selection'. *Pattern Recognition* **39**, 495–497.

Martinez-Diaz, M., Fierrez, J., Galbally, J., Ortega-Garcia, J., 2008. 'Towards mobile authentication using dynamic signature verification: Useful features and performance evaluation'. *Proceedings of the IAPR International Conference on Pattern Recognition (ICPR)*. Tampa, FL.

Martinez-Diaz, M., Fierrez, J., Hangai, J., 2009a. *Encyclopedia of Biometrics*. Springer, Ch. Signature features, Springer, New York, USA; pp. 1185–1192.

Martinez-Diaz, M., Fierrez, J., Hangai, S., 2009b. *Encyclopedia of Biometrics*. Springer, Ch. Signature matching, Springer, New York, NY; pp. 1192–1196.

Modi, S. K., Elliott, S. J., 2006. 'Impact of image quality on performance: age comparison of young and elderly fingerprints'. *Proceedings of the International Conference on Recent Advances in Soft Computing (ICRASC)*. Kent, UK; pp. 10–12.

Modi, S. K., Elliott, S. J., Hakil, K., 2007. 'Impact of age groups on fingerprint recognition performance'. *Proceedings of the IEEE Workshop on Automatic Identification Advanced Technologies (AutoID)*. Alghero, Italy; pp. 19–23.

Morgan, M., Bradshaw, J. L., Phillips, J. G., Mattingley, J. B., Iansek, R., Bradshaw, J. A., 1994. 'Effects of hand and age upong abductive and adductive movements – a kinematic analysis'. *Brain Cognition* **25**, 194–206.

Mueller, P. B., 1997. 'The aging voice'. *Seminars in Speech and Language*, **18**, pp. 159–168.

Ortega-Garcia, J., Fierrez, J., Alonso-Fernandez, F., Galbally, J., Freire, M. R., Gonzalez-Rodriguez, J., Garcia-Mateo, C., . . . Savran, A., 2009. 'The multi-scenario multi-environment BioSecure multimodal database (BMDB)'. *IEEE Transactions on Pattern Analysis and Machine Intelligence* **32**, 1097–1111.

Park, U., Tong, Y., Jain, A. K., 2010. 'Age-invariant face recognition'. *IEEE Transactions on Pattern Analysis and Machine Intelligence* **32**, 947–954.

Poh, N., Bengio, S., Ross, A., 2006. 'Revisiting doddington's zoo: A systematic method to assess user-dependent variabilities'. *Proceedings Workshop on Multimodal User Authentication (MMUA)*. Toulouse, France.

Rabiner, L. R., 1989. 'A tutorial on Hidden Markov Models and selected applications in speech recognition'. *Proceedings of the IEEE* **77**(2), 257–286.

Rattani, A., Freni, B., Marcialis, G. L., Roli, F., 2009. 'Template update methods in adaptive biometric systems: A critical review'. *Proceedings of the IAPR/IEEE International Conference on Biometrics (ICB)*. Springer LNCS-5558, Springer-Verlag, Berlin, Germany; pp. 847–856.

Rattani, A., Marcialis, G. L., Roli, F., 2008. 'Biometric template update using the graph mincut: A case study in face verification'. *Proc. IEEE Biometric Symposium (Bsym).*Tampa, FL; pp. 23–28.

Rawls, A. W., Ricanek, K., 2009. MORPH: Development and optimization of a longitudinal age progression database. *Proceedings of the COST 2101 Workshop on Biometrics and Identity Management (BIOID).* Springer LNCS-5707, pp. 17–24.

Roli, F., Didaci, L., Marcialis, G. L., 2007. 'Template co-update in multimodal biometric systems'. *Proceedings of the IAPR/IEEE International Conference on Biometrics (ICB).* Springer LNCS-4642, Springer-Verlag, Berlin, Germany; pp. 1194–1202.

Scheidat, T., Kummel, K., Vielhauer, C., 2012. 'Short term template aging effects on biometric authentication performance'. *Proceedings of the International Conference on Communications and Multimedia Security.* Springer LNCS 7394. Springer-Verlag, Berlin, Germany; pp. 107–116.

Suo, J. L., f. Min, Zhu, S. C., Shan, S. G., Chen, X. L., 2007. 'A multi-resolution dynamic model for face aging'. *Proceedings of the IEEE International Conference on Computer Vision and Pattern Recognition (ICCVPR).* Minneapolis, MN; pp. 1–8.

Uludag, U., Ross, A., Jain, A. K., 2003. 'Biometric template selection and update: A case study in fingerprints'. *Pattern Recognition* **37**, 1533–1542.

Wayman, J., Jain, A., Maltoni, D., Maio, D., 2005. *Biometric Systems: Technology, Design and Performance Evaluation.* Springer. Springer-Verlag, London, UK.

Yeung, D. Y., Chang, H., Xiong, Y., George, S., Kashi, R., Matsumoto, T., Rigoll, G., 2004. SVC2004: 'First international signature verification competition'. In: Zhang, D., Jain, A. K. (Eds.), *Proceedings of the IAPR International Conference on Biometric Authentication (ICBA).* Springer LNCS-3072, Springer-Verlag, Berlin, Germany; pp. 16–22.

Chapter 7

Ageing in iris biometrics

Judith Liu-Jimenez and Raul Sanchez-Reillo

Iris Biometrics is one of the most used biometric modalities nowadays. These systems were initially used for highly secured environments, but thanks to the decrease in sensor cost, low computational requirements and the effective recognition they provide, increased deployment has occurred in the last years.

Iris Biometrics has shown one of the best recognition rates among different modalities. The results obtained in international and competitive challenges have positioned it as one of the most secure modalities. In fact, the results presented in the ICE 2006 challenge provide recognition rates of over 99% when the FAR is 0.001% [1].

This modality deployment started in 2001, when it was used for the first time in a massive recognition scenario: the United Arab Emirates border control. Till that moment most applications of these systems were limited to high security environments or access to particular facilities or devices. However, when the use was extended to border control, the variety and number of users accessing through these systems increased significantly. In some cases, the use of these systems was voluntary, and only the bravest citizens tried them, but in some other cases it was not, e.g. in 2011, India started one of the biggest projects based on this technology [2]: the whole Indian population is being enrolled for census, medical insurance purposes, etc.

When the use of this technology increased, one of the problems to be considered was ageing. A large population or a population that is not going to have access to this system quite often will require the system to be able to recognize any user after a long time. In those cases it is not possible to re-enrol a user in the system each year because of time constraints and associated quite high cost. Therefore, Iris Biometrics should be prepared for ageing.

The ageing problem is related to the probability of people's iris changing with time, which is influenced by physical ageing as well as template ageing. We should consider not only that the iris itself will change with time but also that all the users will change, and therefore, some other factors will influence how the user interacts

Carlos III University of Madrid, Spain

with the system or how the sample he/she provides will change with time. The ageing topic in the case of iris has not received as much research attention as other biometric modalities, such as face. In this chapter we will describe some of the work done related to that as well as some other related works which point out some topics which are worthy of study in relation to long-term iris biometric systems.

This chapter is organized as follows: we start with a brief introduction to the eye and the iris. It is important to stop to see what we are measuring and try to understand from the anatomical organ what we can expect to happen with time. Following this, we will briefly describe the different components which form an iris biometric system as well as each component's functionality. This will help us to understand how age influences each part of the system. After all these basic concepts, we will move to the studies developed by different researchers on this topic, observing that different researchers have focused their efforts on different parts of the iris biometric system.

7.1 Anatomy of the eye

For understanding the iris we should consider not only the iris, but the complete organ where it is located: the eye (Figure 7.1). The eye is the human organ used for obtaining visual information. Its main task is to provide visual information, and for that, the eye allows light to pass through the pupil to the visual nerve; this

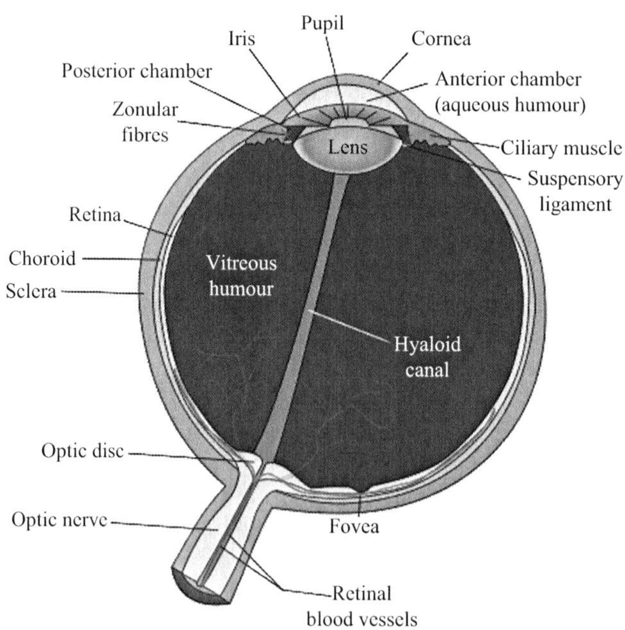

Figure 7.1 Eye anatomy [4]

information is processed in the brain. The eye is basically composed of three different layers [3]:

- The fibrous tunic or tunica fibrosa oculi is the outer layer and is formed by the cornea and the sclera. The sclera is the layer that gives the eye the white colour. The cornea is a transparent tissue which allows light to enter the eye and protects the inner layers. The front part of the cornea gets the name of crystalline lens.
- Vascular tunic or tunica vasculosa oculi is the middle layer which includes the iris, the ciliary body and the choroid.
  - The iris is situated in the middle layer of the eye between the lens and the cornea and aqueous humour. It consists of a pigmented fibrovascular tissue known as the stroma.
  - The stroma connects a sphincter muscle, which is in charge of contracting the pupil, and several dilator muscles, which open it. All these elements form a unique structure of valleys and ridges for each person and are studied in biometrics for identification. To simplify, we can consider that the iris is a muscle which controls the aperture of the pupil, the black part of the iris. The iris is the only muscle which can be seen from outside thanks to the transparent nature of the cornea.
- Nervous tunic or tunica nervosa oculi is the inner layer which includes the retina, composed of photosensitive rods and cone cells which are associated with vision functions.

According to the eye anatomy, several questions arise when thinking of ageing:

- Is the iris structure going to change? Is the iris as a muscle going to change its characteristic ridges and valleys with time?
- Does the iris colour change with time? And if so, will it influence the performance of an iris biometric system?
- Will the iris keep its properties with time? Will the pupil dilation capacity be the same throughout human life? If it is not, would it mean that with time the user will show more iris area or not?
- Does the cornea change with time and therefore make it more difficult to observe the iris?

Probably the reader can solve these questions while reading, and he/she will have some thoughts on these questions. The reader can answer the above questions in the following ways:

- No visual changes have been observed in the iris structure [5], so the first answer for the first question is that we can expect that a biometric system performance will not be influenced because of this reason.
- The question related to the iris colour probably is the easiest to answer. The iris colour changes with time. It is quite common to see babies with bright irises, the colour of which changes with time and becomes darker. The iris usually changes its appearance from bright to darker till a child attains the age of 2–3 years. Then it may undergo a slight change in colour till the teenage years. After being

stable till the age of 55–60 years, the iris changes again to a brighter colour. There are even people whose irises change to a brighter colour in summer due to the exposure to sunlight. Therefore the iris melanin is not stable with time. The question then will be if the iris modality is dependent on the iris melanin.

- Does the iris get lazy with time? Somehow, it does. The iris, as a muscle, loses some of its mobility with time, and the dilation of the iris is slower in older people than in youngsters (19, 20). That's to say, the pupil accommodation requires more time in older population. The pupil dilation depends on several factors. Besides age, illumination, drugs, etc. can influence the pupil size and therefore the usable iris area suitable for a biometric system.
- The cornea can become less transparent with time. This can be due to decay in cornea transparency or because of sediments within the crystalline. In fact cataracts, a quite common eye disease which can appear in older people, translate into a crystalline lens darkening.

7.2 Iris biometric scheme

Considering the iris and its characteristics, in 1993 the first algorithm appeared to recognize persons in this way. Due to the success of that first algorithm, several algorithms have been developed since then. Although there are a wide variety of algorithms, most of them follow the scheme shown in Figure 7.2.

The acquisition is performed using infrared cameras. The use of infrared cameras avoids annoying flashes for the user. In infrared, the images obtained show clearly the ridges and valleys that form the iris, but there is no colour information. Although in iris recognition the colour is not a feature studied, the colour of the iris can affect the recognition because of both the above-mentioned variability of the melanin and the image contrast. In visible images, the iris cannot be clearly appreciated in dark eyes and the segmentation between the iris and the pupil is harder. Conversely bright eyes show lower contrast between the sclera and the iris.

Following this acquisition, the image is processed in order to observe if the image acquired satisfies particular quality features. If the image satisfied them, the system will continue with the segmentation, also known as pre-processing; if not, another image is required, as the initial one is considered not to be of sufficient quality for a correct recognition. Several quality factors have been considered in the

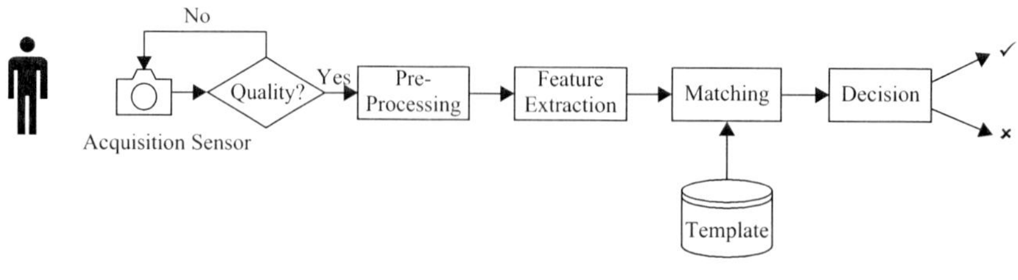

Figure 7.2 Iris biometric system scheme

literature [6, 7] and defined in the International standard ISO/IEC 29794-6, currently under development.

The following block after this quality assessment is the iris pre-processing block. This block prepares the image for the algorithm core: the feature extraction and the matching. Among the different tasks of this block are the enhancing of the iris structure and the more important step of segmenting the iris from the rest of the information in the image, i.e. detecting the pupil-iris boundary, the iris outer boundary, eyelids, eye lashes, etc. This process is necessary to reduce the computational time required by the following steps. Another important goal of this process is to remove external elements of the iris in the image, such as eye lashes. These elements can introduce noise into measurements and lead to erroneous recognitions, as well as making the direct comparison between the iris template and the iris sample impossible.

After the pre-processing, the feature extraction block converts the iris image into a quantifiable vector. In many cases, before obtaining this vector, the iris is normalized from a Polar scale to a Cartesian one. This step is performed to normalize the image and to compensate its size, trying to compensate for possible pupil dilation. Once the iris is normalized, a process to obtain the iris feature vector is performed. Several algorithms can be found in the literature, some of them explore the iris texture using texture tools such as Gabor filters and Gabor wavelet transforms, while others are based on ordinal measures or other kinds of approaches.

Finally the feature vector obtained is compared with the templates already stored in the database for identification purposes, or just compared with the template of the user claiming his/her identity in the case of a verification-based system. This comparison depends on the feature vector obtained, but most of the feature extractor blocks return bit-vectors and the most commonly used comparison algorithms are based on the Hamming distance.

7.3 Iris ageing

Not many studies have been carried out on this topic. Due to the fast spread of biometric systems, not much data related to ageing can be found. However, it is an important matter as we still do not have an answer to the questions: how will these systems behave in the future when the enrolment stage took place a short time (or long time) before identification? Will they be able to recognize people? Are there any difficulties which should be considered when the enrolled user gets older?

The first published iris algorithm papers [8–13] suggested that iris structure does not suffer modification through time, but without a demonstration of this fact. The first test regarding ageing was performed by John Daugman in 2002. In this experiment, Daugman showed the possibility of recognizing a person through time thanks to the iris [14]. Daugman verified the identity of a woman who claimed to be the one whose photo was on the *National Geographic* journal cover in 1984, 18 years after the publication of the original photo.

Daugman performed the following steps: he used his algorithm to check if the applicant is the Afghan girl from the original photo, so he applied his algorithm to both photos and observed the distance obtained when comparing them. He concludes that the distance obtained (0.24 for the left iris and 0.31 for the right iris) is enough to consider that the iris observed belonged to the first girl. However, this experiment just concludes that the comparison result obtained between the initial photo and the current photo is enough to consider her as the same person, but it does not provide information about how the comparison results can change with time and how this possible change will influence a large identification process (i.e. a 1-*N* comparison, instead of 1-1 comparison). Additionally, it is important to note that John Daugman's algorithm usually leads to an intraclass-interclass difference of almost 0.5 in distance in infrared images, when the comparison is made between short-time (a few months) samples. So, although the result obtained in this case, 18 years elapsed samples, is enough for verification, the expected Hamming distance for a database where samples are taken over a long time will get modified from the original one, and probably the difference between intraclass and interclass distances will decrease with time.

As already stated, the comparison of one single person is not enough to predict how a massive biometric system will act. Will the performance rates be affected? Here we will focus on different research works reported by several groups, which point out some of the difficulties and the results they have obtained when considering iris changes with time. For an easy structure we will focus on each of the parts of a biometric system and how age influences each of them.

For understanding the age phenomenon two different aspects will be considered: the template ageing and other age effects. The first effects are related to the iris changes with time, i.e. considering a user enrolled at time T, what will be the effect of an identification or verification process at the time $T + t$. The second ones do not consider a long time between the enrolment and the identification/verification process, but they consider the user's age. Although initially these studies do not seem to be related to the ageing effect, we consider that their results are significant in order to understand collateral effects in an iris biometric system lifetime. They can be a starting point for further research using long interval databases, as the physical effects they point out will influence recognition performance results.

7.4 Sensor

Initially the sensor is just a camera which should acquire the iris image. However, due to the iris size, the positioning of the user is a critical aspect in this step. The iris size usually is 11–14 mm. Acquiring a focused image of an object of this size requires a small range of focal length. Therefore, the user should be positioned at almost an exact distance to the camera. This makes interaction with the device a difficult task for some old people.

The trial carried out in Reference 15 showed that the percentage of errors in the enrolment increases as age increases. As also mentioned in this trial, the main reason of the high failure to enrol (FTE) rate is due to person behaviour, i.e. how the user interacts with the system. The different causes of this behaviour can be generally categorised into one of the following two groups:

- Voluntary reasons: those that the user can control. In this sense, ageing is not directly connected with these results, but indirectly. Older people often have problems to understand new technologies, as people usually reject possible changes and new technologies become a challenge. Usually new technologies, in their design and implementation processes, are focused on a younger population market, for those who already have some knowledge of other technologies, but for older populations, the technology jump is bigger and less understandable.
- Involuntary reasons: those that the user cannot control. In the case of iris recognition, we can find some of these reasons related to age: excessive blinking, impossibility of a correct positioning, diseases such as Parkinson's, etc. These problems can be also related to ageing and the high sensitivity of these sensors to the focusing distance.

In general, this report showed that iris biometric systems are not designed for older people interaction or for their idiosyncrasy; the systems do not probably provide adequate guidance, are not easily usable or adapted for their specific necessities.

7.5 Quality measures

The image obtained by the system should satisfy some quality measures in allowing affordable recognition rates. These quality measures have been defined in References 6 and 7. The IREX II-IQCE report [6] examines how each of these factors influences the algorithm performance. Some of these quality factors are image resolution, focus, occlusion, sensor noise, etc. No studies have been performed regarding ageing influence on quality measures. However, some of these factors can be affected by ageing, although until now, no specific results can be given.

- Pupil-iris dilation: This parameter measures the difference between the pupil size and the iris size. This parameter is clearly influenced by age, and it will be explained in the following section.
- Enough usable iris area: This parameter is determined as one of the most influential factors in an iris recognition algorithm performance. It is defined as the percentage of iris that is not occluded by eyelashes, eyelid, specular reflections (e.g. reflected sunlight, eye-wear reflections, reflections from nearby facial features such as the nose, or glare and reflection from the scene), ambient specular reflections, etc. If the observable and usable iris area decreases, the recognition results also decrease. Older people tend to provide less usable iris area because of the use of glasses.
- Iris-pupil contrast is a measure of the image characteristics of the boundary between the iris region and the pupil. A low iris pupil contrast makes the pupillary boundary detection a difficult task, and it influences the FMR and FNMR [6].

This contrast varies among human irises when images are captured with near infrared illumination. Certain medical conditions, typical in older people, such as cataracts or intraocular lenses, may impact on the iris pupil contrast parameter [34].

– Greyscale: Although all images are obtained in infrared and the colour of the iris is not studied, bright irises lead to infrared images where the contrast between the pupil and the iris, and the iris and the sclera is not as sharp as with dark irises. The image contrast can influence the algorithm, as the texture of the iris cannot be perfectly observed and, depending on the robustness of the algorithm, can decrease the performance results.

7.6 Pre-processing

Initially, the first thought would be that the iris segmentation should not be affected by the age. The segmentation should be initially the same for different ages, but this block has shown important relationships between the results obtained and age. The only study completely dedicated to this topic is reported by Fairhurst and Erbilek [16].

In this study, the authors use the well-known algorithm proposed and developed by Libor Masek in Reference 18. This algorithm, frequently used by different authors for testing purposes, is used to observe how it behaves when considering a user's age. They observed a high influence between the age and the performance result. For testing their ideas the Biosecure database [17] was used, accompanied with the age of the users whose samples formed the database at the time of acquiring them. The image acquisition environment did not change among users or sessions; the lighting was constant and equal for all users. One of the first results they observed is that using the original Masek's algorithm, a significant failure rate in users above 60 years was found. This increase was up to 34% compared with the one obtained in users whose age was below 60 and whose error rate was 11%. This data shows how the age influences the general performance of a biometric system.

Considering these data, in the same paper, Fairhurst and Erbilek proposed a new pre-processing algorithm for more accurate iris segmentation. With the modifications proposed they showed a much better performance results, reducing this 34% error rate to almost 0% in users above 60% and to a less of 1% in other users. Therefore, an effective segmentation is required in order to consider users of different ages.

To find an explanation of this decrease in error rate, the authors look for a reason for this amazing decrease in error in older users. The explanation found joins two ideas already mentioned by other authors: the pupil dilation with age and the size of the usable iris area. Winn *et al.* and Bitsios *et al.* [19, 20] pointed out that with age the pupil size tends to decrease and the iris becomes less reactive. The capacity of modifying the size of the pupil depends on the iris, and according to these authors the adaptation capacity decreases with age. Hollingsworth *et al.* [21] studied the influence of the pupil size in iris recognition systems. Hollingsworth *et al.* using an algorithm based on Daugman [12], Masek [18] and Liu [22] showed that the smaller the pupil is, the better results are obtained, as a higher usable iris area is possible (which is also consistent with the quality measures mentioned earlier).

The results provided by Fairhurst and Erbilek combined those two previous data, showing how as time passes the iris usable area increases as the pupil size decreases and the iris recognition results are better. In such paper, the mean dilation ratios obtained were 0.4317, 0.3816 and 0.3214 for the '<25' age group, the '25–60' age group and the '>60' age group respectively. These results demonstrate very clearly how the pupil dilation changes significantly with increasing age, and consequently, the recognition results.

Although these results are not studied through time and no data regarding how this dilation ratio changes is given, the importance of these results can be extended for a long time systems, and show how important is a proper segmentation to avoid increasing error rates.

7.7 Feature extractor and matching

Template ageing is one of the key points of iris biometrics ageing. With the work now discussed, the authors have tried to understand if the iris structure changes with time and if so, how this influences the performance results of identification systems.

It is important to point out the difficulty that all the following authors have encountered: the lack of long-term databases. As mentioned before, iris biometrics is a quite young technology and therefore, finding databases where the samples are taken in sessions well separated in time is currently nearly impossible.

In this section, the works developed about this topic will be discussed. These works will be explained according to their publication date.

Tome-Gonzalez *et al.* were the first authors in studying this topic. In the paper titled 'On the effects of time variability in Iris recognition' [23], they studied how time influences the performance results of an iris biometric system. For this purpose they used the Biosec [24] and BiosecurID [25] databases and Masek's algorithm without modifications. These databases are formed by iris images obtained in two sessions separated from one to four months. Each user provides 8 (4 images $\times$ 2 eyes) images in each session. They based their study on the different interclass and intraclass curves comparing iris in the same session with other sessions. They observed that as time passes the intraclass curve tends to go nearer the interclass curve, i.e., as time passes the probability of committing an error increases, as if the difference between comparisons among samples from different users shows distances which approaches the distances provided by comparison of samples of the same users. But the interclass curve does not suffer any time impact, i.e. comparing the user's sample with other user's template leads to same results time-independently. These results point out that the FRR is influenced by time, but not FAR.

Regarding time, they observed that this performance degradation just takes place between the first and second acquisition session, but once this time line has passed, the degradation is almost constant, i.e. the degradation between sessions 1 and 3 is similar to that obtained between 1 and 4. This degradation can relate to two possible reasons: a stable iris ageing degradation or user habituation to the system.

This work is significant as it is the first work developed on this topic, and leads to a completely new topic for research. The authors point out their own flaws in this work:

- Libor Masek's algorithm, as pointed out in References 16 and 29, provokes some error in pre-processing. Tome-Gonzalez *et al.* solved this problem by visual inspection of each image and discarded those images which are not well segmented. They check that the pupil dilation is similar between users and sessions.
- As Tome-Gonzalez *et al.* remark, quality measures should be considered in a further study
- Masek's algorithm, although quite useful as a testing reference, does not show the best results in the state of the art and it does not show the same robustness the other algorithms do.

Regarding long-time effects on iris recognition, researchers at the University of Notre Dame have been performing some interesting work for several years.

The first work published about this topic in 2009 was by Baker *et al.* [26]. In this paper they presented the data obtained in a 4-year time period of 13 different users. They were able to re-enrol 13 people who participated in ICE 2005 database preparation 4 years before. They used the same iris camera and environmental conditions. For testing the time influence, IrisBEE algorithm [1] was used. This algorithm is an opensource algorithm, which is based on Masek's algorithm with some segmentation improvements: eliminating edge points (as the Hough transform used for circle detection tends to provoke segmentation errors), reducing the directions in the Hough transform and introducing some constraints in potential edges.

Before observing the variation with time, Baker *et al.* consider the influence of external factors:

- Images of insufficient quality are discarded. This inspection is done visually. Two reasons are considered to discard images: non-correctly segmented and those with not enough quality.
- The number of bits for comparison is desired to be similar among all cases to solve the problem of different usable iris size (and consequently, the pupil dilation ratio). For compensating the iris size observed, a linear normalization is applied (the one suggested by Daugman [9]).
- The presence of contact lenses: the users who had contact lenses in 2004 still have them in 2008, without any change in prescription.

They divide their samples into two possible groups: one group is formed by samples acquired with an elapsed time of six months maximum; the other group is formed by four time-lapse samples. The first group is referred to as short-time sessions, the other as long-time sessions. Two experiments were conducted: one checking the difference in time in one user's iris and the another comparing all irises through time. Between short-time sessions (time similar to the one between Tome-Gonzalez *et al.*'s samples) no such intraclass curve movement was observed. In fact, the authors claimed that no big difference was observed in six months sessions.

However, when comparing short-time recognition with long-time results, a change can be observed: the intraclass curve is shifted to the interclass curve. This shift can be observed in both one iris and all-irises tests with similar characteristics. This approach is translated into an FRR increase of 75% when FAR = 0.01% in four years time.

The authors concluded that initially there seem to be signs that point out that the iris structure changes with time, and also that there are many factors that can affect the results obtained, such as the sensor ageing or the pupil dilation. Regarding this, the authors admit that there are pupil dilation changes, which seem to be smaller in the long-time test, and suggest a further study of it.

Up till now, the works of Tome-Gonzalez *et al.* and Baker *et al.* have reached a similar conclusion (in spite of the fact that the result data from both studies is completely different): with time the intraclass curve varies and approaches to the interclass, i.e. the iris structure seems to change with time.

As claimed by Baker *et al.*, further study should be done and that's what Fenker and Bowyer did in Reference 27. In this paper, some of the issues mentioned in that previous paper are studied more deeply. For this, and considering the lack of previous works, the database used is formed by samples from 43 users (more than in Baker's case) taken over 2 years elapsed time. The sensor is also changed, using a LG4000, opposite to the LG2200 used before (not commercialized anymore, and out of the state of the art). Additionally, two different algorithms were used to observe the algorithm influence. It is important to remark that the two previous studies were based on similar algorithms (although the segmentation process is different, the feature extraction in both previous papers is based on log-Gabor filters), and isolating the algorithm effect should be done to conclude if the iris structure is changing. The algorithms used are IrisBEE [1] and VeriEye SDK developed by Neurotechnology [28]. VeriEye is a Neurotechnology intellectual property algorithm, so its details are unknown. The information shown in Reference 28 points to a quite accurate segmentation algorithm with fast and accurate results in matching.

The database is split into several samples according to:

- The use of contact lenses;
- The pupil dilation: large dilation and small dilation, also called delta.

Tests are performed over short and long time lapses. The short-time tests are performed with samples acquired, never on the same day, between 5 and 51 days apart. Long-time tests are performed for samples acquired in 2 years time period.

Considering the initial database, without subdivisions, IrisEye and VeriEye showed an increase of the FRR, varying approximately from 150 to 350 depending on the decision threshold. This is due to the already mentioned translation of the

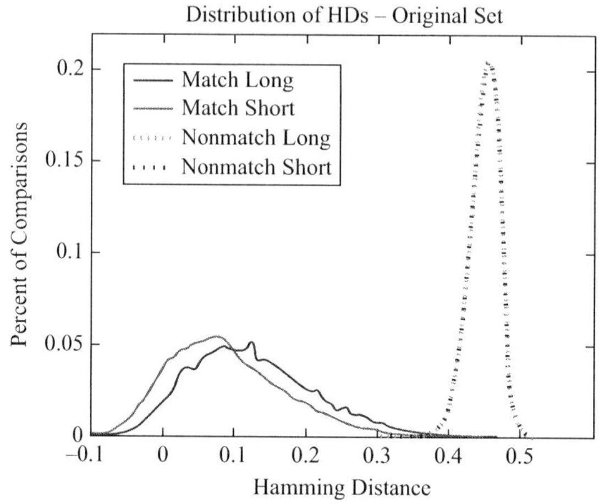

Figure 7.3 IrisBEE algorithm results obtained using short-time and long-time data sets [27]

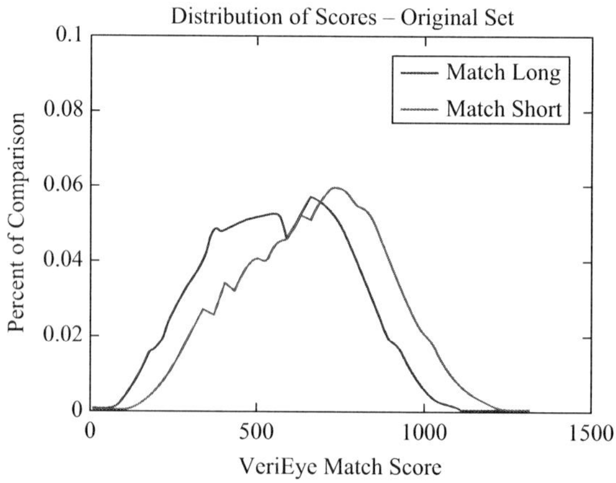

Figure 7.4 VeriEye algorithm result obtained using short-time and long-time data sets. In VeriEye algorithm non-match results are situated around and below 0; this is the way they are not shown [27]

intraclass curve approaching the interclass (see Figures 7.3 and 7.4). As can be observed in Figure 7.3, the tail of the intraclass curve tends to occupy areas where the interclass curve is located and therefore, depending upon where the decision threshold is situated, the probability of considering a user sample as someone else's sample increases with time.

Interesting results were obtained when splitting the database. These results help to explain different factors which affect in iris recognition systems.

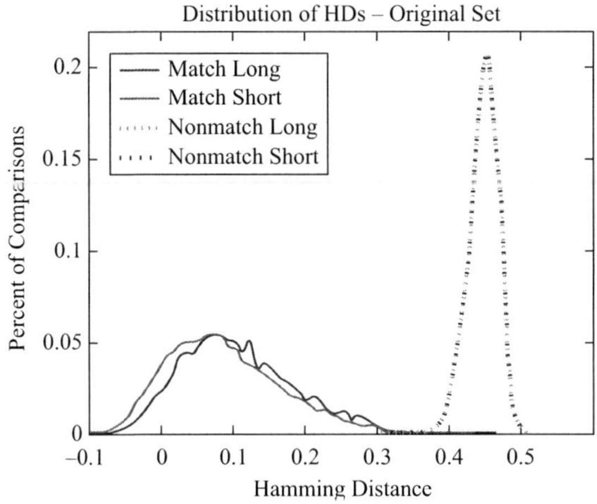

Figure 7.5 Match and non-match distributions of short-time and long-time acquisitions in IrisBEE algorithm when samples show small pupil dilation [27]

Although other works have pointed out that the difference between pupil dilation can lead to bigger error rates, Fenker and Bowyer compared both situations in time, i.e. compared two sets when the difference between pupil dilation between template and sample is bigger and when it is not. The case of IrisBEE algorithm tested on small delta images is shown in Figure 7.5. A small delta means no big difference between pupil dilation. In this case, we can observe in comparison with Figure 7.3 that the matching long-time curve again tends to shift nearer the non-matching. However, this shift is not as dramatic as in the original set. So, although time affects the iris recognition rates, the pupil size and difference between the usable iris area observed between samples are also important and affect these rates.

As can be observed when comparing Figures 7.5 and 7.6, the error rate is not so significant when using VeriEye algorithm. This is probably because of the VeriEye strong and accurate segmentation. This result supports the results obtained by Fairhurst and Erbilek: an accurate segmentation algorithm for pupil dilation is necessary to overcome ageing problems.

The results obtained when studying contact lens influence are quite similar: The FRR increases also with time, but as the authors claimed, this result cannot be considered reliable as for these tests the database is reduced and there are not enough samples to consider them valid.

In 2012, Fenker and Bowyer presented new work on this topic [29]. In this case, the study is based on the difference in the iris with time, but considering more time intervals than in previous papers and four different algorithms. The database used is an extension of their previous paper: it deals with data over three years, acquiring samples from users each year (from 2008 to 2011) [30]. Different data sets were considered: data from one year to the following one, from one year to two years later,

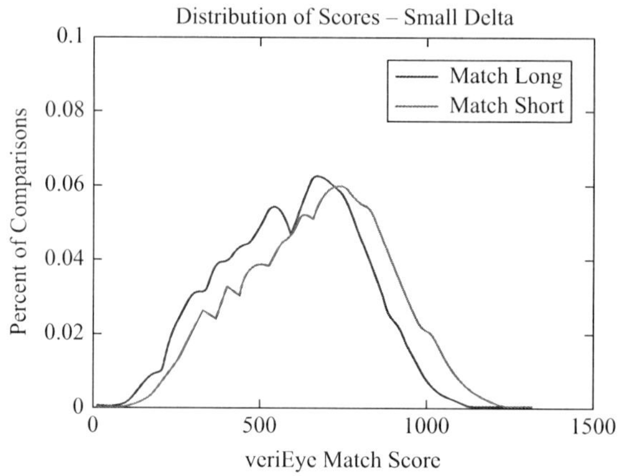

Figure 7.6 Match and non-match distributions of short-time and long-time acquisitions in VeriEye algorithm when samples show small pupil dilation [27]

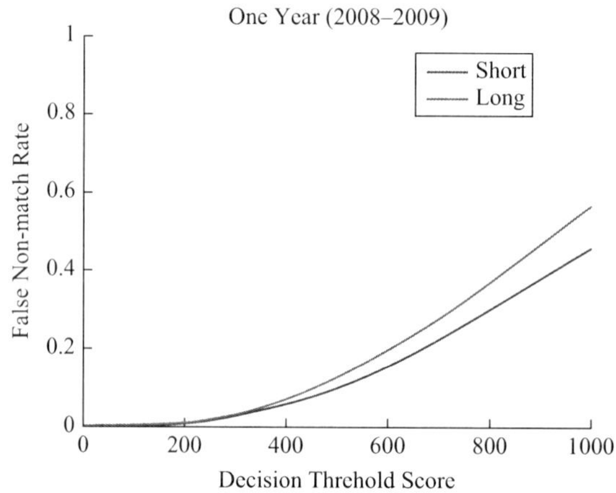

Figure 7.7 Comparison of short and long non-match distribution during one year period [29]

and three years time interval. From this data set, it is expected to get information of the performance degradation through time, something not done before.

Figures 7.7–7.9 show the differences obtained in the FNMR rates in one year (2008–2009), two years (2008–2010) and three years (2008–2011) interval respectively. As can be observed by comparison of these figures, the FNMR increases from one year to the other almost in a linear way.

It is observed that ageing does exist in iris biometrics systems. An increase of the FNMR can be observed in all the four algorithms. This increase is more than 25% through one year to the other and more than 50% in a two years interval.

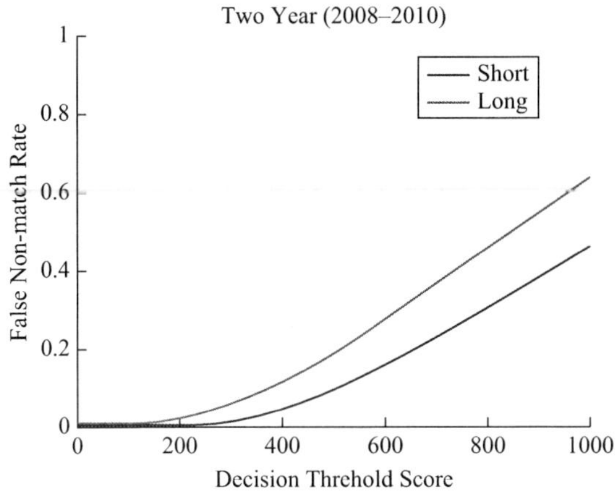

Figure 7.8 Comparison of short and long non-match distribution during two years period [29]

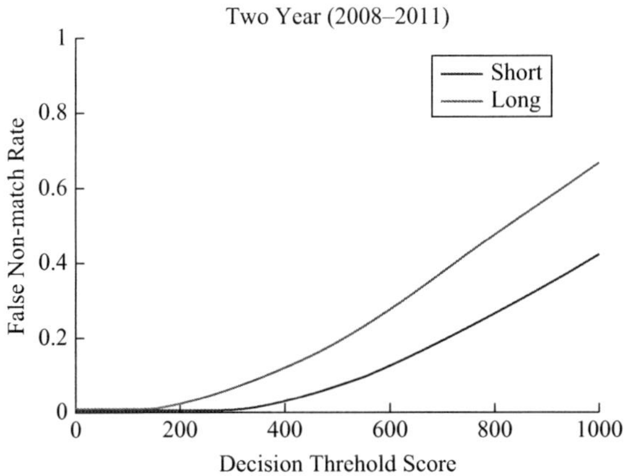

Figure 7.9 Comparison of short and long non-match distribution during three years period [29]

Research at the University of Notre Dame has clearly shown that the iris recognition algorithms are time dependent; in all cases (supporting results obtained by Tome-Gonzalez *et al.*) the intraclass curve suffers a shift with time. This shift depends on time and is independent of the algorithm used, i.e. clearly the iris structure changes with time, although these changes are not visually observed, there seem to be subtle changes.

A completely different point of view was presented by Rankin *et al.* in 2012 [31]. Their study can be considered completely different as the input samples are images obtained in the visible range, which is a fact completely different to most of the iris recognition systems that are based on images obtained in the infrared range. For this purpose, they use a biomicroscopic slit lamp for illumination purposes and a magnifier for the camera.

The database used is formed by samples from 41 users in two sessions with a time interval between them of 3 months. Additionally, 2 users provided their samples in 2 sessions separated by a 6 months interval.

The algorithm employed was based on the work of several authors: for the segmentation they used an integro-differentiator operator such as the one defined in Reference 12, the feature extraction is based on log-Gabor filters as Masek's [18] and dipoles and tripoles are used for non-local texture information as mentioned in References 32 and 33, and finally, comparison is performed using Hamming distance.

The irises are divided into different texture grades. Figure 7.10 shows the division implemented. As can be observed, this division depends on the texture level of each iris, from coarse texture to fine texture.

It is important to remark that the feature extraction is performed using log-Gabor filters, which gives information on local points, and dipoles and tripoles are non-local, and the database division depends on texture grades. These two tools can help to study if the ageing affects more one kind of iris or another.

An interesting result obtained is related to the intraclass curves attained for each set of the database. The intraclass curve attained for fine texture irises (those which belong to grade 6) has a different mean value than the one attained for coarse texture irises (grade 1). When considering time effects, this texture division is also

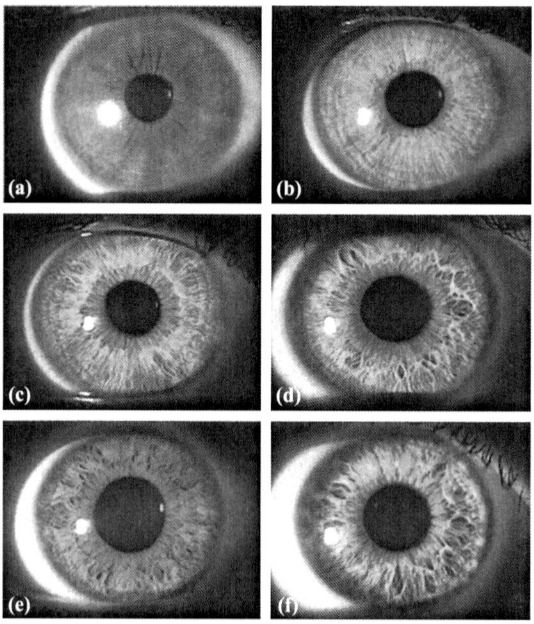

Figure 7.10 Iris texture grades defined in Reference 31: (a) Grade 1, (b) Grade 2, (c) Grade 3, (d) Grade 4, (e) Grade 5 and (f) Grade 6

observed: the irises with more detail show a curve with a smaller deviation than that obtained for coarse texture irises.

The database is not extensive, and also the time interval between sample acquisition is not enough to consider the results reliable. Furthermore the algorithm used does not provide good performance results (21% error rate). But despite all these inconveniences, interesting ideas and a completely different point of view can be observed from this paper: the effects of different texture grades in both algorithm and time.

7.8 Discussion and future work

In this chapter we have observed different works from different authors related to iris ageing. Although the results obtained are different in each work, there is still a consistency among them related to the fact that there is an age influence in iris biometric systems. Against the initial claims that the biometric information used in iris recognition systems does not change with time, several factors seem to influence system performance.

In terms of usability, older users face more problems in providing suitable samples for recognition purposes. Quality measures are highly influenced by the user's age for two main reasons:

- The first reason is related to physical changes in the iris due to age, such as colour change. Regarding this, only more accurate processes to emphasize the contrast can be applied.
- The difficulty of a proper interaction with the system. Many aspects should be improved to provide a time-invariant iris system sensor: The necessity of usable biometric systems is a topic several authors and companies have pointed out as crucial for the expansion of these systems. Regarding age, there is a pressing need to provide sensors with an easier interaction: clear and simple instructions, where the interaction with the sensor is required to be minimal. These instructions should be simple and clear, and suitable for all kinds of users, with different age, study levels, technology knowledge, etc. As one of the main problems of iris sensors is the user position, it is recommended to develop on-the-fly sensors, where the sensor captures the image while the users are moving and are not in a fixed position. Older people have more problems in getting the proper position and maintaining this.

Regarding the pre-processing block, the necessity of an accurate segmentation has been demonstrated. The pupil–iris relation, i.e. the pupil dilation, becomes an important and influential property. When age advances, the pupil size decreases and the recognition rate increases. Most of the algorithms are based on a linear normalization, which converts from a polar to a Cartesian scale. However, the results obtained by several authors [16, 21] point out that this normalization is not enough and therefore, a further study of how the iris muscle deforms should be carried out to model it and propose a better normalization algorithm. This new normalization algorithm can replace the previous one and will have to be considered in the feature extractor and in further steps.

For the extractor: initial ideas and visual inspections [5] pointed out that there was no iris deformation with time and so no template ageing influence was expected. However, work developed in References 23–27, 29, and 31 showed that there is a template deformation and therefore iris structure changes with time, although it is not observed by human visual inspection. Researchers have demonstrated that the performance of iris biometrics decreases with time, and these initial results point out the necessity of continuing to study this phenomenon in order to develop long-time iris systems. For that, some future work can be carried out:

- Study the physical underlying deformation which occurs in the iris with time for establishing a deformation model.
- Once this study is performed, adaptable algorithms can be researched which compensate this deformation according to the time when the template was stored and the identification/verification time.
- Another possible solution is to consider more robust algorithms which will be able to compensate for this time deformation, and, therefore, the performance results will not be altered with time, in spite of the changes due to time in the iris structure.

7.8.1 Other important considerations

Although not mentioned before, it is important to consider other aspects related to age as well as universality. Some diseases are age-related and also quite common. Relatively little work has been reported on this topic, and only Tomeo-Reyes *et al.* have described the effects of some of these diseases in Reference 34. Some of the diseases can influence iris recognition, as they may modify the cornea transparency, pupil shape, etc. Some effects such as losing transparency will result in the image obtained by the sensor not fulfilling the necessary quality requirements and will always be rejected by the system. Among these diseases, we can consider cataract occurrence. This disease is quite common in people older than 50 years and the main effect it has is a crystalline lens darkening, which is the front part of cornea, just before the iris. This leads to images where the iris observed is not completely clear, and a decrease in performance can be expected. This problem is easily corrected by a surgical operation introducing an intraocular lens. This lens produces more reflexes in the pupil, although no modifications in the iris, but these reflexes can make the recognition more difficult. The use of contact lenses can also affect the iris biometric performance, and again, this effect can be emphasized with age. The use of hard contact lenses seems to decrease the performance rates in spite of being transparent [27].

Ageing effects in iris biometrics have not been deeply studied as in other modalities. However, the little work developed about this topic has shown that the initial claims that iris structure does not change with time were wrong. Tests performed showed that the intraclass curve shifts with time, and therefore, there seem to be changes in the iris structure. Not only have iris structure changes been detected but also problems in usability and pre-processing have been pointed out. Iris biometric researchers face a challenge working on this topic, trying to get more usable, accurate and adaptable algorithms to overcome time effects.

References

[1] National Institute of Standards and Technology; NIST ICE database and evaluation results, 2005. Available at: http://iris.nist.gov/ice/

[2] Unique Identification Authority of India: http://uidai.gov.in/

[3] Richard S. *Snell; Clinical Anatomy of the Eye*. Blackwell Science, 1998.

[4] http://en.wikipedia.org/wiki/Eye

[5] K. Hollingsworth; 'Genetically identical irises have texture similarity that is not detected by iris biometrics', *Computer Vision and Image Understanding*, vol. 114, pp. 1493–1502, 2011.

[6] National Institute of Standards and Technology; IREX II-ICQE Report: http://www.nist.gov/itl/iad/ig/irex.cfm

[7] ISO/IEC 4th Working Draft 29794-6, 'Information technology – biometric sample quality – Part 6: Iris image'.

[8] L. Flom and A. Safir; 'Iris recognition system', U.S. Patent 4,641,349, 1987.

[9] J. G. Daugman; 'Biometric personal identification based on iris analysis', U.S. Patent 5,291,560, 1994.

[10] Wildes, R.P.; 'Iris recognition: An emerging biometric technology', *Proceedings of the IEEE*, vol. 85, no. 9, pp. 1348–1363, September 1997.

[11] Y. Zhu, T. Tan, Y. Wang; 'Biometric personal identification based on iris patterns', *Pattern Recognition*, 2000. *Proceedings of the 15th International Conference on*, vol. 2, no., pp. 801–804, Barcelona, Spain, September 2000.

[12] J. Daugman; 'The importance of being random: Statistical principles of iris recognition', *Pattern Recognition*, vol. 36, no. 2, pp. 279–291, February 2003.

[13] D. Monro, S. Rakshit, D. Zhang; 'DCT-based iris recognition', *IEEE Transactions on PAMI*, vol. 29, no. 4, pp. 586–595, April 2007.

[14] http://www.cl.cam.ac.uk/~jgd1000/afghan.html

[15] ATOS Origin Report: 'UK passport service biometrics enrolment trial'. May 2005. Available at: http://hornbeam.cs.ucl.ac.uk/hcs/teaching/ga10/lec3extra/ukpsbiometrics_enrolment_trial_report.pdf

[16] M. Fairhurst and M. Erbilek; 'Analysis of physical ageing effects in iris biometrics', *Future Trends in Biometric Processing, Computer Vision*, IET, vol. 5, pp. 358–366, 2011.

[17] http://biosecure.it-sudparis.eu/

[18] L. Masek and P. Kovesi; 'MATLAB source code for a biometric identification system based on iris patterns'. The School of Computer Science and Software Engineering, University of Western Australia, 2003.

[19] B. Winn, D. Whitaker, D. B. Elliott and N. J. Phillips; 'Factors affecting light-adapted pupil size in normal human subjects', *Investigative Ophthalmology and Visual Science*, vol. 35, pp. 1132–1137, 1994.

[20] P. Bitsios, R. Prettyman and E. Szabadi; 'Changes in autonomic function with age: A study of pupillary kinetics in healthy young and old people', *Age Ageing*, vol. 25, pp. 432–438, November 1, 1996.

[21] K. Hollingsworth, K. W. Bowyer and P. J. Flynn; 'Pupil dilation degrades iris biometric performance', *Computer Vision and Image Understanding*, vol. 113, no. 1, pp. 150–157, January 2009.

[22] X. Liu, K. W. Bowyer, P. J. Flynn; 'Experiments with an improved iris segmentation algorithm', *Fourth IEEE Workshop on Automatic Identification Technologies,* pp. 118–123, Buffalo, New York, NY, October 2005.

[23] P. Tome-Gonzalez, F. Alonso-Fernandez and J. Ortega-Garcia; 'On the effects of time variability in iris recognition', *IEEE International Conference on Biometrics Theory, Applications and Systems,* Arlington, VA, September 2008.

[24] J. Fierrez, J. Ortega-Garcia, D. Torre-Toledano and J. Gonzalez-Rodriguez; 'BioSec baseline corpus: A multimodal biometric database', *Pattern Recognition*, vol. 40, no. 4, pp. 1389–1392, April 2007.

[25] J. Galbally, J. Fierrez, J. Ortega-Garcia, M. Freire, F. Alonso-Fernandez, J. Siguenza, J. Garrido-Salas, ... C. Orrite-Uruuela; 'BiosecurID: A multimodal biometric database', *Proceedings of the MADRINET Workshop*, pp. 68–76, Salamanca, Spain, November 2007.

[26] S. Baker, K. W. Bowyer and P. J. Flynn; 'Empirical evidence for correct iris match score degradation with increased time-lapse between gallery and probe matches', *Proceedings of the International Conference on Biometrics*, pp. 1170–1179, Alghero, Italy, June 2009.

[27] S. Fenker and K. W. Bowyer; 'Experimental evidence of a template aging effect in iris biometrics', *IEEE Computer Society Workshop on Applications of Computer Vision*, Kona (Hawaii), USA, January 2011.

[28] http://www.neurotechnology.com/. Accessed July 2012.

[29] S. P. Fenker and K. W. Bowyer; 'Analysis of template aging in iris biometrics', *IEEE Computer Society Biometrics Workshop*, Providence, RI, June 2012.

[30] http://www.nd.edu/~cvrl/

[31] D. Rankin, B. Scotney, P. Morrow and B. Pierscionek; 'Iris recognition failure over time: The effects of texture', *Pattern Recognition* vol. 45, pp. 145–150, 2012.

[32] B. J. Balas and P. Sinha; 'Dissociated dipoles: Image representation via non-local comparisons', CBCL Paper #229/AIMemo #2003-018, Massachusetts Institute of Technology, Cambridge, MA, 2003.

[33] Z. Sun, T. Tan and Y. Wang; 'Iris recognition based on non-local comparisons', *Lecture Notes on Computer Sciences*, vol. 3338, pp. 67–77, 2004.

[34] I. Tomeo-Reyes, J. Liu-Jimenez, S. Ceballos-Burgos and R. Sanchez-Reillo, 'Spoofing attacks and pathologies: In the search of the border between security and usability'. *12th Biometrics Institute Australia Conference and Exhibition*, Sydney, Australia, May 2011.

Chapter 8
Ageing effects in fingerprint biometrics

Andreas Uhl and Peter Wild

8.1 Introduction

With the enrolment of over 1 million persons every day, 'Aadhaar' issuing a unique identification number to each Indian resident is currently the largest biometric deployment involving fingerprints. The employment of multiple fingerprints and irises of the same person for this project has been an intentional choice. Both modalities have been renowned for their high uniqueness, justified by reported incredibly low 0.14% failure-to-enrol rates, 0.057% false positive identification rates and 0.035% false negative identification rates in the UIDAI programme [20] and acceptance in the market [8]. Francis Galton was the first to study the permanence of the human fingerprint's ridges and furrows in 1892 [5, 7] and it is generally assumed that the fingerprint pattern is fully formed at the gestational age of 24 weeks [7]. Apart from the generally accepted excellent permanence properties of fingerprint biometric systems, it is also known that ageing results in loss of collagen [11] causing skin to be dryer and looser for elderly people. Even behavioural effects like less accurate ability to present fingers to sensors due to arthritis may arise [11]. This change in collectability and appearance may also affect the recognition accuracy of fingerprint-based biometric systems and has attracted many researchers to publish results of ageing effects on recognition performance.

There are two central questions related to ageing: *How does the age of different people affect quality and performance of a biometric system?* and *How does the ageing of templates affect biometric recognition?* In the first case, the application of fingerprint recognition in different user groups (using age as the main characteristic) is investigated – identifying the sensitivity of feature extraction methods with respect to biometric signals from different age groups. While such evaluations require the presence of meta-information (age), the time span between recordings is not considered for this type of ageing effect. However, the template may age itself (i.e. not only the age of the template may be a problem but also the time span between recordings). In the second issue the time-variability of the template is questioned and the subject of investigation, requiring the presence of biometric

University of Salzburg, Austria

samples acquired with longer time spans between recordings. Examples of studies investigating the first type of question are References 7, 11, 12 and 16, but there are also studies related to the second type targeting the modification of the template over time (e.g. References 1, 2 and 9). While existing surveys of the second type concentrate on the degradation effects of age using scanned versions of inked prints in long-term experiments [7], only few experiments on modern optical devices are available. The KFRIA Ageing database [15] with 13,200 fingerprints from 100 persons with a 1-year time span recorded using 3 different sensors is perhaps the only prominent example for fingerprints of very few datasets coming with a systematic large time span between recordings and age information, yet it is – like other datasets – not publicly available. This chapter strives to investigate fingerprint ageing effects from a new full-hand image ageing dataset highlighting template-ageing effects and a separate child vs. adult dataset targeting age-group performance.

8.2 Related work in fingerprint ageing

Skin ageing as change effects in the dermis can be measured by high-frequency skin ultrasonography, prophilometry or skin micro-relief descriptors [6]. By assessing the impact of age on fingerprints we extend our previous own work [19] and studies by colleagues focusing on a range of different topics related to the topic of ageing effects. This chapter tries to put results into context, highlighting the major proved effects of ageing, augmented with new experimental results of a new multimodal single-sensor hand ageing database, acquired at the University of Salzburg. Table 8.1 lists most important ageing studies along with their main observations.

For the first type of questions in ageing concentrating on how different age groups affect performance, the following research papers are important reference

Table 8.1 Survey of ageing studies related to fingerprints

Author	Type	Observation
Modi & Elliott [12]	Age groups	Minutiae count increases with age, but quality decreases
Sickler & Elliott [16]	Age groups	Moisture decreases with age, quality and age correlate
Modi *et al.* [11]	Age groups	Age groups affect recognition accuracy with significance
Uhl & Wild [19]	Age groups	Youth age affects recognition accuracy depending on finger
Gottschlich *et al.* [7]	Age groups	Modelling growth increases accuracy
Bevilacqua & Gherardi [2]	Template age	Skin topography changes can predict age
Arnold *et al.* [1]	Template age	FRR increase of factor 2 for a delta of 10 years for inkprints
Ryu *et al.* [15]	Template age	EER increased for 1-year time lapse for all (3) sensors
Kekre & Bharadi [9]	Template age	Propose adaptive feature set for ageing effects

works: Modi and Elliott [12] analyse the impact of index finger quality in extreme age groups on performance as an important issue in programmes like US Visit using a dataset of 79 aged subjects in the class 18–25 and 60 subjects aged 62+. Results indicated that (1) in age group 62+ more minutiae were present than in the young group – 90 vs. 55 on average – but at the cost of image quality with an average NFIQ index of 4 instead of 1.8 for the young reference group also with performance significantly degraded; (2) young fingerprints exhibited relatively good performance; and (3) a removal of low quality images improved performance. Sickler and Elliott [16] proved empirically a correlation between quality and age for index fingers regardless of device. Furthermore, it was found that young fingerprints (age group 18–25) exhibit more moisture than elderly fingerprints (62+). The hypothesis that young and elderly fingerprints exhibited different quality was verified using a one-way analysis of variance. Modi *et al.* [11] verify an impact of age groups (18–25, 26–39, 40–62 and 62+) on the recognition performance of fingerprints. Using optically acquired fingerprints of 189 users and limiting evaluations to the 90% of best quality fingerprints in each age group the following results were obtained: (1) the 18–25 age group exhibited highest mean quality with smallest variance, followed by 26–39, 40–62 and 62+ confirmed by visual inspections – the hypothesis that all means are equal could not be verified using a Kruskal–Wallis Test with significance 0.05; (2) the performance of groups was shown to be different with an increased error rate for elderly fingerprints; (3) average minutiae counts were similar across ages, but elderly fingerprints exhibited many false minutiae. In Reference 19 we evaluated the impact of children as biometric users on fingerprint recognition accuracy and found that age groups (3–10, 11–18, 19+) exhibit different performance for most fingers and that for the youth age group 11–18 performance differences to the test group 19+ were less pronounced. In a recent study, Gottschlich *et al.* [7] investigate whether special growth models for fingerprints may increase accuracy of nail-to-nail rolled fingerprints using the NIST software and two matchers in Neurotechnology's VeriFinger SDK. They show that for retrieving fingerprints of criminal adolescents (17–34) recorded years before (at ages 6–15), the application of an isotropic rescaling improves recognition accuracy from 10 to 1 error in 48 identification attempts using a database of 3.2 million entries using manually labeled Points of Interests (POIs) and from 11–14% to 5–6% in verification mode (results depend on the matcher employed). Up to 72% reductions in minutiae distances are reported. Experiments revealed that age caused a uniform increase in size only with the scale factor depending on the age at the times of recordings and sex of the person. With rescaling minutiae median distances could be reduced from 0.78 mm to 0.42 mm. But age is not the only property degrading matching performance: Blomeke *et al.* [3] compared fingerprints of a special user group (frontline healthcare workers) with those of general populations in order to verify the hypothesis that due to regular hand washing removing oil fingerprint quality decreases. While in the study of 30 persons using both capacitative and optical sensors fingerprint quality was indeed not decreased significantly, there was a notable difference between the 2 tested user groups with respect to elasticity, oiliness and moisture. While for the optical sensor both groups performed similarly, for the capacitance sensor, the health-care group performed worse for low FAR levels. Besides the effects listed in the study, further quality degrading factors include injuries like scars or (genetical) diseases like dermathopatia pigmentosa. In this case no fingerprint minutiae may be extracted at all. Even environmental factors have been investigated: Stewart *et al.* [17] found that fingerprint recognition performance is not dependent on temperature and humidity (–30° to +20°Celsius), but rugged outdoors and cold weather conditions influence results.

A study trying to determine the age of the template itself is conducted by Bevilacqua and Gherardi [2]. They evaluate skin ageing by analysing skin topography changes from capacitance images (analysing 3D profile). Dermatology assesses skin physiology and dermocosmetics analyses the effect of anti-wrinkle treatments, but there has been no other in vivo measurement analysing skin cells. In capacitance images sensing the ability of a body to store an electrical charge between adjacent plates at 50 micrometers per pixel, grey levels under skin cells are darker than under skin wrinkles – slightly modified by fat or skin hydration. Bevilacqua and Gherardi proposed to (watershed) segment an image from a capacitative fingerprint sensor (resulting in cells as catchment basins and wrinkles as watershed lines) and extract features related to skin cell distribution in order to predict age. They report on ageing causing wrinkles to appear and getting deeper enlarging cells. The extracted feature is the sum of cells with an area smaller than a fixed threshold. In an experiment of 30 people aged 1–94, a strong linear correlation between age and the feature could be observed with most frequent cell area values being 13 pixels for 1 year age to 51 pixels for 91 years age. While age information may be used to derive a degraded expected matching performance, Arnold *et al.* [1] analyse fingerprint recognition systems using data from the German federal criminal police office (BKA) with respect to ageing and other impact factors for application in ID documents. As in the EU both facial images and fingerprints are standardized for inclusion in European ePassports, 1:1 matching performance taking robustness to frequently occurring noise effects into account is an important comparison criterion for the choice of commercially available products to be deployed. The assessment of technologies using a test set of 2,160 fingerprints from 30 people and 11 sensors in 3 sessions highlighted the influence of sensors on accuracy (by a factor of 10) and algorithms (by a factor of 3). In a separate ageing experiment of 183 persons taking fingerprints in intervals of 10–30 years (26 persons for the 30 year experiment) significant performance degradation is shown (same finger experiment). While in theory ageing does not affect structure (which is encoded in the dermis layers and reconstructed in the case of injuries), ageing may affect scaling of ridges and valleys. Experiments reported an FRR increase of factor 2 for a delta of 10 years. They speculate about the effect of area and quality control on recognition accuracy. Ryu *et al.* conclude by employing NIST's fingerprint feature extraction *mindtct*, matcher *bozorth* and quality estimation software *nfiq* on an acquired ageing database (KFRIA) with 1-year time span and using 3 different sensors, that (1) there is ageing degradation by a factor greater than two in terms of EER performance for all 3 sensors, and (2) there are large differences in sample quality over time. Kekre and Bharadi [9] propose an adaptive feature set algorithm to account for ageing adaption in multi-biometrics including fingerprint as a modality. Using a regression-based technique the system assumes that there are independent and dependent features (at least one biometric feature must exhibit low variance) and can be updated using an adaptive gating network.

8.3 System and setup

In order to conduct our own ageing experiments, in our department at the University of Salzburg we collected two ageing datasets of high-resolution human handprints:

- *SBG-KidAdult*: This dataset consists of 2 partitions with 86 users (faculty and students) aged 19+ resulting in 172 hand images in the first partition recorded over 3 days, and 301 children with 498 images (due to some participants not revisiting the recording booth) aged 3–18, again recorded over 3 days in the second partition.
- *SBG-Timespan*: By re-recording faculty members approximately 5 years after the first data acquisition (users were asked to provide their hand image only when they have volunteered 5 years before), this dataset consists of 28 people with 127 hands in the first session recorded in 2007 and 135 hands in the second session recorded in 2012.

Figure 8.1 illustrates the employed system setup consisting of a flatbed scanner with containment and a sample-recorded palmar hand scan. Additional details on preprocessing and system setup may be obtained from Reference 18. Although part of a larger multibiometric system employing geometric and texture features, the experiments described in the following sections target fingerprints only.

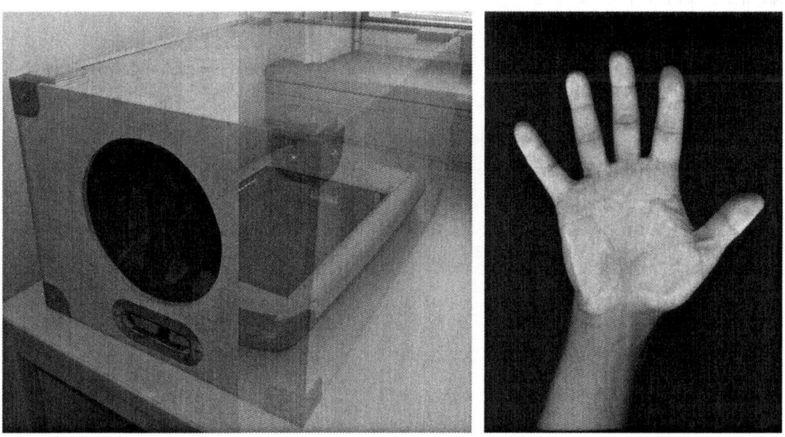

(a) Scanner (b) Sample

Figure 8.1 Employed flatbed scanner solution with containment and corresponding scanned biometric sample

8.3.1 Acquisition and preprocessing

After palmar image acquisition using a flatbed HP 3500c scanner recording a 216 by 297 millimeters area at 500 dpi resolution (resulting in 4250 × 5850 sized images) the input image is segmented and normalized. By allowing for free pegless placement of the hand on the sensor the only limitations were that fingers should not touch each other. Due to the limited illumination capabilities of traditional flatbed scanning devices (enforced by shielding any environmental light by a scanning box with a round hole for hand insertion) the desired object (hand) may easily be segmented using Otsu's thresholding [14] and using best-ellipse fitting to normalize the hand image and individual fingers. Individual minutiae detectors operate on normalized representations of ROIs, i.e. fingerprint images using a fixed fraction of the finger length. Dirt on the glass plate was avoided by regular cleaning of the area (which was done manually but could also be automatized). Due to the recording time needed, some child users experienced problems in keeping their hand motionless during recording.

8.3.2 Fingerprint extraction and comparison

Fingerprint biometrics may extract biometric properties at the three scales: (1) global level, (2) local level, and (3) very fine level [10]. The employed fingerprint feature is a local-level feature extracting minutiae – also called *Galton details* after British researcher Francis Galton – located at tips of fingers, i.e. operating on the 100–300 micrometers large ridge lines in the epidermis. The minutiae extraction software employed tracks the position (x, y) and orientation (θ) of *bifurcations*, i.e. crossings of one ridge breaking into two parts, and *termination* points, i.e. line endings. After generation of a local ridge orientation map assessing pixel sums in oriented windows, minutiae are detected in the binarized image checking for local configurations. In order to extract minutiae positions from fingerprints, NIST's *mindct* feature extraction software [13] has been employed on the normalized finger-axis-aligned fingerprint patches. Additionally, a quality measure is extracted by the NFIS software indicating the quality of the extracted minutiae point. Prior to minutiae extraction using NIST's *mindtct* tool, ROIs are enhanced using CLAHE [21] to cope with low flatbed optical contrast and ridge discontinuities. ROIs are of fixed size being as wide as the finger and one third of its height (one half for the thumb), aligned with respect to the finger axis. Figure 8.2 illustrates a sample. For comparison, we employ NIST's *bozorth3* comparator.

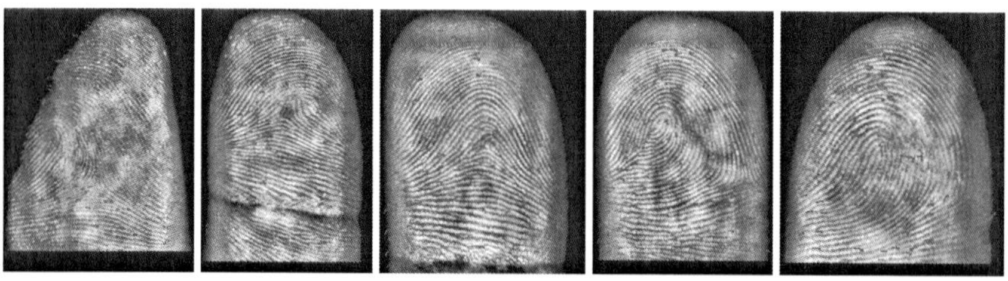

(a) Thumb (b) Index (c) Middle (d) Ring (e) Little

Figure 8.2 Minutiae feature tracking the position of bifurcations and terminations

8.4 Ageing experiments

There are several methods available to assess the effect of age on biometric systems. Certainly, the most important effect of age may be present with respect to biometric system performance evaluated in terms of pairs of false rejection rates (FRR), which represent the rate of incorrectly rejected truthful claims of identity, and false acceptance rates (FAR) being the proportion of accepted imposter attempts. While the choice of biometric system comparison thresholds allows for multiple operating points of a biometric system, permitting for specific choices from high-security systems to high convenience (with corresponding FAR and FRR pairs), we choose the equal error rate (EER), i.e. the operating point with equal FAR and FRR as the main comparison criterion. When useful, we plot the entire range of GAR (Genuine Acceptance Rate being 1-FRR) and FAR pairs in the form of receiver operating characteristics (ROC) to compare different fingers under ageing affects. In the following subsections we focus on specific research questions related to age effects on quality and performance. After reporting youth vs. adult age impact [19] on fingerprints, template ageing effects are assessed.

8.4.1 The effect of age groups on fingerprint performance

In experiments, we tested the performance of 3 different age groups: 3–10, 11–18 and 19+ yielding overall highest error rates for the age group 3–10, followed by 11–18 and 19+ for most of the fingers. Across the fingers tested, performance differences between age groups 11–18 and 19+ were not very pronounced, but observable. As an additional performance-affecting reason for the clearly inferior results for the 3–10 age group, interaction issues were identified with the flatbed scanner requiring over 150 seconds for image acquisition on average causing unintended shifts in the image due to hand movements (Figure 8.3). All obtained EER fingerprint performance rates (see [19] for original results) on the SBG-KidAdult dataset are given in Table 8.2, and Figure 8.4 plots ROC curves for each of the age groups on the tested fingers.

With the thumb yielding higher error rates than in traditional single-finger systems due to the pitched placement on the scanning surface and resulting in a smaller and cropped fingerprint area, no significant changes in performance could be observed (EERs between 2.3% and 2.8%). While being much more accurate, still also for Index, no significant changes (EERs between 1.2% and 1.6%) are observable, at least in a common operating range setting between user convenience and high security. As can be seen from the ROC in Figure 8.4, for high-security applications with requested low FAR, the adult age group performs better than the other two age groups. For Middle, Ring and Little fingers, age differences were more pronounced. For all 3 fingers, age group 3–10 degrades EER performance by a factor greater than 2. The most accurate finger, Middle, degrades from 0.6% adult EER performance to 1.2% for age group 11–18 and 2.5% for age 3–10. Ring degrades from 1.2% to 1.6% and 4.4% for age groups 11–18 and 3–10. Interestingly, Little exhibits a better EER performance for

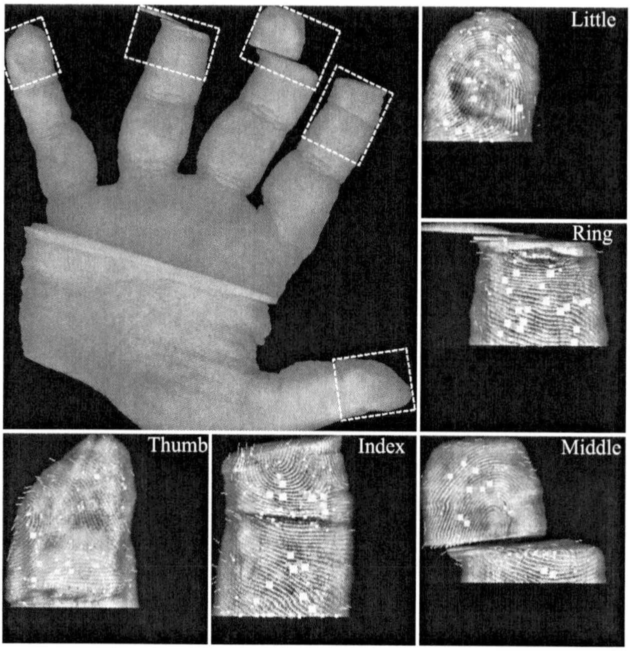

Figure 8.3 Unsuccessful minutiae extraction due to unintended shifts during acquisition

Table 8.2 Children vs. adults age group EER fingerprint performance

Feature	**Age 3–10**	**Age 11–18**	**Age 19+**
Thumb	2.8%	2.4%	2.3%
Index	1.6%	1.6%	1.2%
Middle	2.5%	1.2%	0.6%
Ring	4.4%	1.6%	1.2%
Little	11.2%	4.5%	5.5%

11–18 (4.5%) than for 19+ (5.5%), degraded to 11.2% for age 3–10. Since the minutiae region used for extraction depends on the finger size and performance is most affected for smaller sized ROIs, this value may be critical for performance. Performance differences could potentially be targeted by scale-dependent feature extraction [7].

8.4.2 The effect of template-ageing on fingerprint performance

Since results have shown that for youth fingerprints performance is degraded it is of interest to quantify degradation related to template ageing, i.e. *How much does user*

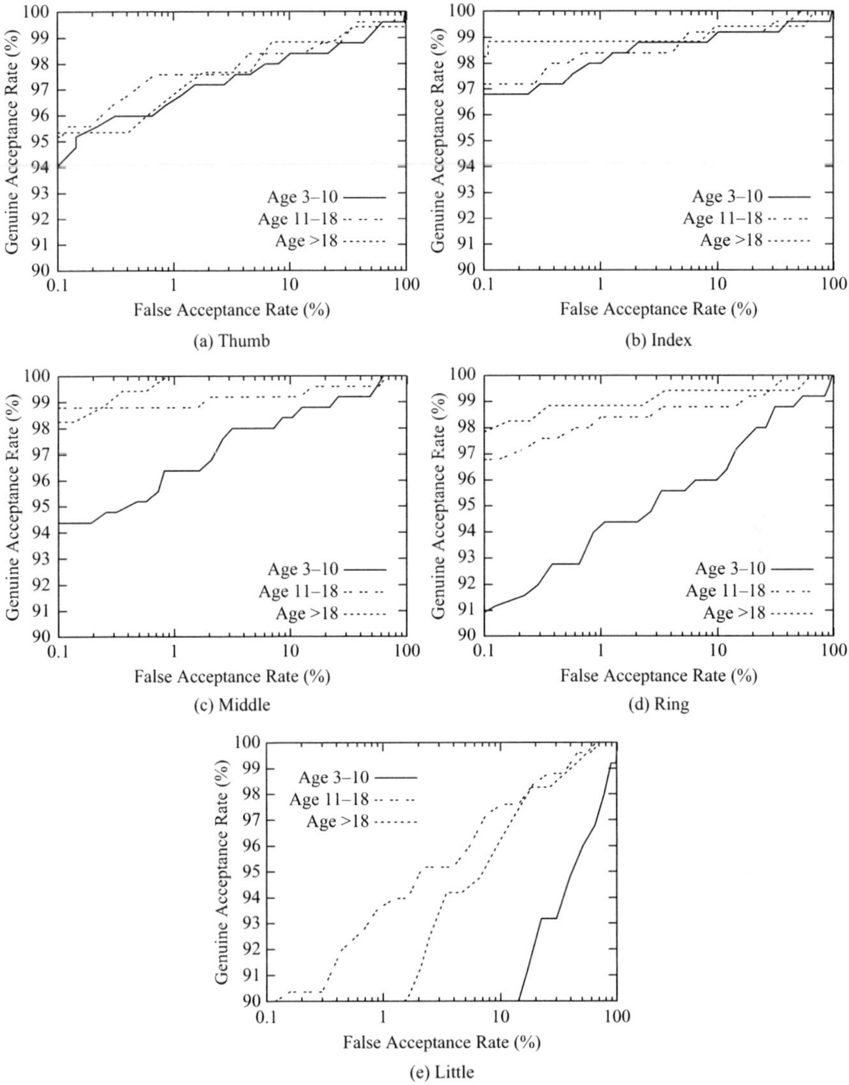

Figure 8.4 ROC curves for kids vs. adults biometric recognition on individual fingers

ageing affect recognition accuracy? While only few studies related to template ageing in fingerprints exist, this experiment contributes an evaluation of all fingers in single-sensor multibiometric configuration (dataset SBG-Timespan with a five years time span between recordings) setting results into context and providing further analysis on effects on comparison scores. In order to assess the impact of template-ageing on recognition performance, we assess each of the two 2007 (Old) and 2012 (New) sessions separately and relate evaluations to the cross-session

Table 8.3 Ageing effects on fingerprint EER recognition accuracy

Feature	EER			Avg.
	Old	New	Old vs. New	Change
Thumb	2.93%	1.18%	7.08%	+321%
Index	0.62%	0.78%	2.74%	+297%
Middle	0.42%	1.76%	3.53%	+421%
Ring	1.65%	3.49%	6.22%	+178%
Little	9.77%	8.30%	15.26%	+70%

(Old vs. New) verification experiment. Confirming results of related studies [1] on single fingerprints, we find increased error rates for all of the tested fingers (see Table 8.3 with EER rates for each of the tested (cross-)sessions).

From the balanced sessions Old and New (with 236 and 255 genuine as well as 7,765 and 8,523 imposter comparisons respectively) we received rather high EER variance between Old and New due to the reduced number of genuine matches for each finger. But degradation effects of the cross-session verification experiments (with 605 genuine and 16,286 imposter comparisons) were even higher yielding 2 to 4 times higher EERs compared to the average EER between Old and New for each finger. The highest degradation was observed for Middle with 3.53% EER for the cross-session experiment vs. 0.42% for Old and 1.76% for New, followed by Thumb with 7.08% EER vs. 2.93% and 1.18%, Index with 2.74% vs. 0.62% and 0.78%, and Ring with 6.22% vs. 1.65% and 3.49% for the respective Old and New sessions. The weakest degradation was observed for Little with only 70% degraded performance; however, in this case the original performance was already quite low (9.77% and 8.3% EER). By trend, degradation is more pronounced for highly accurate fingers (see Figure 8.5).

While the performance change is quite different between different finger types, it is useful to look at the ageing effect on scores directly. Since the employed test set consists of a representative user set (aged 29–54), no variations in imposter scores were expected and indeed imposter score distributions remained almost unaltered. Mean imposter scores lie in the range of 8–12 depending on finger type (by trend, values are higher for larger ROIs corresponding to finger length) and did not change in the cross comparison compared to the average between the 2 intra-sessions Old and New (Table 8.4). For the average genuine scores, however, it is interesting to measure the degree of change and stability with respect to time lapse. All genuine and imposter score distributions for each of the in-session and cross-session experiments are illustrated in Figure 8.6. When comparing images in each row, we can clearly perceive a shift of the genuine score distribution to the left for each of the employed fingers.

Table 8.4 quantifies the shift of the genuine score distributions due to ageing: while the genuine score mean remains almost stable between Old and New (in the order of 65–155 depending on finger type), for the cross-session comparison score means are shifted to the left by approximately one third of the original score mean

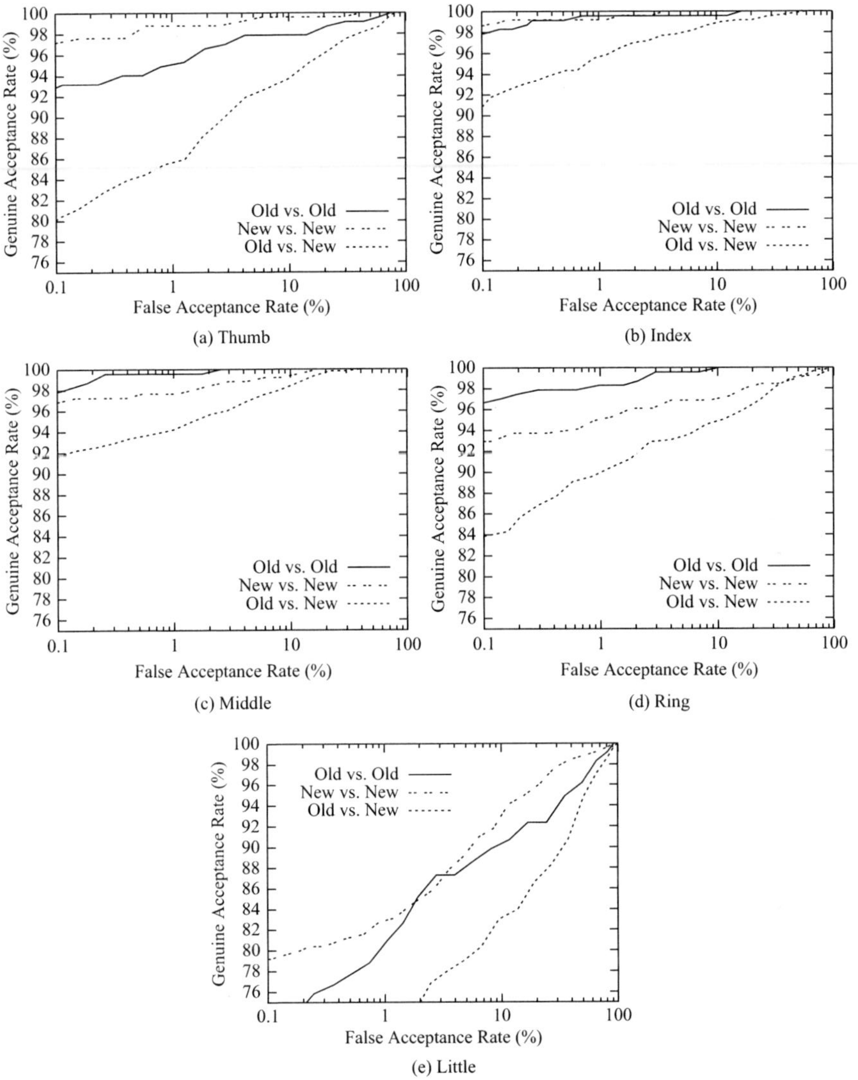

Figure 8.5 ROC curves for template-ageing effects on individual fingers

value across all finger types. The large variation in genuine scores between finger types can be explained by the different size of ROIs.

In order to cope with ageing affects, a system can employ continuous learning from conducted successful verification attempts and perform template updates. If not all features degrade simultaneously or sufficiently ‘slow’, adaptive feature set algorithms [9] can be employed to monitor update decisions. It is therefore desirable to quantify impact, study the effect’s origin (e.g. whether minutiae quality is affected or how the number of minutiae change over time) and identify age

Table 8.4 Ageing effects on fingerprint comparison scores

Feature	Average Genuine Scores				Average Imposter Scores			
	Old	New	Old vs. New	Change	Old	New	Old vs. New	Change
Thumb	99.3	99.6	67.4	−32%	9.6	9.6	9.6	±0%
Index	148.9	143.3	93.8	−36%	11.2	10.8	11.0	
Middle	155.4	136.9	97.2	−33%	11.9	10.7	11.3	
Ring	108.6	100.9	70.5	−33%	9.5	9.7	9.6	
Little	64.1	67.2	41.7	−36%	8.0	8.4	8.2	

predictors (e.g. in Reference 2) to better understand ageing and identify when and where to update templates accordingly. This section targets these investigations.

8.4.3 *Fingerprint ageing goats*

From an analysis of user types in speaker recognition, Doddington [4] identified different categories of users depending on their individual matching performance. He classified users into *sheep* representing typical users with good acceptance, *goats* exhibiting problems in being accepted (i.e. users with low genuine matching scores), *lambs* with exceptional vulnerability to attacks of other users and *wolves* being ideal candidates for the aforementioned attacks. Related to this classification, it is interesting to see whether typical goats in fingerprint recognition in the short time span sessions extend to long time span cross-session experiments or involve completely different users, i.e. separate *ageing goats*. From an analysis of users responsible for the 5% of lowest matching scores listed in Table 8.5, we can see that (1) across different fingers many users recur in the list of goats across different fingers (e.g. users #32 and #30 are responsible for 47% of the 5% of lowest match scores in Old vs. New and 43% in Old and New); (2) almost *all* of the ageing goats are also goats in either of the short time span sessions Old or New; and (3) besides goats across all fingers, there are also specific users with difficulties in extraction of minutiae from certain specific fingers (e.g. #8 exhibits matching scores below average for all fingers except Ring). Finally, although some goats in Old recur in New, the claim that the worst users remain the same after years could not be verified. Figures 8.7 illustrates typical ageing effects resulting in a low genuine matching score.

8.4.4 *Ageing and quality*

Closely related to the question of ageing effects on recognition performance are quality issues. Does ageing affect quality of minutiae and does quality affect recognition accuracy? Doubtless, minutiae extraction software *mindtct* stores a quality attribute in resulting *xyt*-Files of minutiae position and orientation in order to

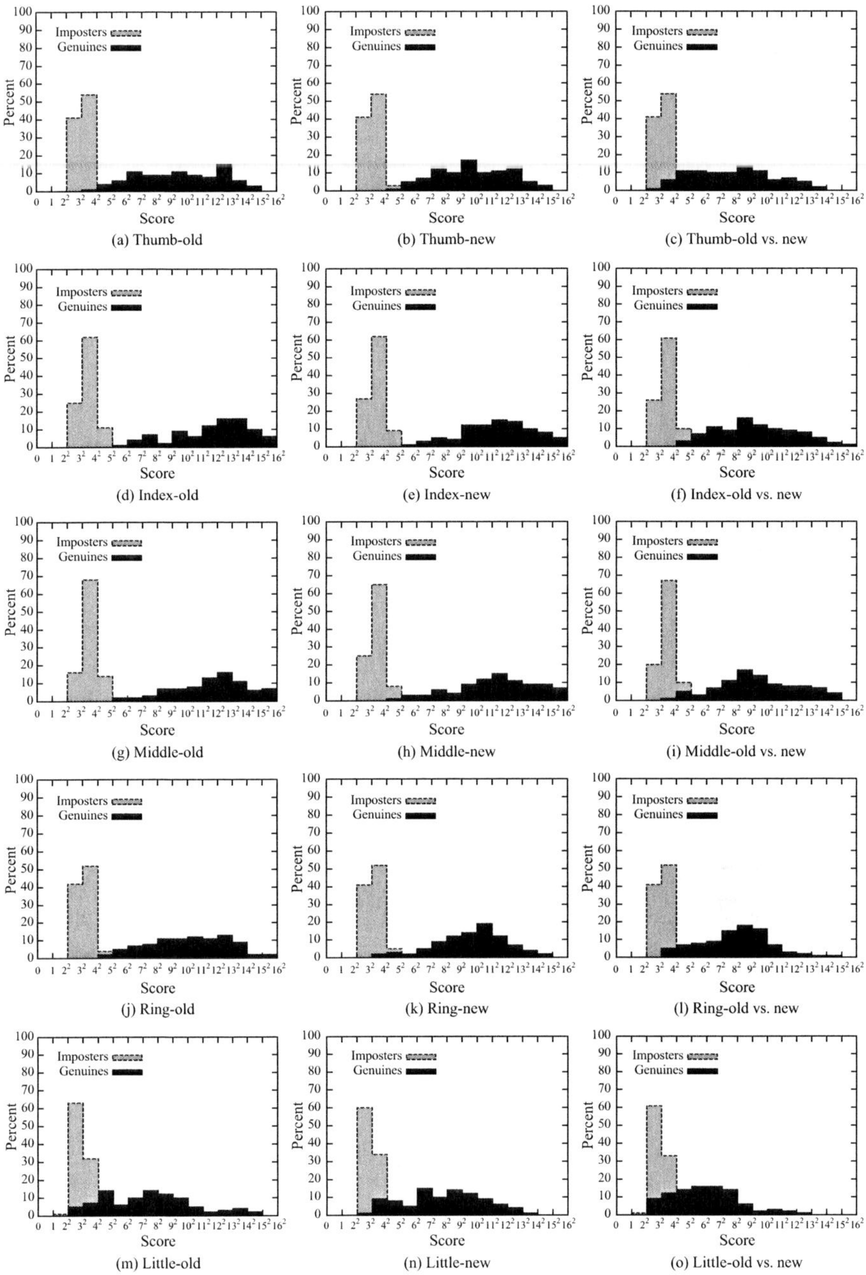

Figure 8.6 Template ageing causing a shift of the genuine score distribution

Table 8.5 Users responsible for 5% of lowest match scores (frequency in parentheses)

	Old vs. New
Thumb	**#32** (7), #8 (6), #14 (5), #11 (5), #24 (3), **#30** (2), #21 (1), #27 (1)
Index	**#32** (14), #24 (8), #14 (6), #13 (1), #12 (1)
Middle	**#32** (16), **#30** (8), #15 (6)
Ring	**#30** (7), #15 (7), **#32** (7), #12 (5), #24 (3), #23 (1)
Little	#12 (13), **#30** (6), **#32** (3), #8 (2), #15 (2), #25 (2), #29 (1), #22 (1)
	Old
Thumb	#12 (6), #11 (2), **#30** (2), #8 (1), #14 (1)
Index	#24 (6), **#30** (3), #15 (2), #14 (1)
Middle	#15 (4), #12 (3), #24 (2), **#32** (2), **#30** (1)
Ring	#15 (5), #12 (4), #24 (3)
Little	#12 (4), **#32** (3), #8 (2), #21 (1), #9 (1), #15 (1)
	New
Thumb	**#32** (5), #24 (3), #11 (2), #8 (2), #27 (1)
Index	**#32** (7), #13 (3), #14 (2), #24 (1)
Middle	**#32** (8), **#30** (4), #15 (1)
Ring	**#30** (9), **#32** (3), #24 (1)
Little	**#30** (6), #25 (3), **#32** (1), #8 (1), #2 (1), #24 (1)

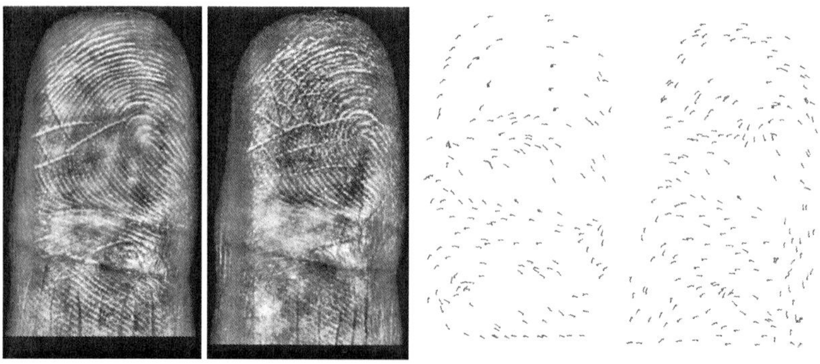

(a) #32 Old (b) #32 New (c) #32 Old (minutiae) (d) #32 New (min.)

Figure 8.7 Visual index fingerprint ageing effect of user #32

be able to conduct a ranking of minutiae points at the comparison stage. Before assessing the effect of ageing on minutiae count and quality it is useful to verify the correlation between high-quality minutiae count and matching scores of genuine comparisons. Figure 8.8 is a scatter-plot of bozorth3 1:1 verification output vs. the average number of high-quality minutiae (quality value $\geq$50) of genuine pairs of *xyt*-Files for the Index and Middle fingers. With few exceptions, the average number of high-quality minutiae is a good indicator of matching performance. Correlation

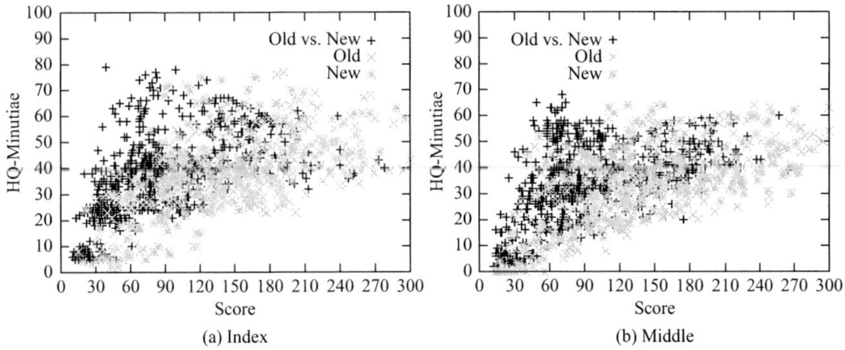

Figure 8.8 Matching score vs. high-quality ($Q \geq 50$) minutiae for Index and Middle finger

Table 8.6 Pearson correlation coefficient between matching score and high-quality minutiae count

Set	Thumb	Index	Middle	Ring	Little
Old	0.70	0.59	0.63	0.74	0.87
New	0.60	0.53	0.63	0.63	0.75
Old vs. New	0.71	0.52	0.52	0.63	0.76

was found to be more pronounced for fingers with less accurate results, like the Little finger with lower average match scores, than for high accuracy fingers Index and Middle. Table 8.6 lists the computed Pearson correlation coefficients of the two samples X_i (matching scores) and Y_i (high-quality minutiae count), i.e.

$$r_{XY} = \frac{\sum_{i=1}^{n}(X_i - \bar{X})(Y_i - \bar{Y})}{\sqrt{\sum_{i=1}^{n}(X_i - \bar{X})^2}\sqrt{\sum_{i=1}^{n}(Y_i - \bar{Y})^2}}. \tag{8.1}$$

Resulting correlation coefficients were in the order of $0.5 < r < 0.9$ for all fingers with strongest obtained correlation for the least performing Little finger and weakest correlation for the best performing Index finger. Now, since the number of high-quality minutiae positively correlates with matching scores, we investigate how ageing affects minutiae quality in terms of this quality measure. Table 8.7 illustrates how minutiae quality degrades with time by tracking the change in minutiae count for high-quality and low-quality minutiae (decision is based on the quality score provided with the corresponding bifurcation or termination point). Note that the quality measure used refers to a single minutiae's quality combining factors from a quality map taking low contrast, low flow and high curvature into consideration and simple

Table 8.7 Old vs. New fingerprint average minutiae count

Feature	Quality <50			Quality ≥50		
	Old	New	Change	Old	New	Change
Thumb	178.9	178.2	–0.4%	26.8	24.9	–7.1%
Index	145.1	151.1	+4.1%	38.2	37.7	–1.3%
Middle	127.4	131.4	+3.1%	35.3	31.7	–10.2%
Ring	120.7	122.8	+1.7%	23.9	22.7	–5.0%
Little	86.0	88.9	+3.4%	12.0	11.4	–5.0%

pixel intensity statistics in the local 11 pixel neighborhood [13] and is provided by *mindtct* as part of NIST's feature extractor, not by *nfiq*. The latter provides a global quality measure, which is related to the entire fingerprint and evaluated in, e.g. Reference 15. Indeed, while the number of low-quality minutiae increases over all fingers except Thumb (with only slight decreases), high-quality minutiae were less observed (minus 1–10%). This behaviour was particularly present for the Middle finger, which at the same time also exhibited the most severe degradation in EER matching performance. Therefore, the number of high-quality minutiae may be a good indicator to judge the need for template updates.

8.5 Conclusion

Ageing in biometrics refers to the accumulated change of physical, behavioural or chemical personal characteristics over time. This chapter examined medium-term stability and youth ageing effects on fingerprint recognition. We found that different age groups of children (3–10, 11–18) affect fingerprint performance with severe degradations for age group 3–10 for all fingers, degradations for 11–18 in Ring and Middle, and even improvement compared to the adult reference group for the Little finger. An analysis of the effects of template ageing on fingerprints resulted in two to four times raised EER performance resulting from roughly 33% decreased genuine matching scores over all fingers. Ageing was shown to increase low-quality and decrease high-quality minutiae strongly correlated with matching scores.

References

[1] Arnold, M., Busch, C., Ihmor, H.: 'Investigating performance and impacts on fingerprint recognition systems'. *Information Assurance Workshop, 2005. IAW '05. Proceedings from the Sixth Annual IEEE SMC*, West Point, NY; pp. 1–7 (2005). DOI: 10.1109/IAW.2005.1495926.

[2] Bevilacqua, A., Gherardi, A.: 'Age-related skin analysis by capacitance images'. *Proceedings of the 17th International Conference on Pattern*

Recognition, 2004. ICPR 2004., vol. 2, Cambridge, UK; pp. 703–706 (2004). DOI: 10.1109/ICPR.2004.1334356.

[3] Blomeke, C., Elliott, S., Senjaya, B., Hales, G.: 'A comparison of fingerprint image quality and matching performance between healthcare and general populations'. *IEEE 3rd International Conference on Biometrics: Theory, Applications, and Systems, 2009. BTAS '09.*, Washington, DC; pp. 1–4 (2009). DOI: 10.1109/BTAS.2009.5339017.

[4] Doddington, G.R., Liggett, W., Martin, A.F., Przybocki, M.A., Reynolds, D.A.: 'SHEEP, GOATS, LAMBS and WOLVES: A statistical analysis of speaker performance in the NIST 1998 speaker recognition evaluation'. *International Conference on Spoken Language Processing (ICSLP)*, Sydney, Australia (1998).

[5] Galton, F.: *Finger Prints*. Macmillan, London (1892).

[6] Gniadecka, M., Jemec, G.: 'Quantitative evaluation of chronological ageing and photoageing in vivo: Studies on skin echogenicity and thickness'. *British Journal of Dermatology* **139**, 815–821 (1998).

[7] Gottschlich, C., Hotz, T., Lorenz, R., Bernhardt, S., Hantschel, M., Munk, A.: 'Modeling the growth of fingerprints improves matching for adolescents'. *IEEE Transactions on Information Forensics and Security,* **6**(3), 1165–1169 (2011). DOI: 10.1109/TIFS.2011.2143406.

[8] International Biometrics Group: *Biometrics Market and Industry Report 2009–2014*, IBG, New York, NY (2009).

[9] Kekre, H., Bharadi, V.: 'Ageing adaptation for multimodal biometrics using adaptive feature set update algorithm'. *IEEE International Advance Computing Conference, 2009. IACC 2009*. pp. 535–540 (2009). DOI: 10.1109/IADCC.2009.4809068.

[10] Maltoni, D., Maio, D., Jain, A., Prabhakar, S.: *Handbook of Fingerprint Recognition*. Springer, London, UK (2009).

[11] Modi, S., Elliott, S., Whetsone, J., Kim, H.: 'Impact of age groups on fingerprint recognition performance'. *IEEE Workshop on Automatic Identification Advanced Technologies*, 2007. Alghero, Italy; pp. 19–23 (2007). DOI: 10.1109/AUTOID.2007.380586.

[12] Modi, S.K., Elliott, S.J.: 'Impact of image quality on performance: Comparison of young and elderly fingerprints'. In: K. Sirlantzis (Ed.), *Proceedings of the 6th International Conference on Recent Advances in Soft Computing (RASC)*, pp. 449–454, Kent, UK (2006).

[13] National Institute of Standards and Technology: 'Fingerprint Image Software 2' (2006). http://fingerprint.nist.gov/NFIS

[14] Otsu, N.: 'A threshold selection method from gray-level histogram'. *IEEE Transactions on Systems, Man, and Cybernetics* **9**(1), 62–66 (1979).

[15] Ryu, J., Jang, J., Kim, H.: Analysis of effect of fingerprint sample quality in template aging'. *NIST Biometric Quality Workshop II*, Gaithersburg, MD (2007).

[16] Sickler, N., Elliott, S.: 'An evaluation of fingerprint image quality across an elderly population vis-a-vis an 18–25 year old population'. *39th Annual*

2005 International Carnahan Conference on Security Technology, 2005. CCST '05. Las Palmas, Gran Canaria, Spain; pp. 68–73 (2005). DOI: 10.1109/CCST.2005.1594817.

[17] Stewart, R., Estevao, M., Adler, A.: Fingerprint recognition performance in rugged outdoors and cold weather conditions. *IEEE 3rd International Conference on Biometrics: Theory, Applications, and Systems, 2009. BTAS '09*. Washington, DC; pp. 1–6 (2009). DOI: 10.1109/BTAS.2009.5339061.

[18] Uhl, A., Wild, P.: 'Personal recognition using single-sensor multimodal hand biometrics'. In: A. Elmoataz, O. Lezoray, F. Nouboud, D. Mammass (Eds.), *Image and Signal Processing. Proceedings of ICISP 2008, LNCS*, Cherbourg, France, vol. 5099, pp. 396–404. Springer-Verlag, Heidelberg, Germany (2008).

[19] Uhl, A., Wild, P.: 'Comparing verification performance of kids and adults for fingerprint, palmprint, hand-geometry and digitprint biometrics'. *IEEE 3rd International Conference on Biometrics: Theory, Applications and Systems,* 2009. BTAS '09. Washington, DC; pp. 1–6. IEEE Press (2009). DOI: 10.1109/BTAS.2009.5339069.

[20] Unique Identification Authority of India: Aadhaar. URL http://uidai.gov.in/. Retrieved May 2012.

[21] Zuiderveld, K.: 'Contrast limited adaptive histogram equalization'. In P. Heckbert (Ed.), *Graphics Gems IV* (pp. 474–485). Academic Press, San Diego, CA (1994).

Chapter 9

The impact of ageing on speech-based biometric systems

Finnian Kelly and Naomi Harte

9.1 Introduction

Biometric authentication relies on the use of biometric templates. A person seeking to access a biometrically controlled system must first identify himself/herself and the system will then verify a match with his/her stored template, created during an enrolment procedure. Such systems are intended for long-term deployment. Hence invariance to human ageing is an important element of any biometric system. The majority of human biometrics such as face, fingerprint, handwriting, iris and voice are subject to ageing effects. Other chapters in this book discuss many of these modalities at length. Here we will focus on speech. Speech is the least intrusive biometric from a user perspective. It can be used independently [1], or combined with other modalities such as face [2, 3] in multimodal systems.

The impact of ageing on speech-based biometric systems has received little attention to date. This does not imply it is unimportant, but rather reflects the lack of longitudinal data to support research on the matter. Lanitis and Tsapatsoulis [4] present a quantitative analysis of the effects of ageing on biometric templates. The importance of ageing for all modalities is discussed but no results on speech data are presented. Kinnunen and Li [1] also recognize ageing as an important element in future horizons for speaker recognition.

9.2 Speech changes with age

It is well established that the human voice changes with age. The initial period of great change in the voice takes place up to the age of approximately 20, or until the larynx and vocal cord tissues are fully mature. The next major period of change is in old age, generally above 60 years of age. Physiological studies [5–7] have identified changes in all parts of the adult vocal system with ageing: The subglottal system (lungs, diaphragm, trachea) is affected by a stiffening of the thorax and a decreased rate and

Trinity College Dublin, Ireland

strength of respiratory muscle contraction. In the larynx, vocal folds thin in males and become thicker in females. The supralaryngeal (oral and nasal cavities) system's primary changes are decreased tongue and facial muscle functionality. The effect of these physiological changes is manifested in the voice in numerous ways [6, 8–10]: A downward shift in fundamental frequency of speech occurs throughout adulthood for both males and females. Around the age of 60 an upward shift occurs in males, continuing into advanced old age. Women's fundamental frequency also rises slightly again, but not until 80 years of age. Common for males and females is a change in timbre or tone quality, along with an increase in instability of both pitch and intensity of speech. A decrease in the rate of speech is typically observed. Other changes include increased breathiness and decreased articulatory precision.

Although the latter vocal attributes are typically most noticeable in an elderly voice, progressive changes occur throughout the lifespan [8, 10]. It should be noted that the extent of ageing-related change is variable [7]. This is particularly noticeable in elderly speakers, and is compounded by the increased likelihood of elderly speakers to experience pathological vocal change due to illness, effects of alcohol and tobacco use, or cognitive decline [7, 11]. An important observation is that the higher level of variability in a range of acoustic measures is consistent with these changes in anatomic structure, physiological mechanisms and motor control. This higher level of variability is in turn reflected in the spectro-temporal features typically employed in speech-based biometric systems.

9.3 Age in speaker verification

To date, research into the relationship between voice and age for speech technology applications has included age perception [12], age estimation and synthesis [13], the effect of absolute speaker age on speaker verification [14] and speech recognition [15]. Harnsberger *et al.* [12] showed that voice quality, incorporating tremor and noise, plus speaker rate are perceptually relevant cues for speaker age. Schötz [13] showed that the most important acoustic correlates of speaker age were speech rate (segment duration) and intensity range. However, F0 (fundamental frequency) and some spectral measures (e.g. F1 and F2 (first and second formants)) may also, if used in combination with other features, be important correlates of age. Other features frequently associated with ageing (e.g. jitter, shimmer and spectral noise) did not yield as much age-related variation as had been found in several earlier studies. Hansen and Lei [14] demonstrated that careful attention in developing the age sensitive models can result in an improved overall speaker verification system using factor analysis. They used just 2 age factors. Younger was defined as 20–30 years and older as 40–50 years. Vipperla *et al.* [15] demonstrated a 10% increase in word-error rate in automatic speech recognition for older voices. Older here indicates male speakers of 60–85 years, with an average age of 68.4. Their analysis of several voice source parameters found that the parameter values of fundamental frequency, jitter and shimmer measures show statistically significant differences in adult and older voices. Interestingly, even though older voices show increased jitter and shimmer, these measures did not appear to affect the speech recognition performance significantly. Acoustic model mismatch was the main problem, with mid vowels, nasals, and phonemes requiring constriction with the tongue tip most affected by ageing.

Actual longitudinal studies in speaker verification are scarce in the literature. An attempt to uncover an ageing effect on speaker verification over a 3 year period was presented in Reference 16. There, Lawson *et al.* found that the impact of ageing on speaker verification over a 3 year period was no more significant that general inter-session variability. Thirty-two speakers were recorded in 21 sessions over a 3 year period. This MARP Corpus [17] is not publicly shared, so the age of the actual speakers used is unclear. More recent work by the current authors [18] confirmed that a 3 year period may not show significant ageing trends, but that such trends are certainly apparent over 5–10 years and more significantly in older speakers. A major obstacle in reaching conclusions on the impact of age on speaker verification is a lack of longitudinal data. The Greybeard Corpus [19], compiled by NIST for their Speaker Recognition Evaluation (SRE) 2010, contains longitudinal data over a 2–12 year period. This corpus is currently only available to SRE 2010 participants, and the authors are unaware of publications reporting results on longitudinal verification on this dataset. More recently though, Doddington [20] reported on how the age difference between target and non-target speakers affects the performance of a speaker recognition system using the extended evaluation results from SRE 2010. The age difference was shown to have a significant impact on the false alarm probability of a system, with that impact increasing with greater age differences. This reflects a growing awareness of the impact of ageing on speaker verification systems.

9.4 Speaker ageing data

An ideal speaker ageing database would contain data collected over several decades, with regular recordings for each speaker in controlled conditions across the entire time span [21]. Due to the evolution of technology for recording, transmission and storage of media over the last few decades, the prospect of retrospectively obtaining data in controlled, consistent conditions is unrealistic. Notwithstanding this reality, we have collected the Trinity College Dublin Speaker Ageing (TCDSA) database (first presented in Reference 22). The primary motivation in collecting the database was to build a useful longitudinal dataset for speaker verification research with limited (or as limited as possible) non-ageing-related variability.

The TCDSA database content is detailed in Figure 9.1. The database contains 18 speakers (9 male, 9 female). As is evident, the quantity and range of the speech samples vary between speakers. There are between 2 and 35 recordings for each speaker spanning a range of between 30 and 60 years. The recordings vary between approximately 1 and 30 minutes in length. The database contains 31 hours of speech content in total.

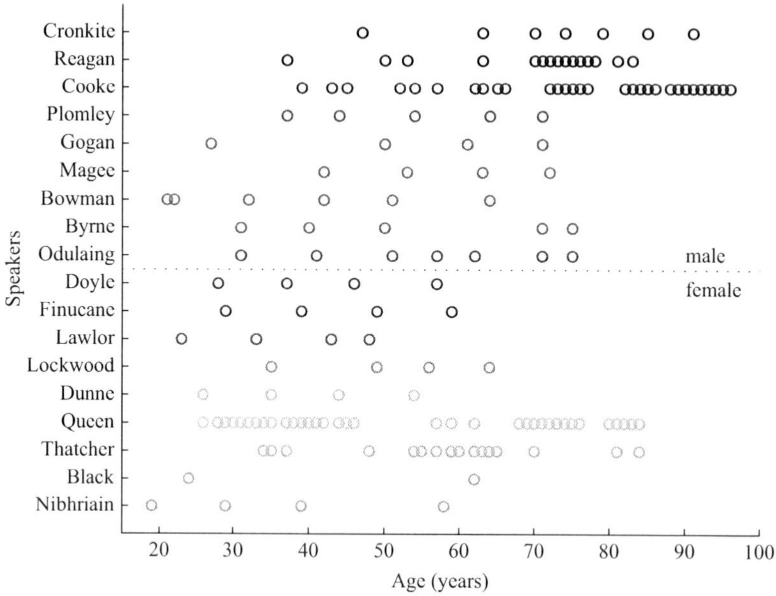

Figure 9.1 A schematic of the TCDSA database. A circle indicates a year where a recording is available for a speaker, with the speaker surnames given on the y-axis and their age on the x-axis. The speakers are divided by gender, with males and females on the upper and lower halves of the plot respectively

The data originates from the national broadcasters of the United Kingdom and Ireland, the BBC (British Broadcasting Corporation) and RTÉ (Raidió Teilifís Éireann), as well as publicly available samples, obtained from YouTube and the Miller Center [23]. Ten of the 18 speakers were sourced from the RTÉ archives (Anne Doyle, Marianne Finnucane, Aine Lawlor, Eileen Dunne, Doireann NiBhrian, Larry Gogan, Jimmy Magee, John Bowman, Gay Byrne, Donncha O'Dulaing); 5 from the BBC archives (The Queen, Margaret Thatcher, Margaret Lockwood, Alistair Cooke, Roy Plomley); 1 from the Miller Center (Ronald Reagan); and 2 exclusively from YouTube (Cilla Black, Walter Cronkite). In addition, the data of many BBC speakers has been supplemented with additional data from YouTube. The content of the recordings is primarily interviews and speeches, thus there is a mix of conversational and read speech. There is a range of Irish, English and American accented speakers in the database. All recordings are in the English language.

As expected for an uncontrolled database, the recording conditions are highly varied. Even in the case of the RTÉ and BBC broadcast data, the specifics of the recording setup and equipment are unavailable. To limit non-ageing-related variability, the initial database material was screened for quality. Any audibly noisy recordings, or segments of recordings, were discarded. The spectral content of the recordings was examined, and any with significant frequency artefacts were removed.

As an objective measure of quality, the likelihood of the recordings given an age-balanced Universal Background Model (UBM) [24] was calculated. This approach has previously been applied for quality measurement [25]. The motivation for this measure can be understood by viewing the UBM as a representation of the distribution of speaker features common to all of the test population. In the case of a degraded test recording, its likelihood given the UBM will fall outside the range of scores from typical 'clean' recordings. In this way, the UBM score can be used as an indicator of recording quality. The notion of quality is addressed in depth in section 9.6.

This was applied to the database material by dividing every recording into 30 second segments, extracting features (as described in section 9.5.1) and calculating the likelihood of each segment given the UBM. Any recording whose complete set of likelihood scores fell outside 1.5 times the interquartile range of the set of all scores was deemed an outlier, and the associated recording was discarded from the database. The 'cleaned-up' database is depicted in Figure 9.1.

9.4.1 Ageing UBM database

An additional dataset was compiled to create a suitable UBM for use with the TCDSA database. The University of Florida Vocal Aging Database (UF-VAD) (extemporaneous set) [26] was combined with data sourced from YouTube and the Miller Center [23]. The resulting dataset contains 1 hour of speech from 120 speakers (30 seconds each). The speakers are evenly balanced across gender, age and accent, and a variety of recording conditions are present. The ages of the speakers are spread evenly across 3 age profiles: under 35, 36–55 and over 55. English, Irish and American accented speakers are included. The dataset was composed in this way in an effort to ensure it was a close representation of the population of the TCDSA database.

9.5 Ageing speaker verification experiment

While there are other systems for speaker verification that have come to prominence recently, we will present this ageing experiment using a GMM-UBM (Gaussian Mixture Model-Universal Background Model) system. This is the first work to explore speaker verification over such an extended time span. As the extent of the ageing effect in this domain is not fully understood, presenting work on a baseline system is therefore in the interest of clarity. Recent speaker verification developments, such as Joint Factor Analysis (JFA) [1, 27], are extensions of the GMM-UBM approach. This work is therefore a basis from which extensions to these methods is possible. It should be mentioned, however, that systems like JFA, while successful in recent NIST SRE events, do not provide an 'out of the box' solution to all applications. In this work, the limited and varied nature of the data introduces extra challenges in training an effective JFA system. The GMM-UBM system will be used to demonstrate the impact of ageing for long-term deployment of a speaker-based biometric system.

9.5.1 Feature extraction and the GMM-UBM system

All samples were pre-processed by downsampling to 16 kHz, removing silences with an energy-based voice activity detector and applying pre-emphasis. Twelve MFCC features were extracted over 20 millisecond windows with 50% overlap, using 26 Mel-spaced filters. Mean and variance normalization were applied to the features followed by RASTA filtering. The MFCC features were appended with delta and acceleration coefficients. These steps are typical in current verification systems [1]. A GMM-UBM verification system was developed by first training a UBM of 1024 mixtures using the contents of the TCDSA UBM dataset (section 9.4.1). To generate speaker-specific models for each speaker in the TCDSA database, the UBM means were adapted with 1 minute of training speech [24]. The standard log-likelihood ratio (LLR) [1] was used for scoring test speech.

9.5.2 Ageing speaker verification evaluation

A model was trained for each speaker using 1 minute of their first year of available data. All recordings in the database taken after the date of the training sample were considered test material. Thus for each speaker, a set of genuine speaker and imposter scores was generated from the set of recordings occurring after the date of their model training recording. The impostor set comprised the remaining (17) speakers. Each test recording was split into 10 segments of 30 seconds duration. The resulting LLR scores were averaged to give one LLR score per speaker per test year. The genuine and impostor LLR scores are shown in Figure 9.2.

It should be noted that this test scenario, whereby all remaining speakers were used as impostors, introduces cross-gender trials. Naturally, male–female

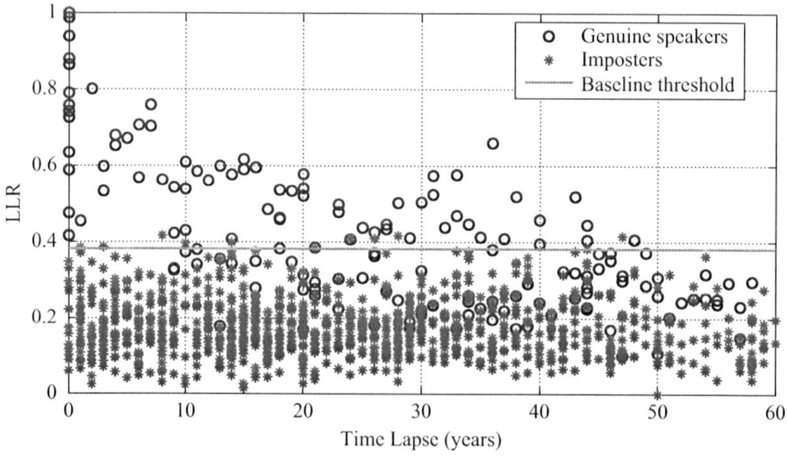

Figure 9.2 LLR scores of genuine speakers and imposters on a 60 year verification evaluation, and a conventional decision threshold, determined from scores at Time Lapse = 0

differences in the extracted acoustic features will result in a lower overall LLR score for this test scenario. While this is not ideal in terms of demonstrating the full potential of the classifier, the main aim of this work is to explore the inter-related effects of ageing and quality. The relative performance of range of classifiers used will not be affected. This comes about purely due to a lack of data to allow fully gender-specific impostor samples.

Errors in a speaker verification system are described by the false acceptance rate (FAR), the percentage of imposters incorrectly accepted, and the false rejection rate (FRR), the percentage of genuine speakers incorrectly rejected. A decision threshold is typically determined by reaching some tradeoff between FAR and FRR using verification scores from the time of enrolment. In our evaluation, in the majority of cases there is only one recording per speaker per year. To set a robust threshold, a global, rather than speaker-specific approach was adopted in which the threshold was found by minimizing the half total error rate (HTER) of the pooled enrolment scores from all speakers. The HTER is given by the average of the FAR and FRR. The enrolment score in this case is the LLR of a test segment usually taken from the same session as, but always distinct from, the training session. A threshold trained in this way is superimposed on the LLR scores in Figure 9.2.

In Figure 9.2, genuine speaker scores decrease progressively until they converge with the set of imposter scores. It is clear that applying the baseline threshold as a verification boundary will result in an increasing error rate as time lapse increases – the effect of ageing on the baseline system is therefore significant.

The ageing issue is of relevance to any verification scenario where there is a large time lapse between enrolment and verification for a user. It is informative to examine this on a per speaker basis, since a threshold in a system will usually be person-specific and set at initial enrolment. LLR scores are plotted against the time between enrolment and verification sessions (i.e. age progression) for two individual speakers, one male, one female, in Figure 9.3.

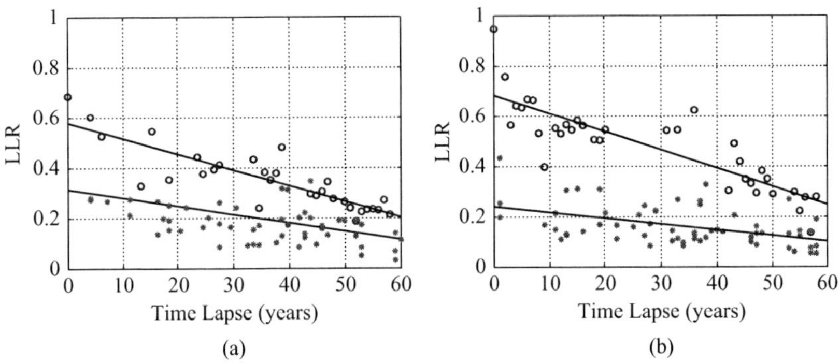

Figure 9.3 Genuine speaker (o) and imposter (*) LLR scores against the time lapse between enrolment and verification for two speakers. A line fit of the data is included to demonstrate the trend of the scores over time: (a) Cooke (male) (b) Queen (female)

It can be observed from the overall LLR trends in Figure 9.2 that there is a general and progressive decrease in LLR score for genuine speakers as the time lapse between enrolment and verification increases. Imposter scores, however, are more stable over the same time period. A similar trend has been observed in the domain of face ageing [28]. The individual speaker examples in Figure 9.3 demonstrate the same trend as the global case. The variability in the scores is also apparent in these examples. It is clear that a fixed threshold determined at the time of enrolment (e.g. where *Time Lapse* = 0, Figure 9.3) will result in an increasing verification error rate as time progresses.

9.6 Effect of quality variation on speaker verification

The effect of sample quality variation has a significant effect on biometric verification [29, 30]. A quality measure that can predict the utility of a sample in a verification system can be jointly modelled with the verification score to reduce error rates [28]. A range of measures were extracted from the TCDSA data based on those measures from previous studies that exhibited promising utility. An additional model-based quality measure developed by the authors is also introduced. The measure is *Wnorm* (weighted norm), which measures quality in a similar way to the UBM score, while retaining independence from the LLR. During verification, for every test speech sample, a GMM is trained by UBM mean-only adaptation. The difference between the mean vectors of this GMM and the UBM is calculated. The difference is multiplied by the corresponding mixture weights and matrix (Frobenius) norm is taken. The motivation is that a recording with a higher quality speech will result in greater 'movement' of important (higher weighted) means in training and will therefore produce a higher Wnorm value.

9.6.1 SNR

Signal to noise ratio (SNR) [25, 31, 32] was calculated using an energy-based voice activity detector. A sample is divided into 20 millisecond frames designated as either speech or non-speech by an energy threshold:

$$SNR = 10 \log \frac{E_s}{E_{ns}} \tag{9.1}$$

where E_s and E_{ns} are the mean energies of the speech and non-speech frames respectively.

9.6.2 Skewness

The skewness [23] of a distribution is a measure of its asymmetry. Since clean speech has a distinctive distribution, higher order statistics like skewness (and kurtosis) can be used to measure quality. Skewness is given by:

$$Skewness = \frac{1}{T} \sum_{t=1}^{T} \left(\frac{s_t - \mu_t}{\sigma_t} \right)^3 \tag{9.2}$$

where s_t is the t th frame of the speech sample and μ_t and σ_t are its mean and variance respectively. The frame length was 20 milliseconds.

9.6.3 Kurtosis

The kurtosis [11, 23] of a distribution is its degree of 'peakedness', i.e. the lower the kurtosis, the flatter the distribution. Kurtosis is given by:

$$Kurtosis = \frac{1}{T}\sum_{t=1}^{T}\left(\frac{s_t - \mu_t}{\sigma_t}\right)^4 \tag{9.3}$$

where s_t is the tth frame of the speech sample and μ_t and σ_t are its mean and variance respectively. The frame length was 20 milliseconds.

9.6.4 UBM likelihood

In Reference 25, the likelihood of the test sample given the UBM was proposed as a quality measure for GMM-based speaker verification. The motivation for this measure is that assuming a UBM represents the acoustic space of the expected operating conditions, then degraded signals are more likely to have a lower likelihood given the UBM than non-degraded signals. The authors themselves acknowledge that the likelihood with respect to the UBM reflects speaker-specific traits as well as quality-related information. Their evaluation of this measure on NIST telephone data suggested promising utility, however. In the GMM-UBM verification framework, given a speech sample O and a GMM λ_s for a particular speaker, the LLR is computed as:

$$LLR = \log p(O|\lambda_s) - \log p(O|\lambda_{UBM}) \tag{9.4}$$

where $p(.|\lambda)$ is the probability density function for any model λ. The second term, $logp(O|\lambda_{UBM})$, is the UBM likelihood.

9.6.5 Wnorm

The proposed Wnorm measure is given by:

$$Wnorm = \sqrt{\sum_{k=1}^{K}\sum_{d=1}^{D}\left(W_k(M_{kd}^S - M_{kd}^{UBM})\right)^2} \tag{9.5}$$

where M^S and M^{UBM} are the $K \times D$ dimensional matrices of means of the test sample GMM and the UBM respectively and W is the vector of mixture weights of length K. D length feature vectors are employed. A 30 second sample is used to train the GMM in this comparison.

A key element in factor analysis for speaker verification is the use of GMM supervectors [27], comprising stacked means from the mixtures. Campbell *et al.* [33] extended the use of GMM supervectors to SVM kernel design in speaker recognition. Distance measures between supervectors are used to design a kernel that is then used in training the SVM-based classifier. The Wnorm measure is closely related to the concept of using distances between supervectors in that the measure is related to a Mahalanobis distance between the mean vectors of the GMM and UBM. The Wnorm is applied only in testing to capture the difference, or distance, between a current test session and the average test session, here the UBM. It is not used in system training.

9.6.6 Quality evaluation

Each of the five quality measures were extracted from the TCDSA database in 30 second segments and a [0, 1] mapping [7] was applied. The [0, 1] mapping was achieved by recording the maximum and minimum value measured on the training data and subsequently scaling between these extremes. In References 25 and 30, utility of quality measures was demonstrated by plotting the LLR scores of genuine speakers and imposters against their corresponding quality measures. A quality measure was deemed useful if, as its value was increased, the scores of genuine speakers increased relative to the imposters. In the case of the TCDSA database, the ageing effect influences the LLR score alongside quality variation, and cannot be controlled for. This considered, the correlation between the LLR of genuine speakers and both ageing and quality measures was examined. The figures are given in Table 9.1. Measures that display stronger correlation with LLR score than with ageing should intuitively be better predictors of non-ageing-related variability.

9.7 Ageing-quality stacked classifier

Having observed that the effects of both ageing and quality influence the LLR score, an effective strategy for long-term verification needs to compensate for both.

9.7.1 Stacked classifier framework

In this approach, the output of several lower level classifiers, together with the original class labels, form the input to a higher level classifier. This is an effective way to combine different sources of 'evidence' in verification. In our previous work [22], this framework was adopted, where the GMM-UBM system and ageing information constituted the lower level classifiers. The higher level classification was then carried out in score-ageing space with a Support Vector Machine (SVM) decision boundary. Here, we extend this framework by supplying the output of each of the proposed quality measures as an additional lower level input. Thus the higher

Table 9.1 Correlation between quality measures, LLR score and ageing progression

	SNR	Skewness	Kurtosis	UBM Likelihood	Wnorm
LLR	0.26	0.11	0.19	0.01	0.37
Ageing	−0.19	−0.13	−0.10	−0.15	−0.11

level classification will now involve a three-dimensional decision boundary in score-ageing-quality space.

In the proposed framework, each of LLR scores from the baseline evaluation, section 9.5.1, together with their associated ageing (time lapse) and quality information are concatenated into a vector. A *Z*-normalization [1] is applied to the LLR score, based on the statistics of the training data. The normalized LLR score and ageing information are then scaled to the range [0, 1] based on the extreme values in the training set. The scaled vector forms the input to a linear SVM classifier. The SVM parameters are set such that the HTER on the training data is minimized.

9.7.2 Stacked classifier experimental evaluation

An evaluation of the proposed framework was designed such that the performance could be directly compared to that of the baseline classifier described in section 9.5.1. For each speaker, the set of all other (17) speakers over their complete 30–60 year time span constituted the training set. Pooling the training data, a score-ageing boundary was trained. Applying this boundary to the test data of the speaker, the HTER was determined. An example of the trained boundary in score-ageing and score-quality planes is shown in Figure 9.4(a) and (b). An application of the

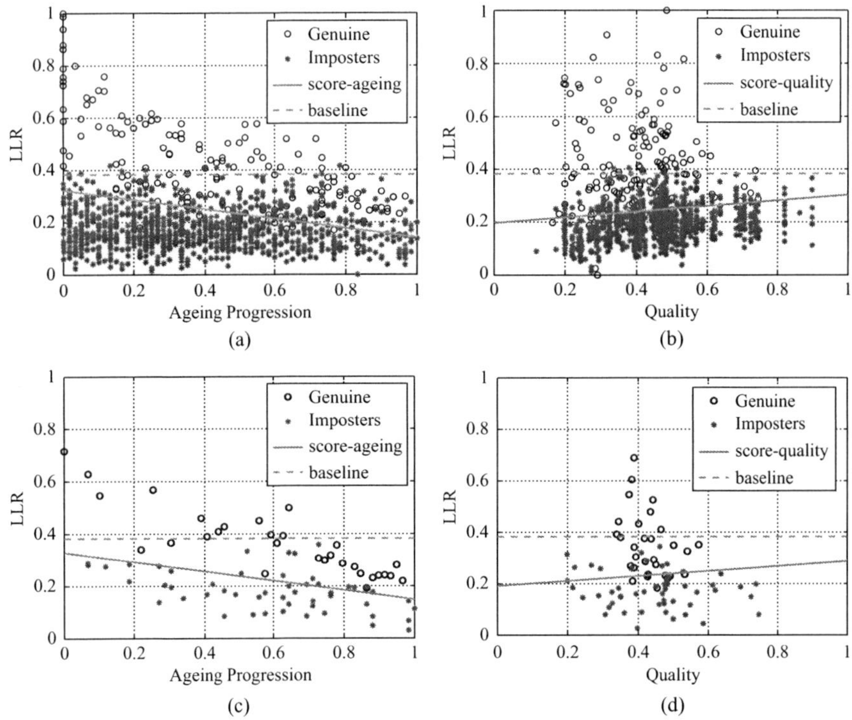

Figure 9.4 Stacked classifier example: (a) training, score-ageing; (b) training, score-quality; (c) testing, score-ageing; (d) testing, score-quality

Table 9.2 Average HTER (%) for all 18 speakers for the baseline GMM-UBM system and the stacked classifier. '+Ageing' denotes a score-ageing boundary, '+Quality' denotes a score-quality boundary and '+Ageing+Quality' denotes a score-ageing-quality boundary

			HTER(%)		
GMM-UBM	23.85				
+Ageing	16.31				
+Quality	SNR	Skew.	Kurt.	UBML	Wnorm
	18.39	19.08	18.93	19.99	17.51
+Ageing+Quality	SNR	Skew.	Kurt.	UBML	Wnorm
	15.66	16.27	16.08	16.29	13.47

boundary to test data, in both cases, is given in Figure 9.4(c) and (d). It is evident that the ageing and quality dependent boundaries track the progression of genuine speaker scores more closely than the baseline linear classifier. The average HTER for all speakers is shown in Table 9.2. Incorporating ageing and quality into the framework independently reduces HTER. The combination of both brings best performance – a score-ageing–SNR combination reduces HTER by 34% relative to the baseline, and by 4% compared to score and ageing alone. The relative performance of the quality measures aligns with the utility predicted in Table 9.1 – SNR and Wnorm, the best performing measures, have a greater correlation with LLR score than with ageing progression.

9.8 Conclusions

The human voice ages. The TCDSA database has allowed us to clearly demonstrate the impact of ageing on decision boundaries fixed at initial enrolment in a speech biometric system (Figure 9.2). At 18 speakers, the dataset remains too small to draw hard conclusions about exactly how many years an original speaker template continues to reflect speaker traits, but certainly beyond 5–10 years the system is unreliable. The issue of quality variation is tightly linked to ageing. To exploit the dependencies between ageing progression, quality variation and LLR score on the TCDSA database, a stacked classifier framework has been used to develop a verification boundary in score-ageing-quality space. The novel use of this framework in a long-term speaker verification evaluation was successful in significantly reducing HTER. This framework provides an effective and flexible approach to the challenging problem of speaker verification in the simultaneous presence of ageing progression and quality variation.

References

[1] T. Kinnunen and H. Li. 'An overview of text-independent speaker recognition: From features to supervectors'. *Speech Communication*, **52**(1):12–40, 2010.

[2] R. M. Jiang, A. H. Sadka, and D. Crookes. 'Multimodal biometric human recognition for perceptual human computer interaction'. *IEEE Transactions on Systems, Man, and Cybernetics, Part C: Applications and Reviews*, **40**(6):676–681, Novmber 2010.

[3] A. Kumar, V. Kanhangad, and D. Zhang. 'A new framework for adaptive multimodal biometrics management'. *IEEE Transactions on Information Forensics and Security*, **5**(1):92–102, March 2010.

[4] A. Lanitis and N. Tsapatsoulis. 'Quantitative evaluation of the effects of aging on biometric templates'. *Computer Vision, IET*, **5**(6):338–347, November 2011.

[5] S. E. Linville. 'Vocal aging'. *Current Opinion in Otolaryngology & Head and Neck Surgery*, **3**(3):183–187, 1995.

[6] S. E. Linville. 'The aging voice'. *The ASHA Leader*, **9**(19):12–21, 2004.

[7] P. B. Mueller. 'The aging voice'. *Seminars in Speech and Language*, **18** (02):159–169, 1997.

[8] E. T. Stathopoulos, J. E. Huber and J. E. Sussman. 'Changes in acoustic characteristics of the voice across the life span: Measures from individuals 4–93 years of age'. *Journal of Speech, Language and Hearing Research*, **54**(4):1011, 2011.

[9] U. Reubold, J. Harrington and F. Kleber. 'Vocal aging effects on F0 and the first formant: A longitudinal analysis in adult speakers'. *Speech Communication*, **52**(7):638–651, 2010.

[10] R. Rhodes. *Changes in the Voice across the Adult Lifespan*. The International Association of Forensic Phonetics and Acoustics, IAFPA, University of South Florida, 2011.

[11] B. J. Benjamin. 'Speech production of normally aging adults'. *Seminars in Speech and Language*, **18**:135–141, 1997.

[12] J. D. Harnsberger, W. S. Brown Jr, R. Shrivastav, and H. Rothman. 'Noise and tremor in the perception of vocal aging in males'. *Journal of Voice*, **24**(5):523–530, 2010.

[13] S. Schötz. *Perception, Analysis and Synthesis of Speaker Age*. PhD thesis, Lund University, Sweden, 2006.

[14] J. H. L. Hansen and Yun Lei. 'The role of age in factor analysis for speaker identication'. *Interspeeech 2009*, Brighton, 2009.

[15] R. Vipperla, S. Renals, and J. Frankel. 'Ageing voices: The effect of changes in voice parameters on ASR performance'. *EURASIP Journal on Audio, Speech, and Music Processing*, 2010. doi:10.1155/2010/525783.

[16] A. D. Lawson, A. R. Stauffer, B. Y. Smolenski, B. B. Pokines, M. Leonard, and E. J. Cupples. 'Long-term examination of intra-session and inter-session speaker variability'. *INTERSPEECH 2009*, Brighton, 2009.

[17] A. D. Lawson, A. R. Stauffer, E. J. Cupples, S. J. Wenndt, W. P. Bray, and Grieco. J. J. 'The Multi-Session Audio Research Project (MARP) Corpus: Goals, design and initial findings'. *INTERSPEECH 2009*, Brighton, 2009.

[18] F. Kelly and N. Harte. 'Effects of long-term ageing on speaker verification'. In C. Vielhauer, J. Dittmann, A. Drygajlo, N. Juul, and M. Fairhurst, editors,

Biometrics and ID Management, volume 6583 of *Lecture Notes in Computer Science*, pp. 113–124. Springer Berlin/Heidelberg, 2011.

[19] L. Brandschain, D. Graff, C. Cieri, K. Walker, C. Caruso and A. Neely. 'Greybeard – voice and aging'. *Seventh Conference on International Language Resources and Evaluation (LREC '10)*, Malta, 2010.

[20] G. Doddington. 'The effect of target/non-target age difference on speaker recognition performance'. *Odyssey 2012: The Speaker and Language Recognition Workshop*, Singapore, 2012.

[21] A. Lanitis. 'A survey of the effects of aging on biometric identity verification'. *International Journal of Biometrics*, **2**(1):34–52, 2010.

[22] F. Kelly, A. Drygajlo, and N. Harte. 'Speaker verification with long-term ageing data'. *International Conference on Biometrics (ICB), 2012*, pp. 478–483. New Delhi, 2012.

[23] Presidential Speech Archive, Miller Center, University of Virginia. http://millercenter.org/scripps/archive/speeches, 2011.

[24] D. A. Reynolds, T. F. Quatieri, and R. B. Dunn. 'Speaker verification using adapted Gaussian mixture models'. *Digital Signal Processing*, **10**(1–3):19–41, 2000.

[25] A. Harriero, D. Ramos, J. Gonzalez-Rodriguez, and J. Fierrez. 'Analysis of the utility of classical and novel speech quality measures for speaker verification'. *Third International Conference on Advances in Biometrics*, pp. 434–442. Springer-Verlag, Alghero, Italy, 2009.

[26] J. D. Harnsberger, R. Shrivastav, and W. S. Brown. 'Modeling perceived vocal age in American English'. In *Interspeech 2010*, Makuhari, Chiba, Japan 2010.

[27] P. Kenny, G. Boulianne, P. Ouellet, and P. Dumouchel. 'Joint factor analysis versus eigenchannels in speaker recognition'. *IEEE Transactions on Audio, Speech, and Language Processing*, **15**(4):1435–1447, 2007.

[28] A. Drygajlo, W. Li, and K. Zhu. 'Q-stack Aging model for face verification'. *EUSIPCO 2009*, Glasgow, Scotland, 2009.

[29] F. Alonso-Fernandez, J. Fierrez, and J. Ortega-Garcia. 'Quality measures in biometric systems'. *IEEE Security & Privacy*, Preprint, 2011.

[30] P. Grother and E. Tabassi. 'Performance of biometric quality measures'. *IEEE Transactions on Pattern Analysis and Machine Intelligence*, **29**(4):531–543, 2007.

[31] D. Garcia-Romero, J. Fierrez-Aguilar, J. Gonzalez-Rodriguez, and J. Ortega-Garcia. 'Using Quality measures for multilevel speaker recognition'. *Computer Speech & Language*, **20**(2–3):192–209, 2005.

[32] J. Richiardi and A. Drygajlo. 'Evaluation of speech quality measures for the purpose of speaker verification'. *Odyssey 2008: The Speaker and Language Recognition Workshop*, South Africa, 2008.

[33] W. M. Campbell, D. E. Sturim, and D. A. Reynolds. 'Support vector machines using gmm supervectors for speaker verification'. *Signal Processing Letters, IEEE*, **13**(5):308–311, May 2006.

Chapter 10

An analysis of biometric performance change over time: a multimodal perspective

Norman Poh, Josef Kittler, Chi-Ho Chan and Medha Pandit

10.1 Introduction

With the increasing need for identity authentication in our daily life – from making purchases online and unlocking personal devices to accessing secure premises – biometric technology certainly has an important role to contribute. Although physical lock, PIN and gesture (which is common for mobile devices) offer convenient solutions, only biometric-based authentication can provide the ultimate means of validating the identity credential. The use of biometric authentication entails many questions, especially when it is used over a long period of time and in different locations: Will the system performance degrade over time? What kind of metrics can best describe the performance over time? How will its performance be affected by different acquisition environments, such as the office environment, public locations, and outdoors?

In this study, we attempt to gain an understanding of whether or not a change in performance of a biometric system is *subject-dependent*. We will not attempt to explain the cause of performance change; we know that it is not only age-related, but is also dependent on biometric sample quality and habituation. As will become clear, measuring the change in biometric performance is challenging, let alone explaining the causes of this change. However, we would like to challenge the common assertion that the performance of a biometric system systematically degrades overtime.

In fact we will show that the performance change is subject-dependent. For some individuals it degrades, but for other it improves and for some it remains stable. Our approach to answering the above question is, first of all, to model the performance change in a subject-dependent manner. Thus, for each enrollee, we attempt to model his/her performance change. Then, the performance change is classified as upward, stable, or downward trend. Once the subjects have been classified, it is then possible to aggregate their matching scores at the beginning and

University of Surrey, UK

at the end of a study period and compare the performance metrics derived from these two points in time.

We validated our approach on the MOBIO database. It contains both talking face and speech biometrics of 150 enrollees collected using a Nokia mobile phone, covering over 600 days of recording. The 600-day time span is sub-divided into 3 time periods each having a length of 200 days. They constitute the 'initial' (<200 days), 'middle' (between 200 and 400 days), and 'end' (>400 days) time periods. Our experimental results show that for some subjects, the system does better in the initial period than in the end period. This implies that the system gets worse over time. This is consistent with the literature [1, 2], including our own work [3]. However, we also observed that there exists a set of subjects for whom the system does better in the end period than the initial period. This means that the system improves with use. To our knowledge, this is rarely discussed or highlighted in the literature. A logical explanation of this is habituation. As subjects get used to the system, they can provide biometric samples of higher quality or in a consistent way.

Whatever the underlying causes of performance change, that is ageing, biometric sample quality, or familiarity with a particular biometric device or mode of acquisition, our study entails several implications. First, a subject-dependent strategy has to be adopted when maintaining a biometric system. Second, the study of biometrics ageing demands more research as any degradation in performance may be compensated by the subject's familiarity. Furthermore, it appears to be important to separately study the effect of ageing and the effect of degradation due to mismatched acquisition conditions, and/or devices. Finally, more longitudinal biometric databases are required to understand the effect of ageing.

The rest of the chapter is organised as follows: Section 10.2 motivates the need of subject-specific performance characterisation. Section 10.3 then extends the notion of subject-specific performance to grouping the subjects by their performance evolution over time. A novel algorithm known as 'homomorphic user grouping algorithm' will be introduced. The remaining sections present the MOBIO database (section 10.4) and the experimental results (section 10.5). This is followed by conclusions in section 10.6.

10.2 User-specific performance characterisation

In order to motivate the need for user-specific performance characterisation, we fitted a Gaussian distribution to the match scores produced by a template (or reference model) *of each subject* for genuine or impostor comparisons respectively. Therefore, if there are 20 enrolled subjects or enrollees in the gallery, there will be 20 pairs of score distributions. The choice of Gaussian distribution is dictated by the small sample size of the data, especially the genuine matching scores (Figure 10.1).

In order to avoid cluttering the figure, we show only the distributions associated with 20 randomly selected enrolled identities (enrollees) out of 200. These scores are taken from the XM2VTS benchmark database [4]. Since there is one pair of distributions per enrollee, there is a total of 40 distributions.

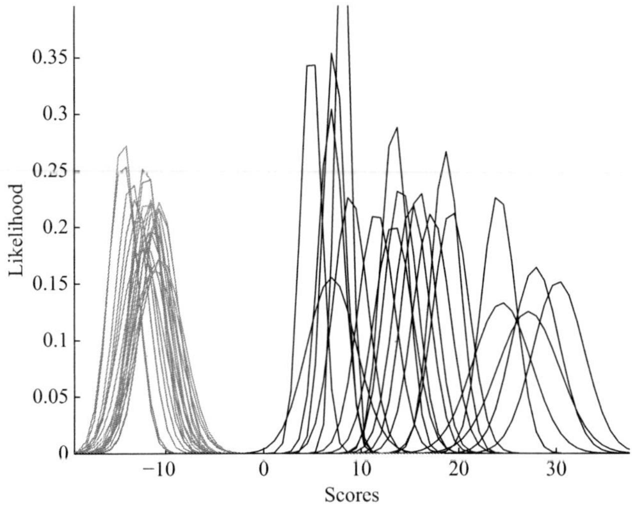

Figure 10.1 *User-specific class-conditional score distributions of a typical speech verification system. Shown here are the distributions of 20 enrollees. The right clusters (in light colour) are for the genuine class, whereas the left ones (in dark colour) are for the impostor class*

Doddington *et al.* [5] defined four categories of users based on the mean of the subject's genuine or impostor scores; they are:

- *sheep*: subjects who can easily be recognised – they have high mean genuine scores;
- *goats*: subjects who are particularly difficult to recognise – they have low mean genuine scores;
- *lambs*: subjects who are easy to imitate – they have high mean impostor scores;
- *wolves*: subjects who are particularly successful at imitating others – they consistently produce high impostor scores for all enrollees.

Goats contribute significantly to the false reject rate (FRR) of a system while wolves and lambs increase its false acceptance rate (FAR). A more recent work [6] further distinguishes four other semantic categories of users by considering both the genuine and impostor match scores for the same claimed identity simultaneously.

Hicklin and Ulery [7] argued that the presence of these "animals" may be due to the poor quality of the template. However, Wittman *et al.* [8] showed that even in a very controlled scenario, subjects still show some tendency towards a particular animal species. Teli *et al.* [9] further suggested that goats and lambs always exist for a given data set but a subject is likely to change animal membership when the matching algorithm changes.

The above literature suggests that there is definitely an interplay between subject, biometric sample quality and matching algorithms that affect biometric performance. The subject-dependency of biometric performance also suggests that tailoring a biometric system by adjusting its decision threshold to each subject, or otherwise normalising the score distribution using subject-specific score normalisation can improve the system performance [10].

In the next section, we will augment the subject-specific characterisation of matching scores with the time notion.

10.3 Our framework: a homomorphic users grouping algorithm

In the previous section, a subject's biometric performance is characterised by four parameters, namely, the mean and standard deviation of his/her impostor and genuine scores, i.e. $\{\mu_j^{\omega}, \sigma_j^{\omega}\}$ for $\omega \in \{0, 1\}$ and each and every user $j \in \mathcal{J}$. In order to consider the time domain, we fit a regression line to a time-series of matching scores for genuine ($\omega = 1$) or impostor ($\omega = 0$) matching. Let the score time-series for subject j be composed of samples $\{y_{j,t}^{\omega}\}$ collected at time $t = 1, \ldots, T$. This time-series is approximated by:

$$y_{j,t}^{\omega} \approx f(\mathbf{b}_j, t)$$

where $f(\mathbf{b}_j, t)$ is a polynomial regression function of degree D for each of the two possible classes of matching scores, ω; it has the following form:

$$f(\mathbf{b}_j, t) = [b_D^{(j)}, b_{D-1}^{(j)}, \ldots, b_0^{(j)}] \quad [t^D, t^{D-1}, \ldots, t^0]^T$$

where $\mathbf{b}_j$ is a vector of coefficients with $D + 1$ elements indexed by $\{b_d^{(j)}\}$ and the symbol T, denotes a vector transpose operation.

The use of regression implies a number of important assumptions, that is (1) linearity in the parameter space $[t^D, t^{D-1}, \ldots, t^0]$, (2) independence of the observations, (3) constant variance across the period observed $\{t\}$ and (4) normality of the distribution of fitting errors. These assumptions are worth commenting here. First, the use of polynomial degree D relaxes the assumption that the scores $\{y_{j,t}^{\omega}\}$ change linearly over time. However, the use of higher D is simply unfeasible when one has very few observations $\{y_{j,t}^{\omega}\}$. This imposes on us a requirement to use a realistic choice of relatively small degree of freedom.

The second assumption implies that knowing one particular value of $y_{j,t}^{\omega}$ does not allow us to predict the value of another. This assumption is not entirely satisfied considering that if two matching scores come from biometric samples obtained from the same session (collected seconds apart), they are likely to be correlated. However, if two biometric samples are obtained from two sessions that are weeks apart, then their matching scores are likely to be less correlated. Hence, there is some weak temporal correlation among matching scores.

Given the few observations in $\{y_{j,t}^{\omega}\}$, the time-dependent variance is hard to estimate. However, even if it could be estimated, the variance is unlikely to be constant over time. Consider a biometric authentication application for mobile devices. Since the acquisition environments are likely to be different from one session to another, for instance, in two different locations, the noise as observed in the matching scores is likely to be different too. Hence, in general, the assumption of constant variance is likely to be violated.

Finally, as the above example shows, the normality assumption of the distribution of error (with zero mean and constant variance) for the observed $\{y^{\omega}_{j,t}\}$ is unlikely to be true. However, this assumption will be required in order to derive confidence intervals around the fitted curve.

All the above assumptions appear to suggest that regression is an unlikely candidate for our choice. However, we have to adopt a rather pragmatic view of data modelling that is best expressed by George Box who famously quoted, 'All models are wrong; some models are useful'. Our view is that any regression models are likely to violate one or more of the assumptions above. We need to understand the limitations of the model and at the same time recognise that the paucity of data does not allow us to use complicated models. Our approach is to use a linear regression model to approximate the score trend which we then use to approximate the evolution of biometric error over time for each subject. The final use of the model is merely to identify if a subject's performance increases, decreases, or remains stable over time. The fitted models are not used to quantify performance, for instance. Therefore, for our purpose of estimating biometric performance trend, an approximate model is deemed to be acceptable.

The estimated regression line $f(\mathbf{b}_j, t)$ gives us the expected value $\mu^{\omega}_{j,t}$ as well as its spread $\sigma^{\omega}_{j,t}$ (thanks to assumption (4) above) at time t for each of the matching classes $\omega \in \{0, 1\}$[1]. From these four time-varying parameters, an instantaneous EER (at time t) can be calculated as [11]:

$$\mathrm{EER}_{j,t} = \frac{1}{2} - \frac{1}{2}\mathrm{erf}\left(\frac{\text{F-ratio}_{j,t}}{\sqrt{2}}\right) \tag{10.1}$$

where

$$\text{F-ratio}_{j,t} = \frac{\mu^{1}_{j,t} - \mu^{0}_{j,t}}{\sigma^{1}_{j,t} + \sigma^{0}_{j,t}} \tag{10.2}$$

and the error function is defined as:

$$\mathrm{erf}(z) = \frac{2}{\sqrt{\pi}}\int_{0}^{z} exp\left[-x^2\right] dx \tag{10.3}$$

Throughout this chapter, we will use polynomial regression. A polynomial function of degrees 1, 2 and 3 has been fitted to a subject's score time-series and the results are shown in Figure 10.2. As can be observed, while the different choices of the degree of freedom can affect the magnitude of EER, it has little impact on the overall trend which is an upward trend in this case. Since our ultimate aim is the

[1] The spread here is taken as the spread of the predicted mean value which can be obtained from any standard implementation of regression. This requires an estimate of the constant noise as well as an estimate of the covariance of the input observations [12].

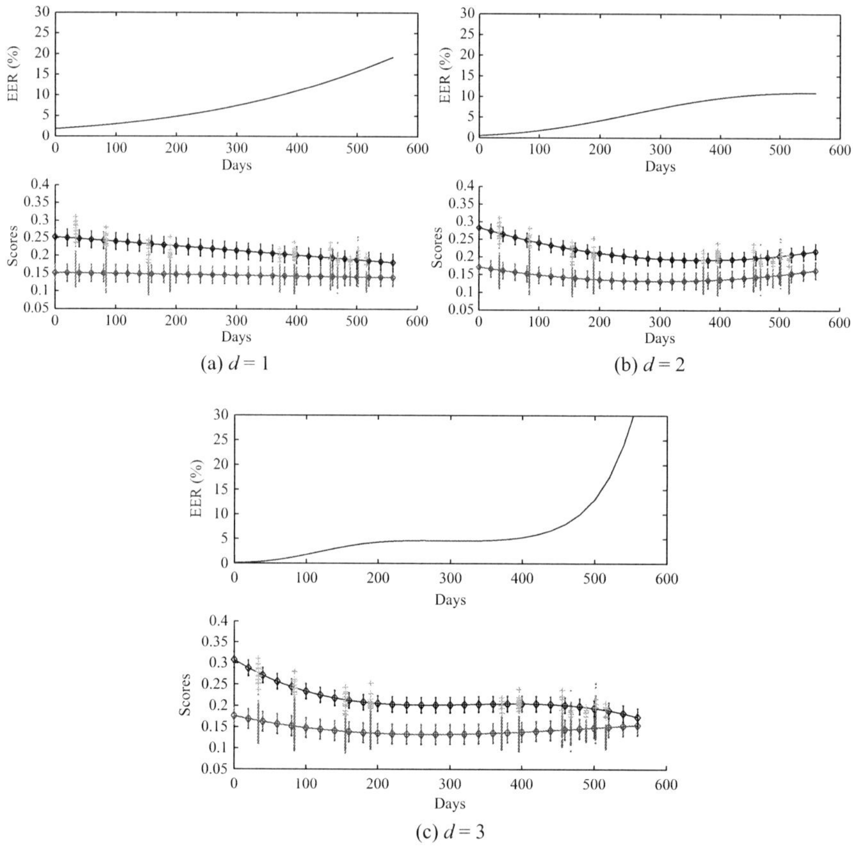

Figure 10.2 Fitting with different degrees of freedom. While the choice of degree of freedom is difficult, it is obvious that EER increases over time for this subject

latter, that is to detect upward, downward, or stable trends, we will use polynomial degree 1. This reduces any unnecessary risk of overfitting while at the same time addresses our need.

The derived subject-specific F-ratio time-series is then further subject to a linear regression fitting so that the gradient so obtained is used to characterise the performance trend.

$$\text{F-ratio}_{j,t} \approx f(\mathbf{a}_j, t)$$

where

$$f(\mathbf{a}_j, t) = a_j^1 t + a_j^0 \tag{10.4}$$

and a_j^i for $i \in \{0, 1\}$ are elements of $\mathbf{a}$, and j is the index of each subject j. Since F-ratio is inversely and non-linearly proportional to EER, an increasing F-ratio time-series actually implies better performance (see Figure 10.3). To detect the

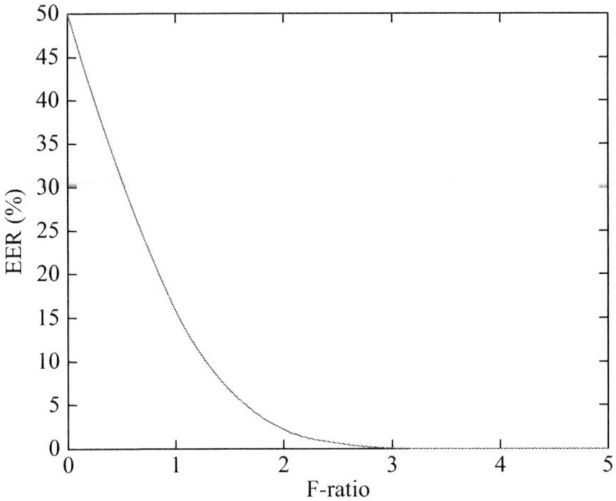

Figure 10.3 The relationship between F-ratio (in the x-axis) and EER (y-axis)

trend of subject-specific F-ratio over time, we need only to infer the value of a_j^1, and we do so for all subjects $j \in \mathcal{J}$.

The next step consists of partitioning the subjects using $\{a_j^1\}$. A simple way is to partition them at a fixed percentile. For this purpose, we assign the subjects who have a_j^1 between 0th and 20th percentiles to the first partition, those who have a_j^1 between 20th and 40th percentiles to the second partition, and so on, until the fifth partition of subjects who have a_j^1 between 80th and 100th percentiles, noting that the 0th percentile corresponds to the minimum observed value whereas the 100th percentile corresponds to the maximum observed value. Therefore, we expect the first partition of subjects should have *increasing* EER over time because they all have negative a_j^1 values. Conversely, we expect the last partition of subjects to have *decreasing* EER over time because they all have positive a_j^1 values.

In order to validate the partitioned subjects, we need to plot the EER of subjects in the first and the fifth partition at different time periods. The MOBIO database that we used covers a period of roughly 600 days. It is convenient to divide the time into 3 periods, each lasting 200 days. Let us define these 3 time periods as 'initial', 'middle', and 'end' periods as discussed before.

For each partition and each time period, one can calculate the EER values. We would expect that for the first partition of subjects, the 'end' period EER is higher than the 'initial' period, hence providing an indication of performance degradation. This would be consistent with the findings in the literature.

However, for the fifth partition, the findings that the EER calculated for the 'end' period is *lower* than the 'initial' period is rather 'surprising'. The results of our findings will further be discussed in section 10.5.

The algorithm above is called 'homomorphic user grouping algorithm' (HUGA) because all the subjects are grouped in a common *parametric score space*, and the final output of the algorithm is a number of subject partitions. The scores associated with these subject partitions are aggregated so that coarsely quantised time-dependent EERs

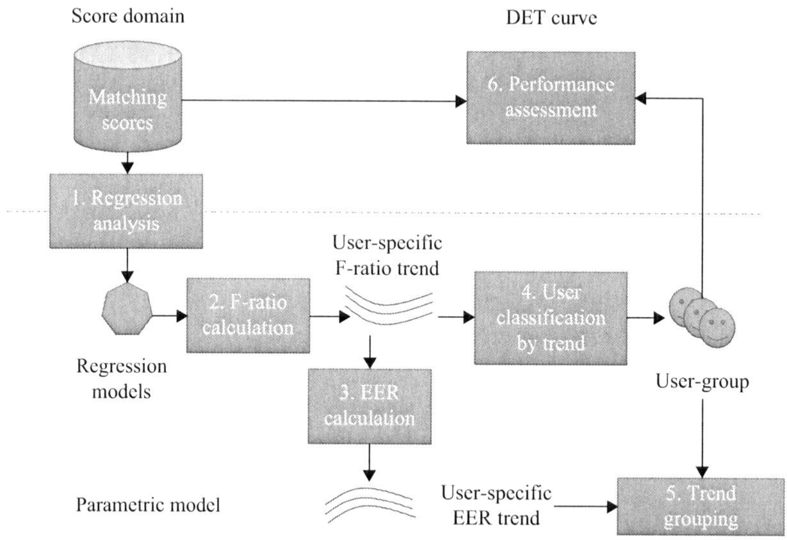

Figure 10.4 An overview of the proposed process in grouping the subjects by their changes in biometric performance

can be estimated. The proposed system architecture is summarised in Figure 10.4. The numbers in each box show the sequence of operations. The algorithms can be stated as:

- For each subject $j \in \mathcal{J}$:
  - Fit the regression function f to $\{y^{\omega}, t\}$ for each class of matching, ω.
  - Calculate the F-ratio over time for subject j.
  - Calculate the gradient of F-ratio, a_j^1.
- Cluster $\{a_j^1|\forall_j\}$ into 5 partitions by their percentiles in a regular intervals of 20% percentiles.
- Estimate the error for subjects in partitions 1 and 5 in the 'initial' and 'end' periods.

10.4 Experiment setup

10.4.1 Database and experimental protocol

The MOBIO database [13], a bimodal face and speech database, is used for evaluating our approach. It consists of roughly 600 days of speaking face and speech data covered by 12 distinct sessions, recorded using a mobile phone (Nokia N900) at different locations simulating a typical office environment. In total, there are 192 unique video samples for each of the 150 subjects. This data was captured at 6 different sites with people speaking English.

Following the Phase II protocol, the database is divided into training, development and evaluation non-overlapping sections. Each section contains 50, 42 and 59 subjects, respectively. While the training section is used to derive background models for the speech classifier, the development section is used to train the fusion classifier. Finally, the evaluation section is used uniquely to measure the system performance. The enrolment data is the first session of the MOBIO database, while the probe (query) data is constituted by the remaining sessions of the MOBIO database.

10.4.2 Face classifiers

A face image is represented by two descriptors: multiscale local binary pattern, MLBP (denoted as F1 in our experiment) and multiscale local phase quantisation, MLPQ (F2). Two types of classifiers have been implemented, each based on one type of descriptor. While the former descriptor attempts to capture variation in local image texture, the latter exploits the phase information and is, therefore, expected to be robust to image blurring. For both features, an image is encoded as histogram bins defined by a statistical summary of these features derived at different image resolutions. The back-end classifier for both descriptors attempts to compare a pair of histograms. This is done first by linearly projecting them into a more compact space via linear discriminant analysis (LDA). The distance between a pair of histograms is defined as a normalised correlation metric in the LDA space. More details about these two classifiers can be found in Appendix A.

10.4.3 Speech classifier

A long-standing state-of-the-art classifier in speaker verification is based on Gaussian Mixture Models (GMMs). A pair of GMMs is used: one to model the distribution of the speech features for the target user/subject and another to model the distribution of these features for an arbitrary number of non-target subjects. While the former model is subject-dependent, the latter model, also called the 'world model' or 'universal background model', is common to all enrollees. In order to use them to verify whether a speaker is the target subject or not, a standard approach is to compute the likelihood ratio of the observed speech features given the two distributions. Modern speaker verification classifiers are still based on the two GMMs but involve more complex classification rules.

Features that have been used for speech recognition also find their use for speaker verification. This is unexpected (but works extremely well) because in speaker verification, it is desirable to retain as much speaker variation as possible, whereas in speech recognition it is better to suppress the variation. For our experiment, linear-frequency cepstral coefficients (LFCCs) that have evenly spaced frequency bands are used [14].

An important assumption when using a GMM is that the LFCC features (of roughly 13 dimensions) from a speech sequence are independent of each other. Because this assumption is violated for features in close temporal proximity, it is common to use first-order time-difference features (called 'delta' features) as well as second order ones ('delta-delta'). All these features are used in our system.

10.4.4 Fusion classifier

The fusion classifier used throughout this chapter is logistic regression. It can be viewed as a weighted sum rule fusion whose weights are tuned in the framework of logistic regression, using the 'gradient ascent' algorithm [15]. The performance of this classifier has been shown to give good results [16].

10.5 Results

10.5.1 Results I: Model fitting

This section presents the intermediate results that show the fitted score models. Regression is extensively used at two different levels: modelling the score density over time, and modelling the F-ratio time series.

Figure 10.5 shows the fitted linear regression lines conditioned on each subject and each matching type (genuine or impostor), using a face authentication system and a speech authentication system as examples. Each continuous line (plotted in darker colour) represents the mean score trend of the match comparison, whereas each dashed line (lighter colour) represents that of the non-match comparison. A linear regression line is used in order to coarsely model the trend as upwards, downwards, or stable. As can be observed, the fitted mean score trends for the match (genuine) comparison exhibit much higher variation than their non-match comparison counterparts.

Figure 10.6 shows the EER curves so obtained by applying (10.1) to the parameters $\{\mu^{\omega}_{j,t}, \sigma^{\omega}_{j,t}\}$ for all t associated with each subject j. Each curve in this figure is an EER time-series for each subject. It is obvious from this figure that the fusion system, as shown in Figure 10.6(d), has better performance on day 1 as none of the subjects have EER exceeding 30%. However, the multimodal fusion system still degrades in performance just like its constituent systems.

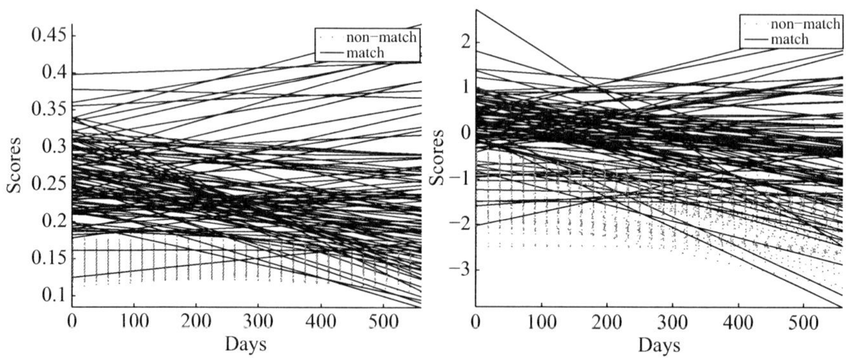

Figure 10.5 The fitted linear curves across all the subjects for the face classifier MLPQ as well as the speaker verification classifier based on GMM. A dark continuous line indicates the genuine mean curve of each subject over time, whereas a light dashed line represents the expected impostor score of each subject over time. It can be observed that the mean genuine lines vary much more within themselves than those of the mean impostor lines

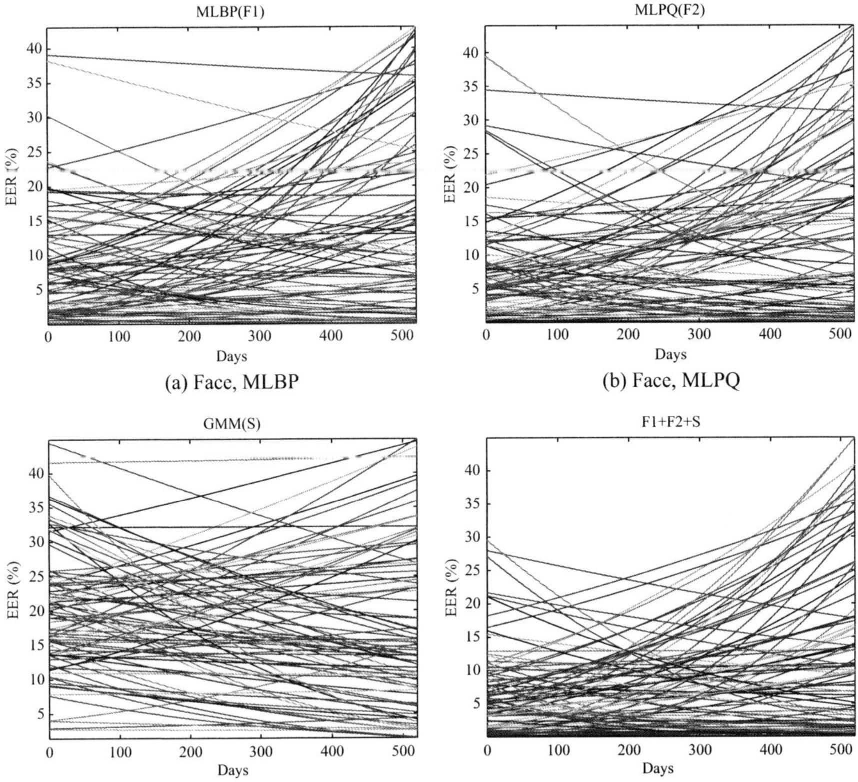

(a) Face, MLBP (b) Face, MLPQ (c) Speech, LPPC-GMM (d) Fusion

Figure 10.6 The evolution of EER curves over time of different systems. (a) and (b) are two face biometric systems, (c) is a speaker verification system and (d) is the fusion of all three systems

Based on the gradient of F-ratio, the HUGA algorithm groups the subjects into a number of partitions in ascending order. Hence, the first partition of subjects have negative gradients (decreasing performance or increasing EER), whereas the last partition of subjects have positive gradient (increasing performance or decreasing EER). Figure 10.7 shows how the EER trends are clustered sensibly for the MLBP classifier. The rest of the classifiers behave similarly.

10.5.2 Results II: Partitioned subjects

The time-dependent EERs of the three unimodal systems, as well as their combined systems, are shown in Figure 10.8. As can be observed, our subject-dependent analysis can identify subjects whose performance improves over time as well as the subjects whose performance degrades over time.

The fitted subject-specific EER series for four different systems, that is, two face systems, one speech system, and their combined system, are shown in Figure 10.6.

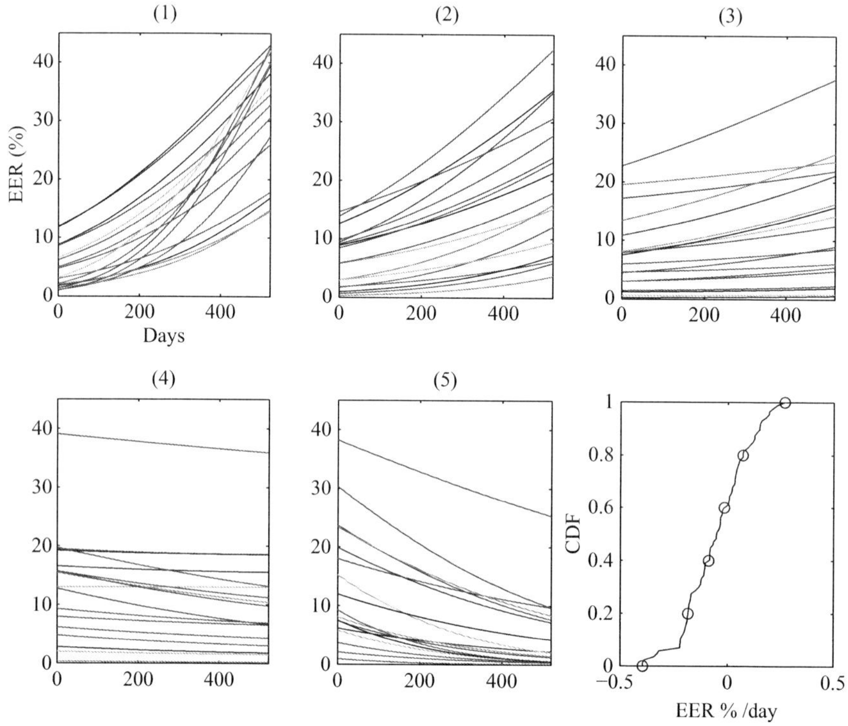

Figure 10.7 Categorisation of the EER trends for the MLBP classifier. Panels 1–5 show the five partitions of users. These partitions are found by dividing the subjects' F-ratio gradient, a_j^1 (10.4) into five partitions such that the number of subjects are balanced. The lower right panel shows the cumulative density function of a_j^1 for all j and the resultant partition boundaries marked by "○"

As can be observed, for all systems, including the fusion systems (with different constituent sub-systems), the subjects in partition 1 exhibit increasing EER over the three time periods. Conversely, the subjects in partition 5 have decreasing EER.

10.6 Conclusions

This chapter presents a new methodology for estimating biometric performance over time. A novel model called 'homomorphic user grouping algorithm' (HUGA) was introduced in order to cluster the subject-specific performance over time. Although the modelled performance trend is approximate, it allows us to detect changes in the score trend and to improve our understanding of the impact of time on biometric performance. Our analysis suggests that biometric performance does change over time. In the literature, it is widely claimed that biometric performance degrades over time. Our experiments support this observation only partially. A rather surprising finding is that the biometric performance can also improve with

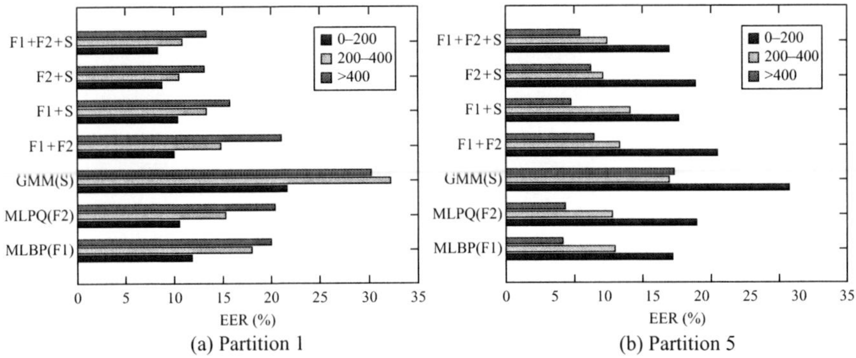

Figure 10.8 EER of different partitions for different systems in the 3 time periods, namely 0–200 days, 200–400 days, and more than 400 days. F1+F2+S denotes the fusion system that consists of the 3 subsystems, namely the 2 face classifiers, F1 (MLBP), F2 (MLPQ), and 1 speaker verification system, S (GMM). The left panel shows the users from partition 1 who have increasing EER over time across the different systems. The right panel shows the users in partition 5 who have decreasing EER over the 3 time periods

use. A logical explanation of this is that as the subjects get used to the system, they learn to use the device in an optimal way. However, it is not possible to quantify the degree of familiarity with the system, nor separate the influence of ageing from the influence of the acquisition environment. The MOBIO database that we used was designed to be as realistic as possible. As a consequence, there was no effort to control for these factors. However, despite such a large source of variation, the proposed methodology, HUGA, can still sensibly cluster the subject-specific performance trends.

Appendix A: The face classifier

A.1 Local binary pattern

The LBP operator, shown in (A.1), extracts information which is invariant to local monotonic grey-scale variations of the image. During the LBP coding, the value of current pixel, f_c, is applied as a threshold to each of the neighbours $f_p(p = 0, .P - 1)$ to obtain a binary number. A local binary pattern is obtained by first concatenating these binary numbers and then converting the sequence into the decimal number. Using circular neighbourhoods and linearly interpolating the pixel values allows the choice of any radius, R, and number of pixels in the neighbourhood, P, to form an operator.

$$LBP_{P,R}(\mathbf{x}) = \sum_{p=0}^{P-1} s(f_p - f_c)2^P \,|\, s(v) = \begin{cases} 1 & v \geq 0 \\ 0 & v < 0 \end{cases} \tag{A.1}$$

A.2 Local phase quantisation pattern

The local phase quantisation (LPQ) [17] pattern is robust to blur effects. The phase information of LPQ can be extracted using the two-dimensional windowed Fourier transform (2DWFT):

$$\breve{f}_{\mathbf{u},\mathbf{m}} = \sum_{\mathbf{x}} \kappa(\mathbf{x}-\mathbf{m}) f_{\mathbf{x}} e^{-j2\pi \mathbf{u}^{\acute{u}} \mathbf{x}} \tag{A.2}$$

where κ is a kernel function with rectangular shape, $f_{\mathbf{x}}$ is an image, $\mathbf{u}$ is a matrix with four column vectors, $\mathbf{u} = [\mathbf{u}_0, \mathbf{u}_1, \mathbf{u}_2, \mathbf{u}_3]$, each specifying one of four frequency points; they correspond to four exponential complex functions that act collectively as a filter bank. Equation (A.2) can be written in a vector form as:

$$\breve{f}_{\mathbf{u},\mathbf{m}} = \mathbf{e}_{\mathbf{u}} \mathbf{f}_{\mathbf{m}} \tag{A.3}$$

where $\mathbf{e}_{u}$, size $= 1 \times z^2$, is a basis vector of 2DWFT. $\mathbf{f}_{\mathbf{m}}$, size $= z^2 \times 1$, is a vector of image pixel values in the window. According to Reference 18, it is sufficient for the transform in (A.3) to be computed at four frequency points, $\mathbf{u}_0 = [a, 0]^{\acute{u}}$, $\mathbf{u}_1 = [0, a]^{\acute{u}}$, $\mathbf{u}_2 = [a, a]^{\acute{u}}$ and $\mathbf{u}_3 = [a, -a]^{\acute{u}}$ where a is the bandwidth of the kernel function. This filter bank yields four resultant complex images consisting of four filtered images of the real part and four images of the imaginary part of the transform, i.e.:

$$\begin{aligned} \breve{\mathbf{f}}_{\mathbf{m}} &= [\breve{\mathbf{f}}_{\mathbf{u}_0,\mathbf{m}}, \breve{\mathbf{f}}_{\mathbf{u}_1,\mathbf{m}}, \breve{\mathbf{f}}_{\mathbf{u}_2,\mathbf{m}}, \breve{\mathbf{f}}_{\mathbf{u}_3,\mathbf{m}}]^{\acute{u}} \\ \breve{\mathbf{f}}_{\mathbf{u}_i,\mathbf{m}} &= \left[\breve{f}^{Re}_{\mathbf{u}_i,\mathbf{m}}, \breve{f}^{Im}_{\mathbf{u}_i,\mathbf{m}}\right] \end{aligned} \tag{A.4}$$

As the real and imaginary parts of these filters have zero mean [19], LPQ is invariant to changes in the mean level of illumination and in illumination contrast. A whitening transform [17] is applied to decorrelate the filtered complex images to improve the system performance. Each pixel of the resultant complex image can be encoded into a binary value as shown in (A.5) by applying the quadrant bit coding.

$$\begin{aligned} \mathbf{B}^{Re}_{\mathbf{u}_i}(\mathbf{x}) &= \begin{cases} 1 & \text{if } \mathbf{F}^{Re}_{\mathbf{u}_i}(\mathbf{x}) > 0 \\ 0 & \text{if } \mathbf{F}^{Re}_{\mathbf{u}_i}(\mathbf{x}) \le 0 \end{cases} \\ \mathbf{B}^{Im}_{\mathbf{u}_i}(\mathbf{x}) &= \begin{cases} 1 & \text{if } \mathbf{F}^{Im}_{\mathbf{u}_i}(\mathbf{x}) > 0 \\ 0 & \text{if } \mathbf{F}^{Im}_{\mathbf{u}_i}(\mathbf{x}) \le 0 \end{cases} \end{aligned} \tag{A.5}$$

This coding method assigns two bits for every pixel to represent the quadrant in which the phase angle lies. In fact, it also provides the quantisation of the Fourier phase feature. LPQ is a binary string obtained for each pixel by concatenating the real and imaginary quadrant-bit codes of the eight Fourier coefficients of $\mathbf{u}_i$. The binary string is then converted to the decimal number by (A.6) to produce an LPQ pattern.

$$\begin{aligned}\mathbf{LPQ}(\mathbf{x}) = \mathbf{B}^{Re}_{\mathbf{u}_0}(\mathbf{x}) + \mathbf{B}^{Im}_{\mathbf{u}_0}(\mathbf{x}) \times 2^1 + \cdots \\ + \mathbf{B}^{Re}_{\mathbf{u}_3}(\mathbf{x}) \times 2^{k-1} + \mathbf{B}^{Im}_{\mathbf{u}_3}(\mathbf{x}) \times 2^k\end{aligned} \tag{A.6}$$

A.3 Multiscale pattern histogram

A multiresolution representation can be obtained by varying the filter size, $z \times z$, and combining the pattern images. This representation [17, 20] has been suggested for face recognition and the results reported for this application show that its accuracy is better than that of the single scale pattern method. A multiresolution representation derived by a set of pattern operators of different filter size may give an unstable result because of noise, but this problem can be minimised by using aggregate statistics, such as the histogram. There are several advantages in summarising the patterns in the form of the histogram. First, the statistical summary can reduce the feature dimension from the image size to the number of histogram bins. Second, using the histogram as a set of features is robust to image translation and rotation to a certain extent and therefore the sensitivity to mis-registration is reduced. Finally, although the effect of unstable pattern responses due to noise is attenuated by histogram binning, it can further be reduced by controlling the number of histogram bins and/or projecting the histogram to other spaces.

In our approach, pattern operators defined on a neighbourhood Q, for both LBP and LPQ, at R scales are first applied to a face image. This generates a grey level code for each pixel at every resolution. The resulting pattern images are cropped to the same size and divided into non-overlapping sub-regions, $\mathbf{M}_0, \mathbf{M}_1, \ldots \mathbf{M}_{J-1}$. The regional pattern histogram for each scale is computed as:

$$\begin{aligned}&\mathbf{h}_{r,j}(i) = \sum_{x \in \mathbf{M}_j} E(Q_r(\mathbf{x}) = i) \\ &\mid i \in [0, L-1], j \in [0, J-1], \\ &\quad r \in [1, R], z = r \times 2 + 1 \\ &E(v) = \begin{cases} 1 & \text{when } v \text{ is true} \\ 0 & \text{otherwise} \end{cases}\end{aligned} \tag{A.7}$$

$E(v)$ is a Boolean indicator. r is the scale index and z is the width or height of the pattern filter. L is the number of histogram bins. The set of histograms computed at different scales for each region $\mathbf{M}_j$ provides regional information.

By concatenating these histograms into a single vector, we obtain the final multiresolution regional face descriptor.

$$\mathbf{k}_j = [\mathbf{h}_{1,j}, \mathbf{h}_{2,j}, \cdots, \mathbf{h}_{R,j}] \tag{A.8}$$

A.4 Image frame matching

This regional facial descriptor can directly be used to measure the face similarity of a pair of images. However, as multiscale LBP and LPQ histograms are of high dimensionality and contain redundant information, better performance can be achieved by projecting the histogram information, $\mathbf{k}_j$, into linear discriminant analysis (LDA) space $\mathbf{W}_j^{lda}$ [20–22] to yield a regional discriminative facial descriptor, $\mathbf{d}_j$

$$\mathbf{d}_j = (\mathbf{W}_j^{lda})^{\text{ú}} \mathbf{k}_j \tag{A.9}$$

After the projection, the similarity between an incoming image, $\mathbf{I}'$, and the template image, $\mathbf{I}$, is measured by summing the similarities, i.e. normalised correlation coefficients of the regional discriminative descriptors:

$$Sim(\mathbf{I}, \mathbf{I}') = \sum_j \frac{\mathbf{d}_j^{\text{ú}} \mathbf{d}'_j}{\| \mathbf{d}_j \| \| \mathbf{d}'_j \|} \tag{A.10}$$

A.5 Video face matching

In the first 200 frames of each video, 10 frames are chosen based on the rank result of the confidence score of a face detector. In each frame, face image is geometrical and photometrically normalised, and divided into 5×5 regions for extracting the multiresolution discriminative regional descriptor. Then, given a video pair for matching, the frame-level similarity scores are computed using 13 and the maximum of these is used as a final score.

Acknowledgement

This work was supported by the Biometrics Evaluation and Testing (BEAT), an EU FP7 project with grant no. 284989.

References

[1] P. J. Flynn, K. W. Bowyer, and P. J. Phillips, 'Assessment of time dependency in face recognition: An initial study', *LNCS 2688, 4th International Conference Audio- and Video-Based Biometric Person Authentication (AVBPA 2003)*, Guildford, 2003, pp. 44–51.

[2] H. Wang and P. J. Flynn, 'Experimental evaluation of eye location accuracies and time-lapse effects on face recognition systems', *LNCS 3546, 5th International Conference Audio- and Video-Based Biometric Person Authentication (AVBPA 2005)*, New York, 2005, pp. 627–636.

[3] N. Poh and J. Kittler, 'A method for estimating authentication performance over time, with applications to face biometrics', *12th IAPR Iberoamerican Congress on Pattern Recognition (CIARP)*, Via del Mar-Valparaiso, Chile, 2007, pp. 360–369.

[4] N. Poh and S. Bengio, 'Database, protocol and tools for evaluating score-level fusion algorithms in biometric authentication', *Pattern Recognition*, vol. 39, no. 2, pp. 223–233, February 2005.

[5] G. Doddington, W. Liggett, A. Martin, M. Przybocki, and D. Reynolds, 'Sheep, goats, lambs and wolves: A statistical analysis of speaker performance in the NIST 1998 speaker recognition evaluation', *International Conference Spoken Language Processing (ICSLP)*, Sydney, 1998.

[6] N. Yager and T. Dunstone, 'Worms, chameleons, phantoms and doves: New additions to the biometric menagerie', *Automatic Identification Advanced Technologies. 2007 IEEE Workshop on Automatic Identification Advanced Technologies*, pp. 1–6, 7–8, June 2007. doi:10.1109/AUTOID.2007.380583

[7] A. Hicklin and B. Ulery, 'The myth of goats: How many people have fingerprints that are hard to match?', Tech. Rep. NISTIR 7271, National Institute of Standards and Technology, 2005.

[8] M. Wittman, P. Davis, and P. J. Flynn, 'Empirical studies of the existence of the biometric menagerie in the frgc 2.0 color image corpus', *Conference on Computer Vision and Pattern Recognition Workshop*, New York, NY, pp. 33–33, June 2006.

[9] M. N. Teli, J. R. Beveridge, P. J. Phillips, G. H. Givens, D. S. Bolme, and B. A. Draper, 'Biometric zoos: Theory and experimental evidence', *International Conference on Biometrics: Theory, Applications, and Systems, 2009. BTAS'09. IEEE 3rd*, Washington, D.C., USA, 2011, pp. 1–8.

[10] N. Poh and J. Kittler, 'Incorporating variation of model-specific score distribution in speaker verification systems', *IEEE Transactions on Audio, Speech and Language Processing*, vol. 16, no. 3, pp. 594–606, 2008.

[11] N. Poh and S. Bengio, 'Why do multi-stream, multi-band and multi-modal approaches work on biometric user authentication tasks?', *IEEE International Conference on Acoustics, Speech, and Signal Processing (ICASSP)*, Montreal, 2004, vol. V, pp. 893–896.

[12] Sara A. Van De Geer, 'Least squares estimation', In: Brian S. Everitt and David C. Howell (Eds.), *Encyclopedia of Statistics in Behavioral Science*, Vol. 2, John Wiley & Sons, Ltd, Chichester, 2005, pp. 1041–1045.

[13] C. McCool, S. Marcel, A. Hadid, M. Pietikainen, P. Matejka, J. Cernocky, N. Poh,...T. Cootes, 'Bi-modal person recognition on a mobile phone: using mobile phone data', *IEEE ICME Workshop on Hot Topics in Mobile Multimedia*, July 2012.

[14] Q. Le and S. Bengio, 'Client dependent GMM-SVM models for speaker verification', *ICANN*, 2003, pp. 443–451.

[15] T. Hastie, R. Tibshirani, and J. Friedman, *The Elements of Statistical Learning*, Springer-Verlag, New York, 2001.

[16] N. Poh, T. Bourlai, J. Kittler, L. Allano, F. Alonso-Fernandez, O. Ambekar, J. Baker, ... C. Vielhauer, 'Benchmarking quality-dependent and cost-sensitive score-level multimodal biometric fusion algorithms', *IEEE Transactions on Information Forensics and Security (TIFS)*, vol. 4, no. 4, pp. 849–866, 2009.

[17] C-H. Chan, J. Kittler, N. Poh, T. Ahonen, and M. Pietikinen, '(Multiscale) local phase quantisation histogram discriminant analysis with score normalisation for robust face recognition', *Workshop on Video-oriented Object and Event Classification (VOEC), in conjunction with the 12th IEEE International Conference on Computer Vision (ICCV)*, Kyoto, Japan, 2009, pp. 633–640, 978-1-4244-4441-0.

[18] J. Heikkilâ and V. Ojansivu, 'Methods for local phase quantisation in blur-insensitive image analysis', *LNLA*, 2009, pp. 104–111.

[19] J. G. Daugman, 'High confidence visual recognition of persons by a test of statistical independence', *IEEE Transactions on Pattern Analysis and Machine Intelligence*, vol. 15, no. 11, pp. 1148–1161, November 1993.

[20] C-H. Chan, J. Kittler, and K. Messer, 'Multi-scale local binary pattern histograms for face recognition', In *ICB*, S.-W. Lee and S. Z. Li (Eds.), 2007, vol. 4642 of *Lecture Notes in Computer Science*, Springer, Seoul, pp. 809–818.

[21] P. N. Belhumeur, J. P. Hespanha, and D. J. Kriegman, 'Eigenfaces vs. fisherfaces: Recognition using class specific linear projection', *Transactions on Pattern Analysis and Machine Intelligence*, vol. 19, no. 7, pp. 711–720, July 1997.

[22] S. Shan, W. Zhang, Y. Su, X. Chen, and W. Gao, 'Ensemble of piecewise FDA based on spatial histograms of local (Gabor) binary patterns for face recognition', In *ICPR*, Hong Kong, vol. 4, 2006, pp. 606–609.

Section 3

Applications and implications for practical applications

Chapter 11

An industrial perspective on biometric age factors

Alastair Partington[1]

11.1 Industrial implications of biometric ageing – a risk-based view

First, a few words of introduction. This chapter does not present a theoretical hypothesis or any new findings from the world of biometric academia, unlike many others in this book. Rather, it draws on the author's practical experience gained through delivering trial and production biometric systems around the globe to address a fundamental question – how is an understanding of biometric ageing relevant to the world of industry, which includes governmental use for immigration and identity management, and commerce?

The following discussion and conclusions are therefore intended to help to set biometric ageing in an industrial context, and may be of interest to those engaged in the topic, regardless of their relative alignment to academia, industrial research and development, or commercial delivery.

To start with, we approach the topic of biometric ageing in the broadest sense, considering what impact it might have on today's industrial and commercial systems. Following this initial 'sanity check', we proceed to consider some of the key factors that might tend to exacerbate the impact of biometric ageing, and present an approach to identify those systems most exposed to this impact, illustrated by some case studies. We then consider some mitigation approaches that might serve to reduce the impact that biometric ageing presents, and consider how these might be put into practice. Finally, we close with a look at how academia and industry might collaborate to bring an improved understanding of biometric ageing to market.

11.2 Biometric ageing – does it matter?

To set us on the right track, it's worth taking a step back and considering two very basic questions – 'Are there any industrial implications of biometric ageing at all,

[1] Alastair Partington leads Accenture's "Unique Identity" practice within the company's Emerging Technologies Innovation group, although the views expressed in this chapter are those of the author on a personal basis

and if so, what are they?' Or, put another way – 'Does industry care about biometric ageing, and if so, why?'

Consider the current macro-level trends in the biometric marketplace. Originally captured from scenes of crime and from suspects, the use of biometrics offered a breakthrough technique for criminal investigators. However, more recently, the benefits of using biometrics as a recognition mechanism to improve security, convenience and inclusion in society, at greater efficiency and reduced cost, have been widely recognised. As a consequence, biometric solutions are increasingly being encountered and used by the general public in both governmental and commercial systems worldwide, in areas as varied as civil ID, travel facilitation, banking and marketing, with a double-digit compound annual growth rate in the number of deployed biometric systems forecast. Accompanying this rapid growth in breadth and reach of biometric systems, we are seeing an increased willingness by both providers and consumers to openly trust biometric identification to ensure the uniqueness of the identity claim – along with a strong trend towards self-service and reduced supervision, with providers especially keen to unlock the efficiency benefits that this can offer. Essentially, then, we are observing a growth in the number and breadth of biometric systems deployed, and in our reliance on them providing correct identity recognitions, like never before.

Biometric identification is not a black-and-white science, but rather one that is designed to function in a probabilistic, greyscale world. Biometric matching systems are assessed by their error rates, among other parameters; they are not 'perfect', and the risk of an incorrect biometric identification – a 'false match' or 'false non-match', in less technical parlance – is always present, albeit slight for a carefully designed and skilfully implemented system. The matching risk is affected by a wide range of factors such as the physical traits being imaged and matched, the quality of the samples captured and the permanence of the physical traits over time (the topic of this book), and is typically associated with impacts to either security (in the case of false matches) or facilitation (in the case of false non-matches). Given the trend described previously for industry's increasing dependence on biometric recognition, the matching risks to which both businesses and users are exposed can also be expected to increase, unless steps are taken to mitigate them.

The fact that the commercial world wants to adopt biometric systems, for the compelling business reasons outlined above, and that consumers are generally ready to use them, is very positive news for all those with an interest in seeing biometric technology thrive. However, it is also clear that some potentially game-changing deployments are not yet viable, due to the high performance demands that these challenging business scenarios would place on their requisite biometric matching systems. Think, for example, of the 'ideal shopping' experience, where you might be recognised by your face image when entering a store, distinguished from perhaps tens of thousands of other regular customers, and automatically billed for your purchases. We have all the technology components required to make a scenario like this work today; however, the risk of falsely identifying a customer means that until the matching technologies can offer low-enough real-world identification accuracy rates, the business case for such a scenario will not stand up. The commercial risk presented and the financial contingency that would have to be accepted by the adopting organisation are, in these kinds of cases, still too great.

This presents industry with a challenge – how to lower the risk that the increased adoption of biometric recognition is bringing, regardless of whether the ultimate goal of that adoption is increased security or improved user facilitation. Fortunately, industry understands risk very well (particularly where it involves a potential financial cost), and so is receptive to discussions on how it might be mitigated or better, lowered.

Commercial risk is frequently viewed as a function of two factors, likelihood and impact – likelihood being the probability that a given issue will occur, and impact representing the damage that the issue might cause to the business if it happened. The highest-risk activities are, of course, the ones that have a high likelihood of occurring and would cause a significant impact if they did. The trends outlined above, of increasing usage of and reliance on biometric systems, serve to increase both of these key risk factors. With the business stakeholders themselves, as well as regulatory bodies and the popular media all alert to the risk that a 'wrong' biometric decision is made by any biometric system, it is fair to say that the collective interest in ensuring accurate biometric matching is stronger than ever.

To drive the risk of false matches down, a strong focus has been accorded to improving real-world matching accuracy and consistency. The last ten years have seen impressive improvements in the performances of deployed biometric systems, with measures such as updated algorithms, strong quality control and multimodal matching being adopted in the quest for improved matching performance – and the accompanying decrease in business risk. The challenge now is that most of the relatively low-cost/high-return gains in this space such as multi-algorithm and/or multi-modal matching – the low-hanging-fruit, if you will – have already percolated through from academia to industrial deployment, and biometric identification, outside of the laboratory at least, is still some way from perfect.

It is reasonable to state, therefore, that industry might have more than a passing interest in the increasing maturity and commercial readiness of 'secondary factors' that can positively influence matching accuracy. The question now is whether biometric ageing is one of those secondary factors worthy of adoption.

While some biometric traits can be readily observed to change with age, such as our faces and voices, there used to exist a general belief that others (such as fingerprint and iris [1]) were largely stable from soon after birth until death. Many systems have been designed on this basis, assuming that these 'stable' traits could be used for identification more or less regardless of how old the reference samples were. However, recent research now indicates that this assumption is invalid; we know that irises do change over a person's lifetime [2], for example. We have come to realise that ageing can indeed negatively affect biometric matching accuracy for most modalities, resulting in increased False Match/False Non-Match or False Positive/Negative Identification rates, and corresponding increases in False Acceptances/False Rejections. Biometric ageing compensation would, therefore, appear to be one area that merits attention – a good understanding of its functioning could contribute meaningful accuracy gains and help drive error rates below their current levels in industrial systems.

We can thus conclude our initial 'sense check' with the suggestion that if an improved academic understanding of biometric ageing can be demonstrated to improve matching performance in the real world, and if this can be effectively harnessed by production systems, then it would likely yield business value. Our basis, therefore, will be that industry *is* interested in biometric ageing, because of the potential it offers to help diminish business risk.

11.3 Industry: at the starting line for biometric ageing

Perhaps we should replace *is* in the previous sentence with *should be*. Despite the fact that the idea of using the age difference between candidate and reference samples in a matching operation is not new – indeed, criminal AFIS/ABIS systems have offered guided, manual 'age progression' features for some time – there is little awareness of biometric ageing in the commercial space, and correspondingly, little understanding of why or how it should be mitigated or allowed for. Consequently, the vast majority of civil biometric matching systems in use today make no allowances for the physical changes to biometric samples that take place due to the human ageing process, due to modifications in structure or shape, general wear and tear, or the possibility of damage.

11.4 Factors exacerbating age-related biometric matching inaccuracies in industrial systems

We shall now explore the factors that might increase the effects of age-related change on biometric matching accuracy, and how accommodating these factors in the matching process might reduce the probability of obtaining false results. We shall attempt to use this set of factors as a mechanism to identify the types of systems that might be most susceptible to inaccuracies caused by biometric ageing, and subsequently, consider how those inaccuracies might best be mitigated.

There are, inevitably, several factors that can influence the likelihood that a reference sample might no longer be fully representative of an individual, due to their biometric traits having changed since the reference was acquired. Without attempting to form a definitive list, these can be said to include:

- the *absolute age difference between the reference and candidate samples*;
- the *susceptibility of the biometric modality (or modalities) to age-related change*;
- the demographics of the user population of the system – in particular, the age distribution (*stage of life*) and *occupation*;
- the *tuning of the matching system* – whether it is skewed towards preferentially rejecting, or accepting, users.

Experience suggests that the first two of these factors are likely to be the most impactful, and thus they are termed 'primary factors' for the purpose of this discussion.

The second pair of factors, while still potentially significant, is believed to be less impactful, and thus is termed 'secondary factors'.

We can consider each of these factors briefly, in turn, to better understand the influence that they might bring to the overall risk picture.

11.4.1 Primary factors

11.4.1.1 Absolute age difference between reference and candidate samples

Generally speaking, for ideal biometric matching accuracy, the smaller the age gap between the reference and candidate samples, the better; if any justification is required for this broad statement, consider the exaggerated case whereby a sample taken yesterday is likely to be considerably more representative of an adult than one taken when he or she was a baby. However, the relationship between the accuracy drop and the age gap is complex – it has been shown to be non-linear [3], and will vary by modality as well as by a multitude of other factors, such as sample quality.

Key to minimising the absolute age difference between reference and candidate samples is the approach taken by system designers to the re-enrolment of users, which can reduce the age variance to zero at a stroke. The choice of re-enrolment approach will be swayed by various factors – user accessibility being key. At one extreme, re-enrolment might be impossible, such as when considering latent images from a crime scene, or an enrolment dataset from a now-uncontactable customer. At the other extreme, re-enrolment might be built into the system as a near-continuous process – think of an access-control system, which benefits from daily contact with the user, and a convenient manual-authentication process. In between these states, re-enrolment may occur at fixed intervals, such as at document renewal, or variable, for example, upon customer or citizen request.

Ultimately, given the importance that the age-variance metric has, minimising it through considered re-enrolment should normally be the first step to minimising the associated ageing risk. Re-enrolment as an ageing-mitigation strategy is discussed in further detail later in this chapter.

11.4.1.2 Susceptibility of biometric modality to age-related change

Aside from DNA, which is considered broadly invariant throughout a person's life, samples of all modalities are likely to undergo a degree of age-related change. The least stable modalities in regular use today are often cited as signature, voice and keystroke [4]; fingerprint and iris patterns, on the other hand, are typically more stable.

A further consideration is that ageing changes in some modalities are better-researched and, consequently, better understood than others. The majority of biometric systems in use today leverage fingerprint or face biometrics [5], with iris, voice and signature also significant. While fingerprint, face and iris ageing are increasingly well-researched, the effects of ageing on other biometric traits are less

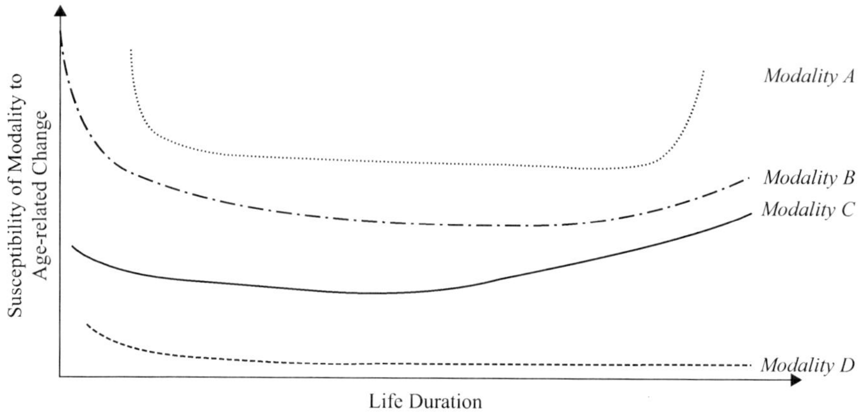

Figure 11.1 Indicative comparison of how the susceptibility of different biometric modalities to age-related change varies with absolute age (not to scale)

well understood – partly, it should be said, due to the lack of readily-available test data for each of these modalities.

Though the full picture of how each biometric modality typically varies with age is still being revealed by current research, we can attempt to illustrate this in an illustrative manner. Figure 11.1 illustrates the relative susceptibility of four arbitrary modalities to age-related change.

Of these four traits, Modality A might represent a behavioural trait, such as signature, that requires a degree of development before it can be captured, and evolves throughout a person's life, before degrading rapidly with advanced age; Modality B, a trait that can be captured from birth, stabilises somewhat into adulthood while continuing to exhibit a significant degree of change year-on-year, and then speeds up its rate of change towards end-of-life, such as facial image; Modality C, a trait that can be captured shortly after birth, again stabilises into adulthood, and then degrades increasingly rapidly from middle-age onwards, such as fingerprint; and Modality D, a highly stable trait that remains broadly unchanged throughout a person's life, such as iris.

This picture is further complicated by the increasing trend for multi-modal systems. Depending on the modalities selected for initial enrolment and subsequent usage, a hybrid susceptibility curve will result. Systems such as India's Unique ID (Aadhaar) scheme are breaking new ground in this respect, pursuing a multi-modal enrol-once, use-forever approach which we discuss in further depth later.

11.4.2 Secondary factors

11.4.2.1 Stage of life

It is well-accepted that the rate of age-related change of biometric samples is not continuous from birth to death, with the greatest change in individuals' faces (for example) taking place between birth and adulthood. Fewer conclusive results are, as yet, available regarding age-related change in different ethnicity populations, or different socio-economic groups, though these may all be further complicating factors. Regardless, the demographics of the user population of the system – in particular, the age distribution – are a significant consideration. Systems that are designed to accommodate very young or very old people (such as those installed in educational establishments, residential homes, or the most extreme cases, birth or death registration) can expect to be early adopters of age-mitigation measures, while those which need to function for any user, regardless of age, perhaps face the greatest challenge.

11.4.2.2 Occupation

The typical occupations of the target population are also a factor to consider – though it is, of course, often difficult to generalise. A system designed for use by a population that has predominantly manual or rural occupations might not want to make use of those biometric traits that tend to be more easily damaged or worn, such as fingerprint or palmprint, and might instead opt for a more resilient characteristic such as iris or vein.

The potential for wear and tear, and accidental damage, is also both age- and modality-dependent. Most people are fortunate enough not to suffer damage to their irises or veins during their lifetime; fingerprints, on the other hand, become increasingly worn throughout a person's life. Practical experience also suggests that finger-vein images can be increasingly hard to acquire consistently from the elderly as their skin consistency (and consequently, finger shape) changes.

11.4.2.3 Tuning of matching system

Biometric matching systems can be optimised for various roles through algorithm and threshold selection. Those tuned for high security typically operate a narrow matching tolerance in order to prevent 'false positive' matches and 'imposter' attacks, either of which would be highly damaging. Systems designed to prioritise convenience, on the other hand, will operate wider tolerances, so as not to inconvenience genuine users. Generally speaking, therefore, high-security systems might be expected to be more sensitive to age-related changes than their convenience equivalents – a small change in a biometric trait might be sufficient to convince the system that the reference biometric sample no longer represents its owner.

It is worth bearing in mind that though the proportion of users impacted by age-related changes might be greater for high-security systems than convenience systems, mass-market convenience systems may still be significantly affected. Enablement systems deployed, for example, in retail scenarios or in the home may often have (a) many times more users and (b) less effective manual processes in place to handle false rejections (perhaps due to operating in self-service mode without supervisors available to provide assistance) than their high-security counterparts. The actual business impact of biometric change due to ageing might thus still be significant for a convenience system, even though proportionally fewer issues might be expected to occur on a 'per user' basis.

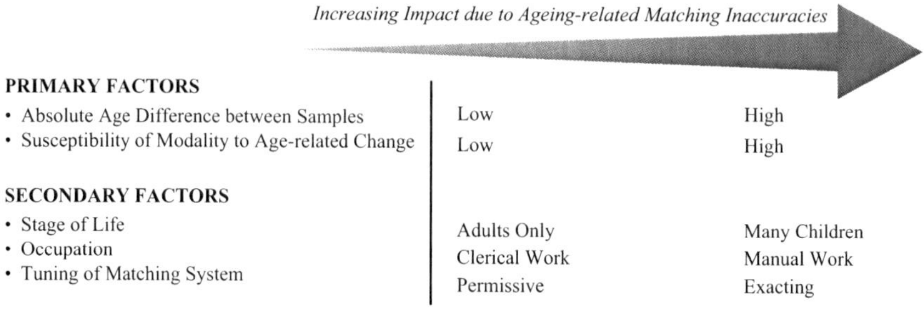

Figure 11.2 Indicative illustration of the impact of age-related change on biometric matching accuracy

This discussion of factors that affect the impact of age-related change on biometric matching accuracy can be summarised in Figure 11.2.

Hopefully, research activities will help us to calibrate this scale with confidence in the near future; for now, it remains qualitative. It can, however, be brought to life with a couple of examples.

11.5 A low-impact scenario: student identification

The educational sector represents a sizeable and largely untapped opportunity for the biometrics industry. Adopting institutions are leveraging biometrics to benefit from streamlined and secure student identification for activities such as daily attendance registration, physical access control, on-campus payments and library book check-out. Most systems in use today use single, flat fingerprints in verification (1:1) mode [6], though others exist that use vein [7] or iris modalities.

11.5.1 Primary factors

11.5.1.1 Absolute age difference between reference and candidate samples

Students would normally be registered when they join their educational institution; courses of study might typically last for three to five years. Reference samples are held in a school-specific database, and so could in theory be updated as required. No guidelines are readily available from governance bodies or the leading technology providers concerning routine student re-enrolment; practice appears to vary between annual re-enrolment, and 'whenever a student can no longer be recognised'. Thus it is probably reasonable to assume that the average age of a sample might be three years, or thereabouts, on average.

11.5.1.2 Susceptibility of biometric modality to age-related change

Fingerprints are generally a stable biometric, as studies [8] have shown. Age variations of a few years should not significantly affect matching accuracy (however, see 'Stage of life' below).

The likely impact of using reference samples with absolute ages averaging a few years for fingerprint modality can thus be judged as Low.

11.5.2 Secondary factors

11.5.2.1 Stage of life

The target user population is children and young adults, including, in some cases, primary-school children. Given the rates of growth of humans, and their fingers, at these ages, age-related changes are likely to be significant. The impact of demographic factors is, therefore, likely to be High.

11.5.2.2 Occupation

Students are unlikely to be involved in activities that expose their fingerprints to significant wear; this factor will therefore be Low.

11.5.2.3 Tuning of matching system

Student identification systems are likely to be tuned for a balance between False Acceptance and False Rejection – they have to be convenient to use, and tolerate a degree of inconsistent sample presentation on behalf of their students, while remaining acceptably secure. The tuning of the matching system will therefore be a Low-impact factor.

11.5.3 Risk assessment

With respect to the discussion above, with primary factors initially suggesting a 'low' risk prior to mixed modification by secondary factors, we can speculate that we might expect age-related factors to have an overall Low to Moderate impact on the overall matching risk to which a student identification system is exposed.

11.6 A moderate-impact scenario: ePassport gates

Since 2010, European countries including the United Kingdom and the Netherlands have operated ePassport gates at their airport terminals. Arriving travellers present themselves and their electronic passports at the gates, which compare each live face with the photograph stored on that traveller's ePassport chip. Once the face match and other checks are successfully met, the gates open automatically and the traveller can enter the country; if any checks are failed, the travellers are directed to a Border Officer for secondary inspection.

11.6.1 Primary factors

11.6.1.1 Absolute age difference between reference and candidate samples

Given the current passport validity duration of up to ten years, we would expect the age of the reference sample (stored in the ePassport chip) to average slightly under five years in the long-term (given a small proportion of within-validity replacements for loss, damage, etc.). Note that today, however, most users will have reference samples of ages significantly below this average, since the ePassport first-issuance cycle is still underway for most countries.

11.6.1.2 Susceptibility of biometric modality to age-related change

The face modality is reasonably susceptible to age-related change, as studies [9] have shown, and age variations of up to ten years (as explained above) may indeed significantly affect matching accuracy.

The likely impact of using reference samples with absolute ages averaging five years for face modality can thus be judged as Moderate.

11.6.2 Secondary factors

11.6.2.1 Stage of life

Eligibility for these systems is typically limited to EU nationals, over the age of 18. Given the exclusion of children from the user population (and our currently limited knowledge about how other demographic factors might affect the impact of biometric ageing), we can characterise this as a Low-impact factor.

11.6.2.2 Occupation

There is currently little evidence relating facial ageing with mainstream occupations; in any case, the travelling population will be drawn from a wide variety of backgrounds, and so it would be difficult to generalise about the proportion of manual vs. clerical workers – we can therefore assume that this factor does not modify the overall matching risk.

11.6.2.3 Tuning of matching system

The ePassport system is tuned for high security, such that False Matches (resulting in travellers being admitted into the country by mistake!) will be minimised. The tolerance of the system to age-related change will therefore be relatively slim, meaning that this is a Moderate- to High-impact factor.

11.6.3 Risk assessment

Given the above moderate primary risk factors, modified by varied secondary factors, we can state that we would expect age-related factors to have a Moderate impact on the overall matching risk to which ePassport systems are exposed.

11.7 A high-impact scenario: anticipated use of India's Unique ID scheme 'Aadhaar'

Conceived in 2006 and with enrolments starting in 2010, India's UID scheme [10] will provide identification for each resident across the country – around 1.2 billion individuals. Aadhaar, which translates into 'foundation', or 'support', is intended to be used primarily as the basis for efficient delivery of welfare services. It will also support service delivery in the commercial sector, with social initiatives such as micropayments prioritised for early adoption.

Aadhaar enrolment is free to all residents, and involves multi-modal biometric capture (fingerprint, face, and iris). Following successful enrolment, identity verification and de-duplication against the Aadhaar database, a unique 12-digit Aadhaar number is issued to the applicant.

Here, we will consider use of the Aadhaar system to deliver identity services in a commercial scenario, 20 years in the future, assuming fingerprint-only modality and verification (1:1) matching, using the Aadhaar number as the key.

11.7.1 Primary factors

11.7.1.1 Absolute age difference between reference and candidate samples

Aadhaar positions itself as an 'enrol once, use many'-type system. Given the challenges with the scale of initial enrolment, and the choice to enrol multiple modalities upfront, it seems reasonable to assume that re-enrolments would be the exception, rather than the norm. Applicants above the age of 3 years are biometrically enrolled; the age of the reference sample could therefore vary from 0 to pretty much a human lifespan. In our scenario, 20 years hence and 22 years after the enrolment programme started, and assuming a constant pace of enrolment, we might consider a typical reference sample age to be 11 years. This age will continue to rise as use of the scheme is extended further into the future.

11.7.1.2 Susceptibility of biometric modality to age-related change

Fingerprints are generally a stable biometric, and even age variations of 20 years are unlikely to greatly affect matching accuracy.

The likely impact of using reference samples with absolute ages averaging 11 years for fingerprint modality can thus be judged as Low.

11.7.2 Secondary factors

11.7.2.1 Stage of life

The target user population includes children and the elderly. Age-related changes to fingerprint stability at these extremities of the age range may be significant. The impact of this factor is, therefore, likely to be Moderate to High.

11.7.2.2 Occupation

A significant proportion of the enrolled population will include rural workers and manual workers, many of whom may suffer fingerprint degradation due to wear. The impact of this factor is likely to be Very High.

11.7.2.3 Tuning of matching system

A commercial banking scenario is likely to see systems tuned towards high security – they have to provide a reliable identity check, and inadvertent rejections can be satisfactorily resolved by manual processes. The impact of a potential increased FNMR due to ageing is likely to be considerable, and this will therefore be a Moderate to High-impact factor.

11.7.3 Risk assessment

With respect to the discussion above, with primary factors initially suggesting a 'low' risk prior to extensive modification by secondary factors, we can speculate that we might expect age-related factors to have a High impact on the overall matching risk to which a commercial adoption of the Aadhaar system is exposed.

The qualitative nature of the scenario analysis above illustrates the space for a formal study to validate and exhaustively identify the pertinent factors affecting the ageing impact, and attempt to calibrate them. A quantitative ageing assessment tool would be a great asset to designers, stakeholders and operators of industrial systems!

11.8 Possible mitigations to age-related biometric matching inaccuracies in industrial systems

Having discussed how age-related factors might affect the matching accuracy of industrial systems, and hence the business risk associated with them, we will now consider how these effects might be mitigated. Two practical approaches to achieving this aim are:

- minimising the system's susceptibility to age-related accuracy degradation through appropriate design decisions;
- correcting (or making allowance for) the changes that are expected to take place due to ageing, as part of the matching process.

These are not exclusive, and indeed it may be desirable to adopt both approaches in certain cases.

11.8.1 Age impact minimisation by design

First, and most simply, we can strive to avoid the problem in the first place – by adopting an approach that we might call "age impact mitigation by design". This involves considering, at the early stages of system scoping, planning and design, how the impact of ageing can be minimised – essentially, how to optimise each of the key factors previously identified. We can have most success by focusing on how we might control the primary factors (Absolute Age Difference between Reference and Candidate Samples, and Susceptibility of Biometric Modality to Age-Related Change); the secondary factors (Stage of Life, Occupation and Tuning of Matching System) are understandably harder to affect, given that they are often defined and fixed for a given system scope. All the choices made at this stage should, of course, be governed by the business scenario that the system will fulfil, and the user population that it will serve.

11.8.1.1 Optimising the absolute age difference between reference and candidate samples

First, when de-duplicating a real-world dataset, we will often end up with a set of multiple reference samples for an individual. Adjudication systems should be designed such that they can guide adjudication operators to take into account sample age, as well as the sample quality, when deciding which to promote as the 'best' reference set. Depending primarily on the modality in use, and (again) the business scenario and user population, the optimal decision between preferring sample quality and sample age will sway towards one or the other. Put another way – we may prefer to prioritise a slightly lower quality biometric sample for an individual, if it is 20 years younger.

Second, we should take every opportunity to re-enrol biometric traits where this can be done in such a way as to not inconvenience the end-users in order to keep the reference set as current as possible. At one extreme, we might see the subjects in an enrolment environment just once – consider a mass-citizen enrolment scheme such as India's UID, which is designed as an enrol-once system (and understandably so, since it would be very expensive to attempt a comprehensive re-enrolment of the 1.2 billion citizens). At the other extreme, systems might seek to achieve 'continuous enrolment', whereby if an individual is biometrically matched during routine system usage, their live sample is of satisfactory quality, and the risk of a biometric 'spoofing' attack is considered negligible; this live sample is stored in their new reference sample – resetting the 'ageing' clock to zero. In between the two are systems that offer periodic re-enrolment, either ad hoc/on-demand, or scheduled, such as the passport renewal or the student re-enrolment processes.

There is often also a convenience and efficiency trade-off to be considered when determining the re-enrolment strategy. Consider, for example, the improved integration between organisations that can drive up efficiency and make people's lives easier. This can manifest itself in benefits such as those offered to UK citizens who can now opt for their driving license to be reissued using the photograph held in the UK passport database. Lowered costs for the department concerned, and very handy for the citizen – no need to worry about having new photos taken! However, this integration – and decreased customer contact – has the effect of driving the re-enrolment period up, rather than down – in this example, some driving licenses are now being produced with face photos that are already several years out of date.

The storage mechanism for the reference sample or gallery is also a pertinent factor to be considered in determining the re-enrolment approach – token-based systems, such as Automated Border Clearance systems that compare a traveller's face with that held on their passport's chip, typically have limited or no opportunities for re-enrolment due to the complexity of updating the physical tokens themselves – whereas those backed by database repositories offer more flexible options for re-enrolment.

So, the decision of which re-enrolment scheme is most appropriate to adopt will be driven by the business scenario, the user population and answers to a set of practical questions, such as 'How often do our end-users biometrically interact with our system?', 'How present is the re-enrolment infrastructure?', 'Where are the reference samples stored and are there any technical barriers to updating them?', 'What legal and privacy constraints are in place regarding biometric sample capture and transmission?' and so forth. The discussion is not a simple one, but it is certainly worthwhile – addressing these questions at design-time will assuredly be cheaper than fixing issues regarding an overly aged biometric reference dataset, later!

11.8.1.2 Optimising the susceptibility of biometric modality to age-related change

While system designers cannot do much to prevent the ageing of biometric traits in their expected user population, they certainly can (and already do) invest significant effort in selecting optimal biometric modalities at design-time. This normally involves consideration of various established biometric performance factors – such as uniqueness, usability, acceptability and, crucially to this discussion, permanence – in light of the business scenario to be fulfilled. According an appropriate weighting to the trait's stability – its tendency to remain unchanged with time – will certainly help to allay the impacts of biometric ageing on a system; indeed, this may be *the* easiest and most significant age-mitigation action that can be taken.

For systems that are expected to remain operational for many years, such as civil ID schemes, multi-modality might be appropriate. Enrolling a stable biometric trait alongside less stable, yet perhaps more usable, modalities can help to de-risk future (perhaps unplanned) re-enrolment activities. While this has its costs, it might represent worthwhile insurance, as well as giving access to the other benefits that multi-modal biometric systems offer, such as improved verification flexibility, improved resilience to fraud and improved de-duplication performance.

11.8.2 Age-related change compensation

For already-established systems, or for those in which a high degree of matching accuracy is required, it may be appropriate to apply technical 'fixes' to compensate for the effects of biometric ageing. This has been a topic of interest in the criminal AFIS space for many years, with techniques being developed in order to artificially age fingerprint and face samples such that the ages of probe and reference images can be matched. These techniques are not yet in use outside the world of criminal investigation and forensics, though as discussed previously, there may be an increasing market for them, if they can be appropriately developed and made usable. The commercial availability of modality-specific ageing algorithms that can be deployed independent of the matching system (so, acting on images rather than templates) would support effective cross-industry adoption. Academic/commercial collaboration can help to migrate these algorithms from lab, through real-world trials and commercial pilots, to readiness for full-scale deployment.

An alternative approach might be to adjust the matching thresholds and/or confidence levels of a system, taking the age variation of the probe and reference samples as a parameter. For higher-age-variation samples, this approach would degrade matching accuracy to some degree – for lower-age-variations, it might drive accuracy rates up (through removing unnecessary tolerance). So, while such an approach would certainly require careful calibration if acceptable performance is to be maintained, it might offer a relatively cheap and simple way to mitigate the effects of age-related biometric change. Again, academia and the commercial world should combine forces to calibrate and test this idea to assess its real-world viability.

Clearly, much research remains to be done in this area, and to achieve this most effectively, academia and industry should collaborate closely. Researchers carry much of the burden of understanding the fundamentals of how the ageing process affects various biometric traits – industry offers the opportunity to test these ideas against large, varied datasets, and on real-world systems. Working together, the two could hope to develop new models that can identify and mitigate age-related change, and understand how to calibrate and commercialise these into algorithms that can help deliver efficient matching systems for the world to use.

It is also conceivable that improvements to our understanding of biometric ageing might deliver additional benefits, and, consequently, ways to drive business value. For example, might further ageing studies bring advances in how we can determine absolute age from biometric samples, and thus increase the value of age as a 'soft biometric'? Might this give another way to reliably detect biometric fraud attempts? This is certainly a rich branch of biometric science waiting to be further developed.

11.9 Closing thoughts

At the start of this chapter, we recognised a challenge – the need to lower the risk that industry's increased dependence on biometric recognition is bringing.

We have subsequently discussed how ageing impacts this commercial risk, considered those factors that tend to make a system more, or less, susceptible to biometric age-related changes, proposed how to identify those systems most at risk, and considered how this ageing risk might be mitigated in practical terms.

Clearly, there are no silver-bullet solutions on the immediate horizon that will remove all exposure to biometric age-related change. People will continue to age, and the nature of ageing is such that the changes it will bring cannot be predicted with total accuracy. The opportunity is there for academia and industry to collaborate in order to better understand the ageing process and develop appropriate compensation strategies; in the meantime, biometric system implementers can adopt an 'ageing-mitigation-by-design' approach that aims to minimise ageing impact through appropriate re-enrolment and modality optimisation. Together, these parallel undertakings should yield biometric systems that offer lower susceptibility to age-related changes, resulting in improved matching accuracy and accordingly, achieve the goal of decreased commercial risk.

References

[1] J. Daugman's 1994 US patent: 'The iris of every human eye has a unique texture of high complexity, which proves to be essentially immutable over a person's life'

[2] S. P. Fenker and K. W. Bowyer, *Analysis of Template Aging in Iris Biometrics*, Department of Computer Science and Engineering, University of Notre Dame

[3] H. Ling, S. Soatto, N. Ramanathan, and D. W. Jacobs, 'Face verification across age progression using discriminative methods', *IEEE Transactions on Information Forensics and Security*, 5, 88 (Figure 9), 2010

[4] N. Belgacem, F. Bereksi-Reguig, A. Nait-Ali, and R. Fournier, 'Person identification system based on electrocardiogram signal using LabVIEW', *International Journal on Computer Science and Engineering*, *4*, 975 (Table 1), 2012

[5] International Biometric Group, 'Biometrics market and industry report 2009–2014'. Retrieved from https://ibgweb.com/products/reports/bmir-2009-2014

[6] http://www.zdnet.com/blog/igeneration/the-students-who-pay-for-school-lunch-by-fingerprint/16331

[7] http://12160.info/profiles/blogs/palm-vein-scanners-rolled-out-at-elementary-school

[8] S. K. Modi, S. J. Elliott, and J. Whetsone, *Impact of Age Groups on Fingerprint Recognition Performance*, Industrial Technology, College of Technology, Purdue University, 2007

[9] Ling, Soatto, Ramanathan, and Jacobs, 'A study of face recognition as people age', *IEEE International Conference on Computer Vision*, *5*, 88 (Figure 9), 2007

[10] http://uidai.gov.in (for further information on Aadhaar)

Chapter 12
Fingerprints and human inspection: a forensics perspective

Clive Reedman

12.1 Introduction

Although this chapter will focus on the effects of ageing related to fingerprints, we do appear to have a need to understand more fully the negative effects on all of the established and emerging biometric modalities and how we may reduce or overcome them. It would seem in many cases that the obvious solution would be to recapture the varying human characteristics from the original donor on a regular basis. This does of course sound ideal, and as long as controls are adopted that ensure that the templates produced are of the same, or of higher quality, than the original enrolment, then all should be fine and the template will age with the donor. But of course this isn't as simple as it may at first sound and in some cases it may not actually be true. A number of factors make this approach difficult to achieve in practice. Clearly, in a situation whereby a donor is regularly also a 'user' of a particular biometric system and uses it with only short- to medium-term intervals then template replacement becomes effective. However, irregular usage will almost certainly reduce the effectiveness of the technique. We also need to understand, for each modality, what effects the ageing process of a human has on that modality. It is fair to say that a fingerprint may be more 'stable' over time than a person's face. A fingerprint will display over time a change in the skin's elasticity levels and overall size as people grow older. Ageing will occur in differing degrees depending on factors such as the occupation of the person, physical climate and their exposure to injury and damage to the ridge detail. Faces almost certainly display less stability over time due to the natural ageing of the skin, changes in bone density and self-imposed variations such as growing a beard and the natural propensity to ageing effects brought about by genetic inheritance, all of which will have a potential effect on the biometric systems effectiveness. A person's iris has been shown to be relatively stable, but illnesses, such as glaucoma, or the development of cataracts can make matching against a template produced prior to the onset of the illness

Previous Chairman of the International Association for Biometrics, Managing Director Identity Solutions Ltd and the Police IT Organisations Head of Biometrics for the UK Police Service - London, UK more difficult, or actually impossible. Indeed, for each modality currently in use today it is possible to identify areas where the natural process of human ageing will have an effect.

It is of course also important to understand that in addition to frequency of biometric system use, other fundamental factors will come into play. Biometric systems are used for diverse purposes and interaction with them correspondingly changes, dependent on their application. A person may be fully cooperative with a system, both during initial enrolment and with subsequent use. However, it is also possible that the person will be uncooperative, particularly with systems used for covert purposes. Training of operators and users also has a dramatic effect on the subsequent accuracy of a biometric system. It may therefore be that despite there being a long time interval between the creation of the original template and subsequent attempts to match that template against a newly created one, the original capture was of a lower quality than that produced subsequently. Simple replacement of the original template would therefore seem to be the solution; however, this is not necessarily the case, particularly if templates are aged using any form of 'age progression' technology, such as that used to identify missing persons after a long period of time. It was interesting and surprising to me that in early investigative work undertaken as part of the UK Identity and Passport Service's initial work on utilising facial recognition technology I often found that higher match scores could result where the original photograph was in fact older than subsequent images of the same person when matched against a present day image!

There is still a long way to go before we achieve a clearer understanding of the effect of ageing on the accuracy and overall effectiveness of biometric systems and there are obstacles that need to be overcome. However, this is not to say that no effort has, or is being, expended in this area, and this chapter will give an overview of the current position, but with a focus on the more mature area of fingerprint biometrics.

In the case of fingerprints and indeed the ridge structure of the palms and soles of the feet we know that the friction ridges form very early on in the development of the foetus and are one of the last recognisable parts of the human body to decay after death. This characteristic has of course been a tremendous boon to the use of them for identifying individual human beings, either as a means of directly confirming, or disproving an identity, or establishing evidence as to the identity of a person who has left latent ridge detail impressions, such as at a crime scene or on an object associated with one. However, in support of a treatise on fingerprint template ageing I give a brief history of the use of ridge detail (from here on collectively referred to as 'fingerprints', but which also encompasses palm and sole ridge detail).

12.2 A brief history of modern human identification methodologies

In 1895 a committee of enquiry was convened in India at the behest of Sir Edward Henry, the then General of Police for the Lower Provinces, Bengal and the future head of the United Kingdom's first fingerprint bureau (1901). The committee was to establish the merits of two systems for establishing human identity and they concluded that fingerprints would be used as the principal method of establishing the identity of human beings. Prior to that reliance was on the anthropometric system that is credited to one Alphonse Bertillon, who after joining the Prefecture of Police in Paris in 1879 as a copyist became somewhat bored and took an interest in how the French Police identified (or failed to identify) criminals. Being the son of an anthropologist, Bertillon had an interest in human behaviour which he expanded into a theory that human beings could be positively identified by measuring a number of the physical characteristics of that person. These included measurements such as length of fingers, arm length and skull circumference. Indeed, the taking of fingerprints to add to the overall record was commonplace, but their use as a definitive means of identification was not included in the anthropometric system as their uniqueness was not understood, or at least not accepted at that point in time. Of course, we all actually perform similar measurements subconsciously and it is how we recognise people that we know or may even just have come into brief contact with. Our brains are particularly adept in making such comparisons against the 'data' it holds in our memories. We make such comparisons many times a day and despite the fact that we make mistakes, we have a tremendous confidence in our ability to recognise people we know, or have seen in the real world, on film, or through other mediums. But the difference was that Bertillon established a system that permitted the recording of these almost subliminal observations in an empirical manner using a methodology and tools that were supposed to make human identification a true science. He also though provided a schema for searching the records for a person to person 'match' and therefore may be credited in some people's eyes as being the true father of modern day biometrics.

Despite a great deal of ridicule and with very little initial cooperation from the authorities, Bertillon was able to prove a deal of success and the system became widely adopted not only in Europe, but in the United Kingdom, America and several other countries. However, he had made a number of critical assumptions regarding the statistical deviation from a 'norm' that human beings displayed. Bertillon had greatly over-estimated the deviation that occurred and his classification system failed therefore to take account of the fact that in the human population differences in the size of any particular finger, or the overall height of a person, or any other measurement, actually deviated to a smaller degree across the whole population than he had at first postulated when developing his system. He had also failed to take into account the fact that human mistakes in the taking of the measurements would cause a high degree of error and also that the effects of ageing were significantly more than he had anticipated.

Although, in defence of Bertillon, his concept was to simply 'narrow the pool' of likely matches, permitting then a positive identification using photographs. But again this concept was proven to be unreliable and unsafe; as an assumption had been made that one human being could reliably confirm a positive match, using a photograph, which from my own personal experience working in this area is not necessarily the case. This assumption and the failings of it were highlighted by the now famous (but sometimes disputed) case of Will West who was sentenced to a term of imprisonment in a US prison based on his 'identification' using the anthropometric system, but was subsequently found not to be the person the Kansas Police thought he was through the use of fingerprints that proved categorically that he was not who they had assumed him to be. A further 'nail in the coffin' of the anthropometric system was the establishment of the first regional fingerprint bureaux, and in 1895 the first of these bureaux was established in Madras and in 1898 the first national bureau was established in Calcutta and the first UK bureau in London in 1901.

So by the early 1900s fingerprints had effectively risen to be the dominant form of human identification, although the anthropometric system, both used as the principal system and used as an adjunct to fingerprints, remained in use for many more years in the United States and other countries throughout the world. But the unique properties of the ridge structure present on the tips of the fingers had been established and accepted as being an almost 'perfect form' of human identification. It is also though interesting to note the fact that the use of fingerprints has not always been to establish the identity of a particular individual or to discover the 'donor' of a latent fingerprint impression. They have been used since ancient times as a form of signature, with examples of such usage dating back many thousands of years and indeed still present to this day. But they have also been adopted in a manner that concerns their classification and study with regard to broad groups of human beings. One person, often cited as 'the father of fingerprints', one Sir Francis Galton, a cousin of Charles Darwin, grew to eminence in the mid-1800s as an English Victorian polymath, anthropologist, tropical explorer, geographer, inventor, meteorologist and statistician. He was without doubt a great man of his day and contributed much to science and he was knighted in recognition of this in 1909. However, we also have to realise that it was Galton that coined the phrase 'Eugenics' and that his interest in human identity was more concerned with identification of 'genetically imperfect' human beings, for the sole reason of being then able to breed out, by forced sterilisation, those that were considered to be adding negatively to the 'true and just' evolution of the human species by 'the study of all agencies under human control which can improve or impair the racial quality of future generations'. It is though somewhat ironic that it was Galton who first pointed out (and gave evidence to the above-mentioned committee in 1895) to the effect that the statistical basis of the anthropometric system, including its lack of understanding of ageing, was flawed, therefore opening the way for fingerprints to rise to their current eminent position within law enforcement and border control.

12.3 Physiology of fingerprints

The friction ridges that are the basis for human identification using fingerprints are visible very early in the developmental cycle of the foetus. At four to five weeks a layer of skin (the ectoderm), which overlays tissue that form a connective layer to the Dermis (the mesenchyme), is present. These cells grow rapidly and form other layers of squamous (flattened) epithelium, the periderm and the basal geminative layer. At eight weeks, the epidermis has grown to be three to four cells in depth and already displays a robust structure. Adjacent cells, which at this stage display smooth structures, have 'anchors', or 'junctions' and at these points structures known as desmosomes are present. As the cell structure grows these desmosomes grow, therefore forcing the cell junctions to form into wavy lines.

The development of epidermal ridges is followed by the formation of volar pads (the swelling of mesenchymal tissue) that first appear as elevations on the palm around 6.5 weeks after conception. The actual fingers of the hand then appear at about 7.5 weeks. These volar pads are at first reasonably consistent in their rounded shape, but by the ninth week they undergo positional and shape changes. Volar pads exhibit rapid growth between 6.5 and 10.5 weeks. Initially the pads appear evenly rounded. However, by the ninth week, the pads begin to vary in both position and shape. It is the volar pads that appear to influence the formation and eventual pattern and ridge structure that we know as a 'fingerprint'. In 1929, Harold Cummins, who is recognised as a father of dermatoglyphics, published a paper, 'The topographic history of the volar pads in the human embryo', in which he postulated that 'two major development circumstances namely, variations in the histology of different regions and differential growth incident to the production of irregular reliefs of the volar surfaces are what the epidermal friction ridge structure is dependent upon'. He also explained his view that 'the various configurations are not determined by self-limited mechanism within the skin. The skin possesses the capacity to form ridges, but the alignments of these ridges are as responsive to stresses in growth as are the alignments of sand to sweeping by wind or wave'. This would explain why, even when a common DNA structure is present between identical twins, there is a distinct difference present in the actual friction ridge detail, due to the presence of random physical effects on each foetus as it develops. This randomness is accepted now to be so distinct in its effect that the practical chance of two human beings having identical friction ridge structures on even one of their fingers is highly improbable, if not impossible.

12.4 Physiological effects of ageing on fingerprints

Throughout the life cycle of a human being the largest organ of the body (the skin) is constantly undergoing changes. And of course the rest of the human body is also constantly changing. Babies grow bigger into children and then adults. The whole structure of the body will alter in terms of shape and size, and external factors such as climate, occupation, illness and injury will have an effect. In relation to the skin the effect of growth and ageing will result in the organ altering its physical composition and appearance. However, we know through over 140 years of collecting fingerprints that, provided the underlying (anchoring) structure of the dermis is not heavily scarred, or destroyed, the ridge structure we are born with will persist, as it is governed by a regenerative process that is dependent on the underlying dermal ridge structure (a kind of negative image where a furrow is mirrored by a ridge on the epidermal layer). Attempts to wear away fingerprints, or destroy the epidermal ridge detail using substances such as acids, may work in the short term, but left to heal, the original ridge structure will return to the epidermal layer. Only deep rooted injuries that destroy the dermal layer detail will be reflected in this regeneration and modern day civil biometric, or AFIS systems are able to remain highly accurate even when deep rooted scars are present that alter the template of a capture made prior to the occurrence of the injury.

Bearing in mind that friction ridge detail is highly stable from before birth until after death, it is a natural assumption to make that as a finger grows under the normal process of ageing, then the fingerprint structure will simply grow with it at a uniform rate across the population. However, it has now been shown that these anticipated patterns of growth are not uniform, resulting in the observation that 'growth patterns of fingerprints as measured (in this study) do not appear to follow a general pattern'. (*Quantifying the dermatoglyphic growth patterns in children through adolescence* –Dr J.K. Schneider Ultra-Scan Corporation – Final Technical Report FR00A178000-1, July 2010). This study was conducted to provide the US National Institute for Justice (NIJ) and the scientific community a means for predicting the shape changes of fingerprints during this period of growth, in order to significantly enhance the probability of matching an individual's adolescent/adult fingerprints to fingerprints acquired when the individual was a child. Schneider went on to conclude that, 'however, the potential of developing a "Minutiae Growth Map" that could be employed to "scale up" the details commonly used to match fingerprints in AFIS and most biometric systems (see current modalities below) of a child's fingerprint to that of an adult was not possible at this time'. A solution was though found that employed a previously developed process to adjust for plastic distortion. Initially, this algorithm was planned to augment the automated Minutiae Extraction Matching Tool (MEMT), but was hypothesised to be of use in compensating for growth patterns.

As we grow older the elasticity of our skin decreases and the ridges present on the epidermal layer get thicker. This results in the gaps (furrows) between the ridges decreasing. It follows therefore that the prominence of the ridges decreases when pressure is applied, i.e. when the finger is pressed onto a surface, as that pressure causes the ridges to merge and produce an image that excludes much of the actual ridge detail present when the finger is not under pressure.

There are two distinct stages of skin ageing that affect the friction ridges present in fingerprints. Intrinsic ageing, which is also often referred to as the 'natural ageing process' is the first. This process sets in around the twentieth year of life. Within the skin, collagen production slows, and elastin, the substance that enables skin to snap back into place, has a bit less spring. Dead skin cells do not shed as quickly and turnover of new skin cells may decrease slightly. This process is largely, if not wholly, subject to the genetic effects of the person's DNA, which will predetermine the start of the stage and the rate at which it takes effect. While these changes usually begin in our early to mid-20s, the signs of intrinsic ageing are typically not visible for decades, but accelerate at an exponential rate as we grow older. The effects of intrinsic ageing on fingerprint friction ridge detail are:

- the appearance of fine wrinkles;
- thin and transparent skin;
- loss of underlying fat, leading to noticeable loss of firmness on the skin covering the hands and fingers;
- bones shrinking away from the skin due to bone loss, which causes sagging skin;
- dry skin that may itch and flake, due to the reduced ability to sweat;
- smoothing of the friction ridges present on the fingers, palms and the soles of the feet.

The second stage is that known as extrinsic ageing, which is predicated by environmental factors, predominantly exposure to the sun, working conditions, smoking or exposure to other external factors that may be unique to an individual, or particular groups of individuals based on geographic, economic, or social conditions.

It follows therefore that the ageing process that affects the accuracy of both manual and automated fingerprint matching systems cannot be seen as a 'predictable constant', as all human beings from their moment of birth will be subject to variable conditions, be they physiological or environmental.

12.5 Ageing effects across the fingerprint matching modalities

At the end of the 19th century when fingerprints rose to their dominant position as the principal method for determining positive human identification the process of collection, classification and matching was a manual one. It wasn't until the 1970s that the power of computing systems came to be utilised to attempt automated matching. Initially these dactyloscopic systems used all of the information available in a manually taken fingerprint image (ink on paper scanned into the system). That is they used both the patterns and a comprehensive set of features. However, this form of matching required a great deal of computing power, which was simply not practicable at the time, resulting in very slow and often inaccurate matching. As an answer to this most biometric systems that were employed as simple access control aids reverted to the extraction of coordinates (minutiae points) related to the point at which a friction ridge either ended or split (branched, or bifurcated). Modern biometric systems, however, may also now employ the use of a filter mask (Gabor Filter) that uses the detail surrounding a particular detail.

There exist today two basic modalities for use in automated fingerprint matching systems, namely the minutiae-based systems that utilise the deviations from 'the normal flow of a ridge', i.e. where the ridge ends, of 'bifurcates', or those that use the pattern formed by the friction ridges. The predominating matching modality for large-scale AFIS systems is minutiae-based and the trend now is for most small-scale (civil) applications to adopt this also, although many systems will employ a combination of the two techniques.

The vast majority of both AFIS and all biometric fingerprint systems now employ the use of electronic sensors to capture the initial and subsequent fingerprints. The prevailing technologies used to achieve this are either optical, utilising an image capture technique, or capacitance-based, whereby an 'image' is formed from the difference in capacitance between a furrow and a ridge, where the ridge has a higher capacitive value than the furrow. Other forms of sensor technology include those that use a high-frequency ultrasound, or an arrangement of prisms that detect changes in light reflectance between the furrows and ridges. One other type of sensor adopts a measurement of temperature over time as a finger is swiped across a small sliver of silicon. Each of these sensor classes have their merits and are employed in such a way as to fit the particular end use of the overall AFIS, or civil biometric system. AFIS systems require much higher quality images and to date only the optical solution provides this, whereas a civil use biometric system may employ any sensor technology, but tend to be constrained by the need for lower overall costs and also the need for a lesser degree of accuracy.

Before moving on to the variation of ageing effects on the various matching and sensor modalities and technologies it is necessary to note a few important points. First, we now have many large-scale AFIS systems used for both criminal and border control use whose original data set was derived from ink on paper capture, with all of the human error factors that are associated with such a form of capture. The electronic capture of fingerprints directly from an individual has been utilised for over 25 years, but where systems have been introduced after the use of such 'livescan' technologies were available, original data sets would have been 'back converted' from paper forms. It also follows that a person may have had their fingerprints taken originally on paper, but subsequently via electronic means, creating databases that contain a mix of both forms of capture. Depending on the laws of a particular country it is also possible that original templates that were converted from paper-based images are now augmented, or completely replaced by images captured via electronic means. In the United Kingdom the law allows for the replacement of poorer quality images with better (subsequently taken) images so as to improve the overall quality of the data set. It may be that one or a few of the images are of higher quality than the originals, or it may be the whole set that is replaced. This applies even if the fingerprints taken following a subsequent arrest appertain to a case that is not proceeded with for whatever reason, or where the defendant is discharged or found not guilty. The second point to note is that there are fundamental differences between the matching techniques used in the diverse AFIS and civil biometric systems in terms of the number of comparisons made between an enrolled 'set' and the subsequent capture, which is made to either search for a previous enrolment and establish a potential 'match', or derive that the donors' fingerprints have not previously been recorded in that particular system. AFIS systems adopt a number of methods for searching against very large databases, i.e. they may adopt a method whereby a single or two fingerprint templates are searched against the complete data set, or all ten. They may also complete a 'one to one' comparison in the first instance where other available information leads an official to believe that they have enough information available (or even personal knowledge) for them to suspect that a person is already known. They may also segment the search parameters based on physical derivations of gender, race, age etc. and only complete a full search of the data set if an initial match is not likely. In AFIS systems used for law enforcement purposes it is usual for a skilled expert to make the final decision as to whether a match has or has not been found.

Civil use biometric systems tend towards limiting the amount of 'database penetration' needed to establish a match, or non-match. This comes about because of either (or a combination of) the fact that the overall database size is smaller, the risks involved in not identifying an individual or falsely identifying a donor against the true donor are not high, or that the computing power available and the limitations of the matching algorithm are such that database penetration of the data set would result in increased false matches (the false match rate), or increased rejections of templates that match but are not accepted as such (the false reject rate). It is therefore quite normal for a civil biometric system to utilise a methodology whereby a 'known' identifier, such as a PIN number, is used to 'call' a particular template against which the initial match is attempted, a 'one to one match'. However, where data sets are small in number even a civil biometric system may employ a 'one to all' matching methodology as the computing power required has little effect on the overall return in respect of result reporting time.

So what effect will ageing have across these diverse modalities and methodologies? I would first say that with all systems we must realise that it is difficult to design accurate algorithms capable of extracting salient features and matching them in a robust way. This problem is often exacerbated, especially in poor quality fingerprint images and when low-cost acquisition devices that capture small areas of the fingerprint are adopted for design or cost reasons. However, where minutiae-based matching is utilised it is possible (and often normal) for the attempted match to be undertaken on only a small area of the overall template, relating to a part of the fingerprint that displays the best characteristics in terms of resistance to ageing. This may be augmented in the matching algorithm by a 'move out' technique, whereby a low match score may require the examination of one or more other portions of the fingerprint template. Such a technique is less efficient where lower amounts of data are available either from the enrolled template or from the template derived from the recent capture. This naturally disadvantages those modalities that derive less initial data, or those that rely on matching across the whole template, without the ability to segment. However, we must also consider the effects of intrinsic and extrinsic ageing and how they affect the diverse capture and matching modalities and how choices of sensor technologies play their part.

12.6 Issues surrounding the study of fingerprint ageing

When investigating and discussing the potential (and real) issues surrounding the ageing of fingerprint templates, there are a number of elements that when combined make meaningful conclusions hard to arrive at. Principal among these are:

- There is a lack of 'aged' data available in many biometric modalities, due to their being relatively young technologies.

- Vendor imperatives for commercial confidence and differing applications of the 'base' modalities reduce the effectiveness of time-based study as comparisons of 'like for like' are not always possible.
- Relatively recent introduction of biometric interoperability standards is exacerbating vendor-specific data sets not being suitable for combining into larger data sets for particular modalities.
- Testing methodologies/frameworks/tools need to be established and agreed.
- Further funding of research needs to be established.
- There is a corresponding lack of data available from children, due to sensitivities surrounding collecting biometric data from them.
- Sensor technologies are always improving and test sets will therefore be captured using different devices.
- Existing data from operational systems is hard to come by due to data privacy laws, with the largest databases being utilised in large-scale national systems associated with criminal activity and citizen identification linked with border controls, or other government systems.
- Securing large-scale databases is difficult as volunteers are not always easily recruited and it is hard to take multiple samples from them over time.

12.7 Conclusions

This chapter has purposely been written with the lay person in mind. However, we need to be aware that whatever techniques are employed to capture and match fingerprint templates a human intervention may well be required, which will vary in its approach depending on the age of the donor, or other factors that may influence the success rate of the particular system being used. I believe that we need to examine the effects of human intervention more robustly in terms of the effects of ageing. Whatever system is used and whatever hardware and software we employ, there is always going to be a trade-off between obtaining the best capture and the best chance of an accurate match or non-match. I have pointed out that human friction ridges develop early in the foetus development and persist long after death when compared against other human features. We understand that humans age in much the same way, but that their genetic make up and the physical conditions they are exposed to vary and therefore make prediction of ageing effects difficult to forecast in any empirical manner. The varying requirements of human identification using fingerprint matching technologies display differing sensibilities to the effects of ageing, and the 'risks' involved also differ according to the function of that system. There are ongoing studies relating to the ageing of fingerprints and these must be embraced and performed in a manner that allows current and future vendors and users of systems (AFIS and civil biometric) to understand fully what effects ageing will have on system performance and how best to maximise that performance. But there are obstacles that need to be overcome before we can arrive at a position whereby our data sets are truly deemed to be adequate and fit for the purposes of study. Above all we need to be aware that this is a complex issue, but one related to a simple premise, i.e. whatever we do, we cannot avoid getting older!

Chapter 13
Age prediction in face images

Guodong Guo

Guodong Guo

13.1 Introduction

The ageing factor is very significant in face images. As shown in Figure 13.1, the facial appearance changes a lot with the person's ageing process. The facial appearance changes could be used to characterize the ages of an individual for the purpose of age prediction. Actually, humans have the capability to perceive the ages of other people by looking at their faces [46]. So, age perception from faces is quite natural for people.

A question is then raised: Can we use a computer to estimate an individual's age? We call this problem "age prediction" or "age estimation." Age prediction is defined as predicting or estimating the age of an individual, based on the individual's face image. Because different individuals may age quite differently, age prediction is actually a quite challenging problem.

Because of many potential applications in practice, age prediction has become an active research topic in recent computer vision studies [13, 45]. Researchers have proposed different approaches to deal with the age prediction or estimation problem, and some significant progress has been made. Now, it is a good time to summarize the recent progress, and discuss some challenging research topics.

In this chapter, we will provide an overview of some existing approaches to age prediction, and discuss some research trends and challenges. There was a book chapter that we wrote in Reference 24, which covers both age estimation and gender classification. In the chapter here, we only focus on age prediction, and give some new updates on more recent work and discuss more challenging problems.

Before presenting different methods, it might be useful to look at the potential applications of age prediction. For readers who are not familiar with the topic of age prediction, these applications may be helpful to inspire some interest.

Department of Computer Science and Electrical Engineering, West Virginia University, Morgantown, WV, USA

Figure 13.1 Face images of the same individual at different ages, containing significant variations

13.1.1 Applications

13.1.1.1 Age-specific HCI

Human computer interaction (HCI) is an emerging research topic. Its goal is to create more intelligent and friendly user interfaces. If the user's age can be predicted by the computer, then an age-specific HCI might be developed [21]. For instance, there are some real applications that need age-specific HCI. One is secure Internet access control for young kids. The access for young kids can be constrained to appropriate materials, while some adult content, such as texts, images, and videos, could be blocked for them. Another example is to use age prediction in a vending machine. By recognizing the user's age, a vending machine can refuse to sell alcohol or cigarettes to young people (e.g., under age 18). In building a smart environment, age prediction might also be useful, such as finding the elderly (who may need special care) in a crowd of people.

13.1.1.2 Business intelligence

Business is related to customers. To promote business, it is necessary to know more about the customers, such as their age, gender, ethnicity, education, and income. Different groups of people may show different interests or buy different products. Hence, advertisements could be adapted to the proper groups of people, rather than broadcast to unrelated people who have no interest. In dividing people into groups, age is one of the most important cues. So age prediction or estimation is very useful for business intelligence [24].

13.1.1.3 Security

Age estimation can be useful for preventing money transfer frauds. In Japan, the money transfer frauds on ATMs are very severe. It was reported that the victims were swindled out of 2.75 billion Japanese Yen across 20,481 cases in 2008 [40]. Based on the statistics, most victims are women over age 60. So, if age (and gender) prediction systems can be developed and installed in the ATMs, a warning signal could be sent to the bank when a woman over age 60 is transferring money. Here age prediction could be used to help avoiding money transfer frauds.

13.1.1.4 Image filtering

Along with the popularity of the Internet and online social networks, a huge number of images and videos are uploaded each day. How to organize the huge data sets is critical for image and video browsing, and search. Age could be an important component in organizing the data. For example, people can be grouped into different age groups based on automated age estimation in face images. Then the computer server can store the data according to age information, which will make the retrieval relatively easier. For example, a query of "Bill Clinton at age sixties" can be responded to by the search engine.

13.1.1.5 Soft biometrics

Age is also a kind one of the so-called soft biometrics, similar to gender, ethnicity, eye color, weight, height, etc. It might be interesting to seek an approach to enhancing a biometric system by using such soft biometrics [32]. Given the progress of age prediction, the cue of age might be helpful for face-based biometric recognition.

13.1.1.6 Social behavior analysis

Humans are social animals. Thus it is important to analyze various social activities in human-centered image and video understanding. Age prediction could be a useful cue for recognizing social behaviors, since people may act with different social roles at different age groups. It could be interesting to study the relation between age groups and social behaviors, and use it to facilitate solving the challenging problem of social behavior understanding.

In summary, age prediction by a computer has many potential applications. We believe that more and more applications could be explored along with the progress and performance improvement in automated human age estimation.

13.1.2 Organization

In the remainder of this chapter, let me overview some of the approaches to age prediction from face images, including face representation and ageing function learning in sections 13.2 and 13.3. Then age group classification is described in section 13.4. Some databases useful for age prediction are introduced in section 13.5. Evaluation criteria and performance measures for age estimation are presented in section 13.6. Some research trends and challenges are discussed in section 13.7, and finally, the concluding remarks are given.

13.2 Face representation

To perform age prediction from face images, the first step is to derive some kind of representation of the face images with ageing variations. In the literature, various methods have been proposed for facial feature extraction for the purpose of age estimation [13, 45]. Some selected methods will be described in the following.

13.2.1 Wrinkle models

It is very intuitive to consider facial wrinkles when we talk about human ageing. When humans become old, skin ageing is more and more significant. Facial muscles lose elasticity and become loose gradually, which causes the appearance of wrinkles, creases, and sagging in facial regions. Based on this observation, Kwon and Lobo [36] presented the earliest work of detecting wrinkles from face images to separate young adults from the more senior. Computationally, facial wrinkles can be computed in several regions, such as on the forehead, next to the eyes, and near the cheek bones. The detection of wrinkles in a facial region is based on the survival test of some randomly dropped snakelets in the specified regions. One example is shown in Figure 13.2.

The wrinkle model can be used to separate old people from young adults and babies, but it cannot predict the specific age of a person, such as 56 or 57 years. In practice, it might be combined with other models for age group classification. For example, the wrinkle model was used with facial anthropometry measures, e.g., the ratios of distance measures from various fiducial points, to classify people into three age groups: babies, young adults, and senior [36].

13.2.2 Anthropometric models

Anthropometry is a branch of anthropology that is about comparative measurements of the human bodies and their parts. Face anthropometry deals with measuring sizes and proportions on human faces. Farkas [11] presented a systematic and comprehensive measurement of facial distances between various fiducial points and angles. Specifically, Farkas [11] reported the measurements over 57 facial landmarks or fiducial points at different ages, ranging from infancy to adulthood. Based on the anthropometric measures, one can develop some computational methods to characterize the faces at different ages. A typical use of facial anthropometry is to compute the distance ratios measured from facial landmarks. For example, Kwon and Lobo [36] computed six ratios of distances on frontal face images, as shown in Figure 13.2, and used them to separate babies from adults.

Another approach was proposed by Ramanathan and Chellappa [44], which uses 8 ratios of distance measures to model age progression in young faces, e.g., 0–18 years, rather than just separating young kids from adults. The purpose is to predict an individual's appearance across age and to perform face recognition across age progression. An illustration of the created age progression process is shown in Figure 13.3. Their experiment was performed on a database of 233 images of 109 individuals, collected partially from the FG-NET ageing database [12] and their own data. Gunay and Nabiyev [22] proposed a variant of Ramanathan and Chellappa's approach to detect facial anthropometric features and then used these for age estimation.

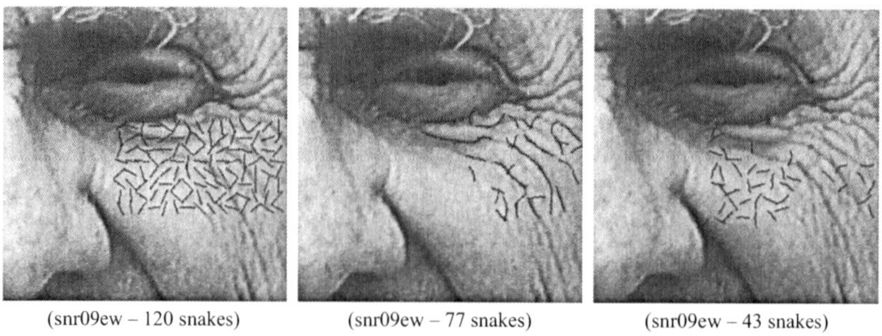

Figure 13.2 Wrinkle analysis process. Left: initial snakelets dropped in an image; middle: wrinkle snakelets (including quantity); right: non-wrinkle snakelets. Originally shown in Reference 36

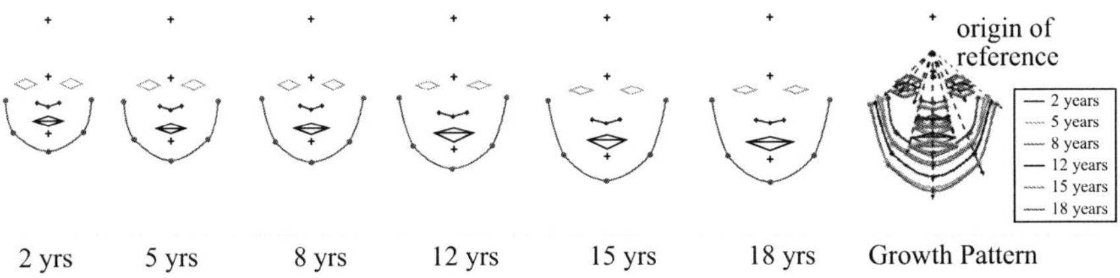

Figure 13.3 Prototype faces created for different ages using anthropometric measures, originally shown in Reference 44

In practice, there are some issues in using the anthropometric models: (1) it is not trivial to localize facial fiducial points accurately, although some progress has been made in recent years, and (2) it is difficult to model facial shape variations among different people at the same age. The anthropometric models could be used to separate young people from adults, although they may not be accurate enough to estimate the exact age. Another thing is that the anthropometric models have not been evaluated on a large database yet.

13.2.3 Active appearance models

The active appearance model (AAM) was proposed by Cootes *et al.* [8] as a statistical model to represent face images. The AAM method begins with a set of manually labeled landmarks to encode the facial structure, such as eyes, nose, mouth, facial contour, and so on. The principal component analysis (PCA) technique is then applied to these landmarks to learn a shape basis. Similarly the PCA method can also be applied to the image intensities given the structure or shape constraint. The learned AAMs can be applied to new face images to obtain the corresponding representation based on the learned basis. One example of the AAM fitting is illustrated in Figure 13.4. The AAM is a general method for face encoding and has to be applied to different tasks of face processing. Lanitis *et al.* [37, 38] adapted the AAMs to represent face images for a facial ageing study. Later on, the AAMs were also used in other approaches, e.g., in References 25–27 and 39, for age estimation.

One should notice that it is not trivial to localize the facial fiducial or landmark points accurately for the AAM models, especially when the faces undergo some

Figure 13.4 One example of the AAM fitting result on a face image

variations, e.g., expression change, or head rotations. Inaccurate point locations may have a significant influence in using the AAMs for age estimation.

13.2.4 Ageing pattern subspace

Geng *et al.* [20, 21] proposed a method called Aging Pattern Subspace (AGES). An ageing pattern is defined as a sequence of face images of the same individual arranged in the temporal order [36]. In the AGES approach, each face image is represented by the AAMs [8] as Lanitis *et al.* [37]. But the AGES method uses all face images of the same individual at different ages. The AGES approach requires all face images at different ages available. If some face images are not available, the AGES method can synthesize the missing faces by using an Expectation-Maximization (EM)-like iterative learning algorithm (Figure 13.5). In practice, the synthesis results might not be good if there are too many missing faces.

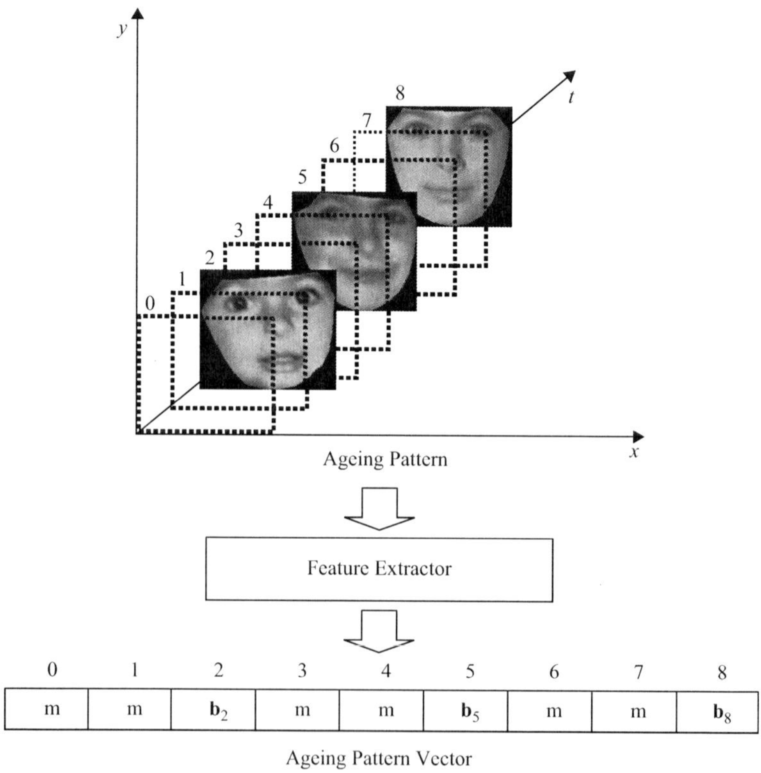

Figure 13.5 Illustration of the AGES model. The ages (0–8) are marked at the top-left to the corresponding positions and above the corresponding feature vectors. The missing parts in the ageing pattern vector are marked by "m." It was originally shown in Reference 21

13.2.5 Biologically inspired models

The biological vision systems are very powerful. It has been a long-term interest to adapt visual cortex models to computational algorithms for image representation and understanding. A recent theory of the feed-forward path of object recognition in the cortex accounts for the first 100–200 milliseconds of processing in the ventral stream of the primate visual cortex [48]. This biologically inspired model consists of simple (S) and complex (C) layers alternately, and has shown good performance in object category recognition [42, 50]. Guo *et al.* [31] adapted the bio-inspired model to human age estimation in face images, using the descriptive term biologically inspired features (BIFs). In previous approaches to object category recognition [41, 42, 50], a number of prototypes (approximately 1,000) are randomly selected from the learning images. These prototypes are stored for template matching in S_2 units. Guo *et al.* [31] found that the S_2 (and then C_2) features using pre-learned prototypes cannot work well for age estimation. They proposed to use a model that has S_1 and C_1 units with a modified C_1 for feature extraction, and show good age estimation performance. Figure 13.6 illustrates the necessity to use the modified C_1 in characterizing the ageing patterns.

13.2.6 Compositional and dynamic model

Face images can be represented at different level of detail in a hierarchical fashion. For example, the first layer is the whole face, the second layer could be facial parts, such as eyes, nose, and mouth, and the third layer could be the skin details and wrinkles. Xu *et al.* [59] investigated this idea in detail and derived a general method for face modeling. In order to combine the different levels and consider different individuals together, they used a so-called AND-OR graph which is a top-down tree

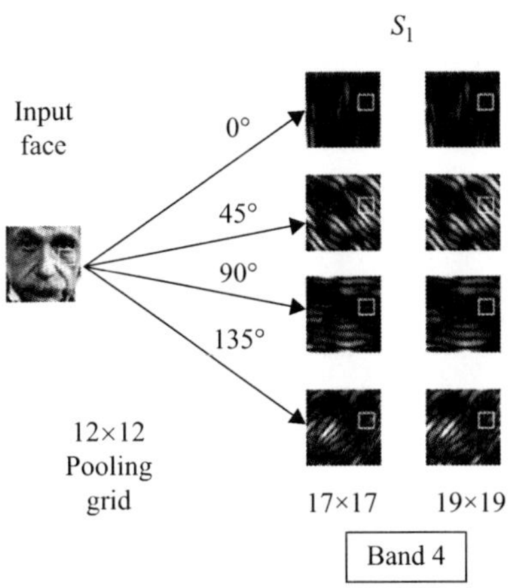

Figure 13.6 Illustrates the usefulness of the "STD" operator in C_1 pooling. The "STD" operator may encode more information than a pure "MAX" pooling in characterizing the ageing patterns (within the indicated box for pooling). It was originally shown in Reference 31

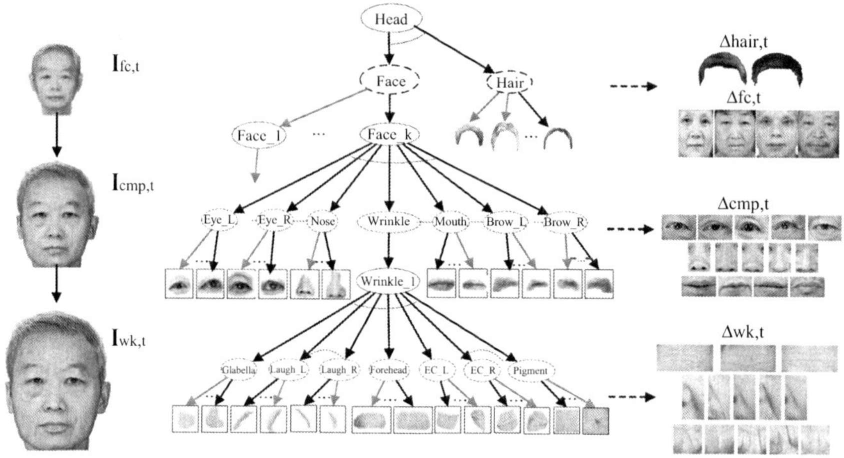

Figure 13.7 The compositional and dynamic model for human ageing, originally shown in Reference 86

structure. The root node of the tree is the whole face, while the second and third layers consist of the local parts and facial details. The AND operation will composite all these nodes to form the whole face. The appearances of different individuals form alternatives for the facial parts or details, and combined with an "OR" operation. Suo *et al.* [54] adapted this hierarchical compositional face model to model face ageing and call it a compositional and dynamic model, especially focusing on the dynamic wrinkle variations with ageing. This model can synthesize new face images at different age groups, as well as age estimation for a given face image. An illustration of the model is shown in Figure 13.7. Notice that the compositional and dynamic model is complex, and it is not easy to implement the whole model without sufficient knowledge in each step. Another issue is the need to localize each facial part precisely, which is not trivial in practice.

13.2.7 Local spatially flexible patches

Rather than using a compositional model as in Reference 54 to organize facial parts and details, local facial patches can be used directly for facial age estimation. Yan *et al.* [60, 64] used facial patches to represent face images for age estimation. They call these local spatially flexible patches, which encode both the facial patch appearance and its location information. An illustration of the process to extract the spatially flexible patches is shown in Figure 13.8. When all facial patches are extracted with overlap, the cost for face representation might be high in comparison with the compositional model [54].

13.2.8 Other methods

In addition to the typical approaches presented above, there are also some other face representations for age estimation. For example, Hayashi *et al.* [34] computed

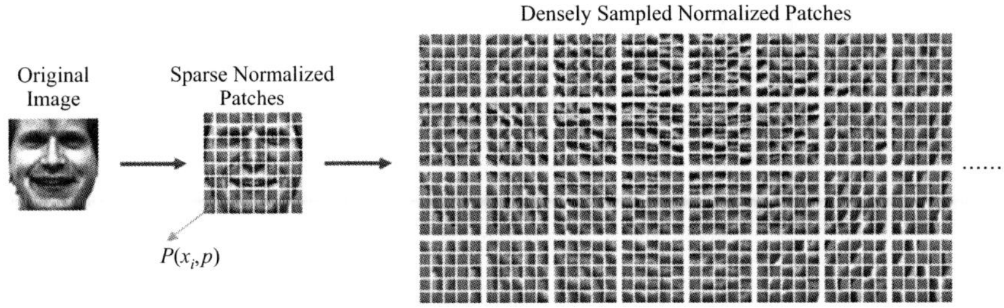

Figure 13.8 The process of extracting the spatially flexible patches, originally shown in Reference 60

wrinkles and facial geometry to represent ageing faces. Fukai *et al.* [16] took the fast Fourier transform (FFT) and genetic algorithm (GA) for feature extraction and selection for ageing face representation. Gunay and Nabiyev [23] and Yang and Ai [65] adopted the popular local binary pattern (LBP) method which has become a standard face encoding method for face recognition [3]. Gao and Ai [19] used the Gabor filters for feature extraction and showed its superb performance over the LBP features in age classification.

Another kind of approach is to use manifold learning for ageing pattern representation. Manifold learning, e.g., References 49 and 55, aims to exploit a low-dimensional embedding of the original high-dimensional input. It can be usually used for non-linear dimensionality reduction or visualization of the proximity of the data. The manifold learning techniques can be used on raw images or combined with other features. For example, Fu *et al.* [14, 15] applied the OLPP method to learn the age manifold on raw face images. Guo *et al.* [25, 27] showed that the age manifold can represent the facial ageing as well as obtain low age estimation errors when a proper age estimation method is used. Guo *et al.* [30] combined the biologically inspired model with various manifold learning techniques for facial ageing representation. Several manifold learning methods are evaluated based on their combination with the BIF, including the orthogonal locality preserving projections (OLPP) [5], marginal Fisher analysis (MFA) [63], and locality sensitive discriminant analysis (LSDA) [6].

More recently, Guo *et al.* [29] proposed a unified framework for estimation of age, gender, and ethnicity altogether based on the kernel partial least squares (KPLS) formulation [68]; Yang *et al.* [67] proposed to estimate age, gender, and ethnicity based on a convolutional neural network (CNN) framework [69]. Their system was even adapted to a mobile platform [66]. Guo and Wang [33] studied the age estimation problem under facial expression changes.

A summary on facial ageing representation: A number of approaches have been developed for facial age representations. Some are adapted from the methods that were originally proposed for object recognition or face recognition, while some others are specifically related to ageing, e.g., wrinkles and facial anthropometry. Depending on the application scenarios, each representation may have its advantages as well as disadvantages, as we discussed above. It is important to perform a comprehensive evaluation of the face ageing representations using *common testing strategies* and *common evaluation protocols*.

Given the facial ageing representation, the next step is to learn the ageing function for age estimation.

13.3 Ageing function

Given a face image I with the corresponding age y, and the representation of the face images using various methods and models, let us now discuss how to learn the ageing functions. Denoting the derived face representation by X, now the problem is how to learn an ageing function that builds the relationship between X and y. Let $y = f(X)$ indicate the relation, and f is the so-called ageing function.

It is difficult to have an explicit form for the ageing function, and we are even not clear what form the ageing function could be. Lanitis *et al.* [37] proposed to use a quadratic function for ageing. But there is no evidence to show why the ageing function is quadratic. We believe that the ageing function is not a simple polynomial or exponential function, considering a large span of ages, e.g., from 0 to 90 years. To alleviate the difficulty in specifying a specific function for ageing, a practical way is to learn the ageing function from data. The learned ageing function is just a mapping from X to y. For example, when the support vector machine (SVM) is used with a radial basis function (RBF) as the kernel, we cannot know the specific function form or the order of ageing function. To learn the ageing function, many different methods can be used, which can be categorized into three broad categories: classification, regression, and hybrid.

13.3.1 Classification

Considering each age label (only integer numbers) as one class, the ageing function can be learned by classifiers. Then age estimation can be viewed as a classification problem. Many classifiers developed in the machine learning community can be adopted, such as the artificial neural networks (ANN) [4] and SVMs [57]. Lanitis *et al.* [37] used the nearest neighbor classifier and ANN for age estimation, based on the AAM face representation. They also used a quadratic function for ageing mapping. In their experiments performed on a relatively small database of 400 images at ages ranging from 0 to 35 years, they reported an average error of 5.04 years using the quadratic ageing function, which is slightly lower than the nearest neighbor classifier, but higher than the ANN. Guo *et al.* [25, 27] applied the SVM for age estimation, based on the age manifold for face image representation. They performed age estimation on a large database of 8,000 images and reported the mean absolute errors (MAEs) of 5.55 and 5.52 years for females and males, respectively. They also reported an MAE of 7.16 years on the FG-NET [12] ageing database using the AAM face representation.

13.3.2 Regression

Another way to deal with age is to consider this as discrete and ordered integer numbers, e.g., 0, 1, 2, ..., and then age estimation can be viewed as a regression problem. Various regressors might be used to learn the ageing function *f*. Lanitis *et al.* [37] used a quadratic function for regression. The face representation X is derived from the AAM models. Fu *et al.* [15] applied a multiple linear regressor to learn the ageing function where X is the age manifold obtained by using the manifold learning method OLPP [5]. Guo *et al.* [25, 27] applied the support vector regression (SVR) [57] method for ageing function learning. The input X is also the age manifold learned by the OLPP method. They performed age estimation on the public FG-NET database [12] and the large YGA database. A nice property of the SVR method is that it can deal with outliers in age labeling. The SVR optimization is based on the non-linear programming technique that has the capability to deal with outliers using slack variables [57]. Later on, Luu *et al.* [39] also used the SVR method combined with the AAMs for age estimation. Yan *et al.* [62] employed a semi-definite programming (SDP) technique to learn the regression function. The SDP formulation is nice but the optimization usually has a high computational cost. Yan *et al.* [61] extended their work by using an EM [9] optimization method to speed up the learning process and reduce age estimation errors. Ni *et al.* [43] proposed to use a regression method called robust multi-instance regression (RMIR) to deal with age label outliers in regression. Their database was collected from the Internet and manually labeled later; the labeled ages may not be accurate compared with the true ages, which can be considered as label outliers. Xiao *et al.* [58] used metric learning to find latent semantic similarity between ages for SDP regression. Chen *et al.* [7] compared different methods among least angle regression (LAR), principal component analysis (PCA), and locality preserving projections (LPP), and selected the best one for age estimation.

13.3.3 Hybrid

So far, we have presented two different ways to learn the ageing functions: classification-based and regression-based. One may ask a question naturally: Which is better for age estimation: classification or regression? It is difficult to answer this question in general and theoretically. But one may use empirical studies to compare the two different approaches on multiple databases. Guo *et al.* [25, 27] studied this problem on two databases: the FG-NET and YGA databases. They chose the SVMs as the classifier and the SVR as the regressor, using the same input X that was derived by manifold learning [5] or AAM [8]. From their experiments, the SVMs perform much better than the SVR on the YGA database (5.55 vs. 7.00 and 5.52 vs. 7.47 years for females and males, respectively), while the SVMs perform much worse than the SVR on the FG-NET database (7.16 vs. 5.16 years). From this experimental comparison on different databases, we can see that there is no unique ageing function that is always the best for age estimation. Depending on the databases used, the ageing function can be learned based on either classification or regression.

Combining the two different approaches, i.e., classification and regression, can also be considered. Guo *et al.* [25, 27] explored this idea and proposed a method, called the locally adjusted robust regressor (LARR). They showed that a consistently better performance can be obtained by combining classifiers with a regressor. Using the LARR scheme, the MAEs can be 5.25 and 5.30 years for females and males on YGA, and 5.07 years on FG-NET, respectively. Specifically, the LARR method first performs a regression using all available data and then uses the regression results to guide the classifiers using a small local search range. Since global regression provides a constraint for local search, the LARR method improves the age estimation performance significantly. Theoretically the LARR method provides a nice way to combine classifiers with regression, but practically it cannot determine how big the local search range could be. To determine the local search range automatically in a data-driven manner, Guo *et al.* [26] proposed a probabilistic approach to combine regression and classification results. The idea is to transform or interpret the regression results into probabilities using a uniform distribution, and then the probabilities from the classification result are cut off by the uniform distribution.

To summarize, the ageing functions can be learned by either classifiers or regressors. So, age estimation is not purely a classification or regression problem. A promising approach is to combine regression and classification methods as demonstrated in References 25–27. It is worth investigating more advanced schemes in combining classifiers and regressors for ageing function learning, and better performance might be expected in future for age estimation.

13.4 Age group classification

In age prediction or estimation, usually we consider to estimate ages in specific years, i.e., integer values for ages, such as 0, 1, 2, ... But in some applications, one may just want to know the age groups, such as babies, adults, or senior people, or the computer system can only predict the age group information. We call this kind of problem "age group classification."

Age group classification can be approached in different ways: (1) a pure group classification problem with a limited number of groups, e.g., young, adult, and senior and (2) performing the regular age estimation and then reporting the age group information based on the estimated age. For pure group classification, the task is usually simpler since there are only a small number of groups to consider, e.g., three or more. The problem can be solved using standard classifiers. Further, the task of age labeling is simpler since there is no need to know the specific age of a face. In practice, if one wants to collect a learning database from the Internet, he or she can collect the face images first, and then label the age for each face by just providing the age group information. On the other hand, if the standard age estimation problem can be solved, the problem of age group classification will be trivial, where the age groups can be determined easily by categorizing the estimated ages into specified groups.

Age prediction or estimation is generally harder and more complex than age group classification. In age group classification, only classifiers can be used, while in age estimation, various methods can be adopted, including classifiers, regressors, or even a combination of classifiers with a regressor [25–27].

On the other hand, age group classification can be useful for the standard age estimation. Guo *et al.* [30] showed that the age estimation accuracy can be improved if the age estimation follows an age and gender group classification module. When gender and age are considered together in group classification, we may have multiple groups, such as "young male," "young female," "male adult," "female adult," "senior male," and "senior female." It was showed in Reference 30 that the features used for age and gender group classification can be different from the features used for age estimation.

In the literature, there are published works that only deal with age group classification. For example, the earliest work by Kwon and Lobo [36] presented a three age group classification system. They used an anthropometric model, i.e., ratios of distances between facial landmarks to separate babies from adults and a wrinkle model to recognize senior people. Ueki *et al.* [56] built 11 Gaussian models in a low-dimensional feature space. A test face image is classified into one of the age groups based on the likelihood with respect to each of the learned Gaussian models. They computed the classification accuracies in terms of different age ranges. Kanno *et al.* [35] dealt with a four-class age group classification problem using the ANN. Recently, Gallagher and Chen [17] analyzed images of groups of people using social context latent in a picture. They investigated both image content and social context for age and gender classification, and collected a database of people's photos for evaluation. The authors manually labeled age group information (7 age groups: 0–2, 3–7, 8–12, 13–19, 20–36, 37–65, 66, or above), since no ground truth of age is available in data collection. Shan [52] used LBP and Gabor features as face representation, AdaBoost for local feature learning, and SVM for age group classification. The database in Reference 17 was used to evaluate the classification performance, and a result was obtained better than the original approach presented in Reference 17. Gao and Ai [18] used Gabor features with a fuzzy version of the linear discrimination analysis (LDA) for facial age classification on consumer images. Four age groups were considered: baby, child, adult, and elder people.

13.5 Databases

It is quite difficult and time-consuming to develop a facial ageing database in comparison with many other face databases. The difficulty comes in two aspects: (1) face image capture with a large span of age variations and (2) age labeling for each face. In face image acquisition, one has to recruit many people with a large range of ages, e.g., from babies to adults and to very old people. In age labeling, it is better to let the person whose faces are captured tell his or her true age at the capture time. Usually it is hard to use human examiners to label the accurate ages after image capture. Further, if the captured person tells a wrong age, it is not trivial to find and correct the mistake after his or her absence. Finally, even if a face database could be collected from the Internet containing a large span of ages, it is still very hard to label the exact, true ages.

In capturing an ageing database, usually one may think to capture each individual's face images at many different ages. Actually, the first public available ageing database, FG-NET [12], was developed based on this consideration.

Because of the difficulty to get a large age span, lots of face images are scanned from past photos with low quality and uncontrolled variations in FG-NET.

Realizing the difficulties to cover a large span of years for each individual, recent ageing databases capture just one face for each individual at the current age or a small range of age span for each individual. The private YGA database [62] captures each individual at a single age, the MORPH database [47] captures many individuals (more than 10,000 in Album II), but each individual has only a small number of face images at the same age or different ages. The LHI face database [54] has only one face image per individual at a single age. It makes the data collection easier by requiring only a small number of ages or a single age for each individual. However, a database captured in this way might not be good to study the age progression for each individual, since different people age differently, and there is no ground truth to verify whether an age progression model is correct or not.

There are other ageing databases available; see Reference 13 for a list of them. Another recent database collects face images from videos containing pose variations that are continuous in the video sequence [53]. Thus the temporal information could be used to enhance the age estimation performance when the pose variations exist. The labeling is offline performed by human examiners.

There are also ageing databases with facial expression variations [10, 41]. Two databases have been collected in the psychology community and introduced to the computer vision community recently [33]. These databases provide the exact labels of both ages (in years) and expression categories, which may be useful to study age estimation under facial expression changes [33].

Even though some ageing databases are available, there are various shortcomings in the existing databases. We hope that new databases could be collected in future that will be better than the current collections, and advance age prediction research to a new level.

For age estimation, a good database should have a balanced number of faces at each age or at least each age group within 5–10 years. Further, the ageing database should have a large span of ages. Currently, FG-NET and MORPH are the most popular databases that are publicly available for age estimation study. The FG-NET database contains faces mostly at young ages although the age range is from 0 to 69 years, while the MORPH database has no face below 16 years or above 68 years. It is still demanding to have a large database with a balanced distribution of individuals, age range, gender, and even ethnicity.

13.6 Evaluation

Two quantitative measures are usually used to evaluate the age estimation performance. One is called the mean absolute error or MAE, and another is called cumulative score or CS [21, 37, 38]. The MAE is defined as the average of the absolute errors between the estimated ages and the ground truth. The CS is defined as $CS(j) = Ne \leq j/N \times 100\%$, where $Ne \leq j$ is the number of test images on which the age estimation makes an absolute error no higher than j years.

These two measures are usually used together to measure the age estimation accuracy. The CS measure draws curves to visualize the age estimation accuracies at different acceptable error levels in years, while the MAE is a specific number of years of estimation error.

Besides the often used MAE and CS curves, there are some other measures that might be used to get more details about the age estimation performance. One is the so-called "MAE per age," which measures the mean absolute error at each specific age, e.g., 5 or 70 years. Usually it is better to draw a figure to show the MAE per age results, rather than using a tabular presentation, since the table could be very large containing many items (one average error per age, given a large span of ages). For example, Guo *et al.* [25] used the MAE per age measure to present more details in age estimation. A nice property of this measure is that it allows a better understanding of an algorithm's performance at each age, not just the overall performance, or it can show a detailed comparison between various methods based on displaying which method performs better at which age.

Another detailed measure is called "MAE per decade," which is something between the overall MAE and "MAE per age." It computes the mean absolute error for every ten years instead of per year. Usually the measure of MAE per decade is presented by using a table, e.g., in References 31 and 64, although a bar graph display might be used as well.

To obtain an appropriate performance measure, one needs to divide the database into training and test sets. The training set may also contain a small part for tuning parameters. Usually, the leave-one-person-out (LOPO) test scheme is used on FG-NET [21, 25, 61], and a fourfold cross-validation scheme on YGA [25, 61]. On MORPH, a random selected set of about 10,000 images are used for training, and the remaining about 40,000 images for testing [28, 29], considering the memory issue.

In age group classification, usually the classification accuracies are measured and reported, since it can be viewed as a standard classification problem in pattern recognition.

13.7 Research trends and challenges

Human ageing is a complex process. Facial ageing is determined by not only the gene but also many external factors, such as health, living style, living location, and weather conditions. Males and females may age differently [30]. People belonging to different ethnic groups may also have different age progression [28]. There are still many challenges in age prediction in face images. Here we present some research trends and challenges based on a personal view of the author.

13.7.1 Combining age with other facial traits

Human faces encode many cues, e.g., age, gender, ethnicity, and facial expression. It is difficult, if not impossible, to decompose theses cues into independent ones, such that we can only study one trait each time, e.g., age only, or expression only. So a promising way might be studying all these cues together. For example, Guo and Mu [29] recently proposed to recognize age, gender, and ethnicity jointly based on a partial least squares (PLS) formulation [68]. The idea is to predict all three cues rather than working on each cue independently.

13.7.2 Age prediction with variations of pose, expression, and illumination

The problem of age prediction is different from the traditional face recognition problem. However, the traditional influential factors for face recognition, such as pose, expression, and illumination, will also affect the age prediction performance. Researchers have expended significant effort to deal with pose, expression, and illumination in face recognition. It is also necessary to deal with these factors in age prediction. There are some research works toward processing these factors. For example, Song *et al.* [53] proposed a method to deal with multi-view age estimation using video contexts. Very recently, Guo and Wang [33] studied and found the significant influence of facial expressions on age estimation. They also proposed some methods to deal with the expression influence.

All these recent works show some preliminary results to deal with pose and expression variations. Further efforts are needed to perform robust age prediction under pose, expression, and illumination changes.

13.7.3 Ageing database collections

As described previously, there is still a lack of good facial ageing databases, although several databases are already available. The progress of age prediction or estimation depends heavily on the available databases. A good ageing database should contain (1) a large number of individuals; (2) a large age span for each individual; (3) a large span of ages in the whole database, e.g., from 0 to above 90 years; (4) and a balanced distribution of gender and ethnicity. To facilitate the study of robust age prediction with respect to pose, expression, and illumination changes, it is also valuable to have an ageing database that contains different poses, expressions, and illumination conditions. Currently there is no database that satisfies all of the listed requirements here. Hopefully either academia researchers or industrial partners can consider the challenge of databases, and several groups of people may work together to contribute a database that can be available to other researchers to advance the state-of-the-art research on age prediction.

13.7.4 Cross-domain age prediction

Usually age estimation is executed on each captured database only, and the database may be captured under the same or similar conditions. In practice, however, age prediction may be executed in different situations. So, there may be a domain difference between the training and testing data. We call this "cross-domain age estimation." This cross-domain requirement comes from the difficulty in ageing database collection. Suppose there is an application scenario of a shopping mall. We may not expect to collect the training database in the same scenario. The training database may be collected in a studio environment by asking participants to visit a lab or studio. Further, the developed age estimation system may be used in different scenarios, such as security entrance, retailer store, or Internet image filtering. So the cross-domain age prediction will be useful for real applications. The challenge is how to adapt the learned age estimator to different domains without losing the recognition capability, as well as being useful in the same domain.

13.7.5 Develop a robust and working system for age prediction

Currently the study of age prediction is mainly within academia research. It needs to see real applications in industry and business. Recently, some commercial products are being developed for age estimation. For instance, NEC develops an ultra-compact sensor that estimates age and gender [2]; Intellio builds a system, called VisiScanner [1], for custom analysis including age and gender. In real applications, however, the performance of currently developed methods is still not good enough. Researchers need to think about how to make the age estimation systems robust and workable in complicated real applications. There is still a long way to go.

13.8 Concluding remarks

In this chapter, we have introduced some potential applications of age prediction in face images. Various approaches to age estimation have been presented, covering some typical approaches to facial ageing representation and ageing function learning. We should say that this chapter is not intended to present a complete list of all existing approaches.

Different measures of the age prediction performance and some available databases have been described. We have also discussed the research trends and challenges in age estimation, although based on a personal view. Hopefully, the materials presented here can help readers to have a basic understanding of the state-of-the-art approaches and inspire new research efforts to move forward the study and development to a new horizon. The ultimate goal is to develop a working system that can estimate or predict human age with a reasonably good accuracy in any application scenarios.

References

[1] 'Intellio visiscanner retail customer analytics solution'. http://www.intellio.eu/visiscanner.php

[2] 'NEC develops an ultra-compact sensor that estimates age and gender'. http://tweetbuzz.jp/entry/56752845/www.nec.co.jp/press/en/1105/3101.html

[3] Ahonen, T., Hadid, A., Pietikainen, M.: 'Face recognition with local binary patterns'. *European Conference on Computer Vision*, Slovansky ostrov, Czech Republic, 469–481 (2004).

[4] Bishop, C.M.: *Neural Networks for Pattern Recognition*. Oxford University Press (1996).

[5] Cai, D., He, X., Han, J., Zhang, H.: 'Orthogonal Laplacianfaces for face recognition'. *IEEE Transactions on Image Processing* 15, 3608–3614 (2006).

[6] Cai, D., He, X., Zhou, K., Han, J., Bao, H.: 'Locality sensitive discriminant analysis'. *Proceedings of the International Joint Conference on Artificial Intelligence* (2007).

[7] Chen, C., Chang, Y., Ricanek, K., Wang, Y.: 'Face age estimation using model selection'. *IEEE CVPR Workshop*, pp. 93–99 (2010).

[8] Cootes, T., Edwards, G., Taylor, C.: 'Active appearance models'. *European Conference on Computer Vision*, pp. 484–498 (1998).

[9] Dempster, A.P., Laird, N.M., Rubin, D.B.: 'Maximum likelihood from incomplete data via the EM algorithm'. *Journal of the Royal Statistical Society, Series B* **39**(1), 1–38 (1977).

[10] Ebner, N., Riediger, M., Lindenberger, U.: 'Faces – A database of facial expressions in young, middle-aged, and older women and men: Development and validation'. *Behavior Research Methods* **42**(1), 351–362 (2010).

[11] Farkas, L.: *Anthropometry of the Head and Face*. Raven Press, New York (1994).

[12] 'FGNET: The FG-NET ageing database'. http://www.fgnet.rsunit.com/ (2002).

[13] Fu, Y., Guo, G.D., Huang, T.S.: 'Age synthesis and estimation via faces: A survey'. *IEEE Transactions on Pattern Analysis and Machine Intelligence* **32**(11), 1955–1976 (2010).

[14] Fu, Y., Huang, T.S.: 'Human age estimation with regression on discriminative ageing manifold'. *IEEE Transactions on Multimedia* **10**(4), 578–584 (2008).

[15] Fu, Y., Xu, Y., Huang, T.S.: 'Estimating human ages by manifold analysis of face pictures and regression on ageing features'. *IEEE Conference on Multimedia and Expo*, pp. 1383–1386 (2007).

[16] Fukai, H., Takimoto, H., Mitsukura, Y., Fukumi, M.: 'Apparent age estimation system based on age perception'. *SICE Annual Conference*, pp. 2808–2812 (2007).

[17] Gallagher, A., Chen, T.: 'Understanding images of groups of people'. *IEEE CVPR Workshop*, pp. 256–263 (2009).

[18] Gao, F., Ai, H.: 'Face age classification on consumer images with Gabor feature and fuzzy lda method'. *The 3rd IAPR International Conference on Biometrics* (2009).

[19] Gao, W., Ai, H.: 'Face gender classification on consumer images in a multiethnic environment'. *International Conference on Biometrics* (2009).

[20] Geng, X., Zhou, Z.H., Smith-Miles, K.: 'Automatic age estimation based on facial ageing patterns'. *IEEE Transactions on Pattern Analysis and Machine Intelligence* **29**(12), 2234–2240 (2007).

[21] Geng, X., Zhou, Z.H., Zhang, Y., Li, G., Dai, H.: Learning from facial ageing patterns for automatic age estimation. *ACM Conference on Multimedia*, pp. 307–316 (2006).

[22] Gunay, A., Nabiyev, V.V.: 'Automatic detection of anthropometric features from facial images'. *IEEE Conference on Signal Processing and Communications Applications* (2007).

[23] Gunay, A., Nabiyev, V.V.: 'Automatic age classification with LBP'. *Proceedings of the International Symposium on Computer and Information Science* (2008).

[24] Guo, G.D.: 'Human age estimation and sex classification'. In: C. Shan, F. Porikli, T. Xiang, S. Gong (eds.), *Video Analytics for Business Intelligence*, pp. 101–131. Springer (2012).

[25] Guo, G.D., Fu, Y., Dyer, C., Huang, T.S.: 'Image-based human age estimation by manifold learning and locally adjusted robust regression'. *IEEE Transactions on Image Processing* **17**(7), 1178–1188 (2008).

[26] Guo, G.D., Fu, Y., Dyer, C., Huang, T.S.: 'A probabilistic fusion approach to human age prediction'. *International Workshop on Semantic Learning Applications in Multimedia* (2008).

[27] Guo, G.D., Fu, Y., Huang, T., Dyer, C.: 'Locally adjusted robust regression for human age estimation'. *IEEE Workshop on Application of Computer Vision* (2008).

[28] Guo, G.D., Mu, G.: 'Human age estimation: What is the influence across race and gender'? *IEEE International Workshop on Analysis and Modeling of Faces and Gestures* (2010).

[29] Guo, G.D., Mu, G.: 'Simultaneous dimensionality reduction and human age estimation via kernel partial least squares regression'. *IEEE Conference on Computer Vision and Pattern Recognition*, pp. 657–664 (2011).

[30] Guo, G.D., Mu, G., Fu, Y., Dyer, C., Huang, T.S.: 'A study on automatic age estimation on a large database'. *IEEE International Conference on Computer Vision*, pp. 1986–1991 (2009).

[31] Guo, G.D., Mu, G., Fu, Y., Huang, T.S.: 'Human age estimation using bio-inspired features'. *IEEE Conference on Computer Vision and Pattern Recognition*, pp. 112–119 (2009).

[32] Guo, G.D., Mu, G., Ricanek, K.: 'Cross-age face recognition on a very large database: the performance versus age intervals and improvement using soft biometric traits'. *International Conference on Pattern Recognition* (2010).

[33] Guo, G.D., Wang, X.: 'A study on human age estimation under facial expression changes'. *IEEE Conference on Computer Vision and Pattern Recognition* (2012).

[34] Hayashi, J., Yasumoto, M., Ito, H., Koshimizu, H.: 'A method for estimating and modeling age and gender using facial image processing'. *Seventh International Conference on Virtual Systems and Multimedia*, pp. 439–448 (2001).

[35] Kanno, T., Akiba, M., Teramachi, Y., Nagahashi, H., Agui, T.: 'Classification of age group based on facial images of young males by using neural networks'. *IEICE Transactions on Information and Systems* E84-D(8), 1094–1101 (2001).

[36] Kwon, Y., Lobo, N.: 'Age classification from facial images'. *Computer Vision and Image Understanding* **74**(1), 1–21 (1999).

[37] Lanitis, A., Draganova, C., Christodoulou, C.: 'Comparing different classifiers for automatic age estimation'. *IEEE Transactions on SMC-B* **24**(4), 621–628 (2002).

[38] Lanitis, A., Taylor, C.J., Cootes, T.F.: 'Toward automatic simulation of ageing effects on face images'. *IEEE Transactions on Pattern Analysis and Machine Intelligence* **34**(1), 442–455 (2002).

[39] Luu, K., Ricanek, K., Bui, T., Suen, C.: 'Age estimation using active appearance models and support vector machine regression'. *IEEE Conference on BTAS*, pp. 1–5 (2009).

[40] Martin, A.: Bank transfer fraudsters have that grift of gab. *The Japanese Times*, http://search.japantimes.co.jp/cgi-bin/nn20090203i1.html (2009).

[41] Minear, M., Park, D.: A lifespan database of adult facial stimuli. *Behavior Research Methods, Instruments, & Computers* **36**(4), 630–633 (2004).

[42] Mutch, J., Lowe, D.: 'Object class recognition and localization using sparse features with limited receptive fields'. *IEEE Conference on Computer Vision and Pattern Recognition*, pp. 11–18 (2006).

[43] Ni, B., Song, Z., Yan, S.: 'Web image mining towards universal age estimator'. *ACM Multimedia* (2009).

[44] Ramanathan, N., Chellappa, R.: 'Modeling age progression in young faces'. *IEEE CVPR Workshop*, pp. 387–394 (2006).

[45] Ramanathan, N., Chellappa, R., Biswas, S.: 'Age progression in human faces: A survey'. *Visual Languages and Computing* (2009).

[46] Rhodes, M.G.: 'Age estimation of faces: A review'. *Applied Cognitive Psychology* 23, 1–12 (2009).

[47] Ricanek, K., Tesafaye, T.: 'Morph: A longitudinal image database of normal adult age progression'. *IEEE Conference on AFGR*, pp. 341–345 (2006).

[48] Riesenhuber, M., Poggio, T.: 'Hierarchical models of object recognition in cortex'. *Nature Neuroscience* **2**(11), 1019–1025 (1999).

[49] Roweis, S.T., Saul, L.K.: 'Nonlinear dimensionality reduction by locally linear embedding'. *Science* 290, 2323–2326 (2000).

[50] Serre, T., Wolf, L., Bileschi, S., Riesenhuber, M., Poggio, T.: 'Robust object recognition with cortex-like mechanisms'. *IEEE Transactions on Pattern Analysis and Machine Intelligence* **29**(3), 411–426 (2007).

[51] Serre, T., Wolf, L., Poggio, T.: 'Object recognition with features inspired by visual cortex'. *IEEE Conference on Computer Vision and Pattern Recognition* (2005).

[52] Shan, C.: 'Learning local features for age estimation on real-life faces'. *ACM International Workshop on Multimodal Pervasive Video Analysis* (2010).

[53] Song, Z., Ni, B., Guo, D., Sim, T., Yan, S.: 'Learning universal multi-view age estimator by video contexts'. *IEEE ICCV Workshop* (2011).

[54] Suo, J., Zhu, S., Shan, S., Chen, X.: 'A compositional and dynamic model for face ageing'. *IEEE Transactions on Pattern Analysis and Machine Intelligence* **32**(3), 385–401 (2010).

[55] Tenenbaum, J.B., de Silva, V., Langford, J.C.: 'A global geometric framework for nonlinear dimensionality reduction'. *Science* 290, 2319–2323 (2000).

[56] Ueki, K., Hayashida, T., Kobayashi, T.: 'Subspace-based age-group classification using facial images under various lighting conditions'. *IEEE Conference on AFGR* (2006).

[57] Vapnik, V.N.: *Statistical Learning Theory*. John Wiley, New York (1998).

[58] Xiao, B., Yang, X., Xu, Y.: 'Learning distance metric for regression by semidefinite programming with application to human age estimation'. *ACM Multimedia* (2009).

[59] Xu, Z., Chen, H., Zhu, S., Luo, J.: A hierarchical compositional model for face representation and sketching. *IEEE Transactions on Pattern Analysis and Machine Intelligence* **30**(6), 955–969 (2008).

[60] Yan, S., Liu, M., Huang, T.: 'Extracting age information from local spatially flexible patches'. *IEEE Conference on ICASSP*, pp. 737–740 (2008).

[61] Yan, S., Wang, H., Huang, T.S., Tang, X.: 'Ranking with uncertain labels'. *IEEE Conference on Multimedia and Expo*, pp. 96–99 (2007).

[62] Yan, S., Wang, H., Tang, X., Huang, T.: 'Learning auto-structured regressor from uncertain nonnegative labels'. *IEEE Conference on ICCV* (2007).

[63] Yan, S., Xu, D., Zhang, B., Zhang, H., Yang, Q., Lin, S.: 'Graph embedding and extensions: A general framework for dimensionality reduction'. *IEEE Transactions on Pattern Analysis and Machine Intelligence* 29, 40–51 (2007).

[64] Yan, S., Zhou, X., Liu, M., Hasegawa-Johnson, M., Huang, T.: 'Regression from patch-kernel'. *IEEE Conference on CVPR* (2008).

[65] Yang, Z., Ai, H.: 'Demographic classification with local binary patterns'. *International Conference on Biometrics*, pp. 464–473 (2007).

[66] Yang, M., Yu, K.: 'Adapting gender and age recognition system for mobile platforms'. *Third Chinese Conference on Intelligent Visual Surveillance* (2011).

[67] Yang, M., Zhu, S., Lv, F., Yu, K.: 'Correspondence driven adaptation for human profile recognition'. *IEEE Conference on CVPR*, pp. 505–512 (2011).

[68] Rosipal, R.: 'Nonlinear partial least squares: An overview'. In: H. Lodhi and Y. Yamanishi (eds.), *Chemoinformatics and Advanced Machine Learning Perspectives: Complex Computational Methods and Collaborative Techniques*, pp. 169–189 (2011).

[69] LeCun, Y., Bottou, L., Bengio, Y., Haffner, P.: 'Gradient-based learning applied to document recognition'. *Proc. IEEE* **86**(11): 2278–2324 (1998).

Chapter 14

Short- and long-time ageing effects in face recognition

Massimo Tistarelli, Daksha Yadav, Mayank Vatsa and Richa Singh

14.1 Introduction

Human faces play an important role in the identification process as well as to describe other information such as age, ethnicity, gender, and expression. Humans have a tremendous capability of recognizing learned faces very fast and with high accuracy, which may be difficult to be achieved by currently available automated systems. A very challenging issue is the definition of the optimal face representation, either holistic (based on global features) or component-based (based on local features) [1]. Regardless of the adopted representation, biometric patterns are always affected by the changes in the face appearance due to ageing. The impact on the intra-class variations can often spoil the distinctiveness of a face template even after a short time from the user's enrollment. It has been reported in several evaluation studies that even a short elapsed time of one or two weeks between acquisition sessions can induce variations in the face appearance which degrade the performance of the recognition system [2]. In order to overcome this problem a quantitative evaluation of the changes in the facial appearance over time, as well as a regular update of the stored facial template, is mandatory.

In the course of time, several algorithms were proposed to overcome several ambiguities arising in face-based identification due to changes occurring in the face appearance. As an example, Figure 14.1 shows face images of a celebrity in different poses, illumination, and facial expressions. Face ageing may change the appearance of the face shape as well as its texture (i.e., the skin pigmentation and wrinkles). The performance of any face recognition algorithm drops just because these changes impair the matching of the captured face image with a previously acquired reference face image, or template.

Ling *et al.* [3] published a study on the performance degradation of a face recognition algorithm due to the presence of age-separated samples in the dataset.

University of Sassari, Computer Vision Laboratory, Sassari, Italy
IIIT-Delhi, Okhla, Phase 3, New Delhi, India

Figure 14.1 Images of a subject from FG-NET Facial Aging Database, with different illuminations, poses, alignment, and facial expressions

As reported in Reference 4, during the formative years of a person, the variations in the shape of the face are more prominent, while in the later stages of life texture variations such as skin wrinkles and the change in pigmentation are more visible. Hence, a face recognition system which takes into account age progression needs to be able to incorporate both types of ageing factors.

Only recently ageing effects on face recognition have been rigorously studied, both for adult and young people's faces[1] [6, 12–15]. In general terms, the analysis of the changes in the face appearance over time can be analyzed over two different time scales: *long-time* as opposed to *short-time* effects. As for the former, it involves gradual and deep modifications in the face structure and texture. These can be accounted for in terms of reduction in the fat tissue, progressive degradation of the skin elasticity, reduction of the tone of the facial muscles, development of the skull bones and uneven pigmentation of the skin [16]. The short-time effects can be due to the same phenomena as well as to other, even transient effects such as the change in facial tan, hair, and beard growth, or small textural changes in the skin. As these are related to finer and sometimes temporary changes in the face appearance, they require a different methodology than the analysis of long-time ageing. First and foremost, long-time changes can be analyzed from coarsely sampled face images over a considerably long-time span (ten years or more), while short-time ageing requires the analysis of face images sampled very closely within a short time (one year or less).

In this chapter a quantitative approach to facial ageing is presented, where the evolution of changes in the face appearance are *measured* over a short-time or a long-time period. Most of the work reported in the literature is related to the analysis of long-time variations. In References 12 and 15 algorithms to simulate the age progression in adult face images are presented. In Reference 14 a computer-based automatic ageing simulation technique in 3D is described, with some qualitative results. The work in Reference 13 describes the textural changes produced by three types of geometrical manipulations on face images, which are comparable to ageing effects on the face appearance.

The most similar study to the one presented in this chapter is reported in Reference 6. In this case a quantitative approach is adopted to establish the identity

[1] Face images of people from childhood into adulthood (18–20 years old).

between two age-separated face images of an individual. The computational framework is based on Bayesian functions to estimate the intrapersonal subspace (differences between two age-separated faces of the same individual) and the extrapersonal subspace (differences between different individuals). The estimated subspaces are used to correctly classify the face image of a given subject.

The estimation framework presented in this chapter does not rely on a probabilistic measurement of the changes in the facial appearance but rather on the direct measurement of the time evolution of the face shape. In order to capture the global changes in the face a holistic approach is adopted. Given the short-time variability and the transient nature of the analyzed changes, the physical phenomena underpinning the ageing effects are not systematically analyzed in this chapter. Rather, a global face map is computed, which allows us to define the face areas showing the same *ageing factor*. This estimation allows us to determine the variability of facial features over time.

The pictures used in the reported experiments on short-time ageing were extracted from a series of publicly available videos which can be downloaded from several public Internet websites.[2] This dataset was chosen because the high temporal sampling in these video streams allows the analysis of the time evolution of facial features over a short time span. Concerning long-time ageing two major publicly available databases have been often used: the MORPH [7] and the FG-NET [8] databases. To analyze the ageing effects in an unconstrained acquisition scenario, a novel dataset has been assembled, which includes 2,618 images of 102 subjects, which are publicly available in the network. The data include age-separated images of the same subjects as well as photometric variations due to changes in illumination, facial expression, and pose.

14.2 Analysis of short-time ageing effects

The primary objective of the proposed approach is to quantitatively measure the face changes, either local or global, within a given time interval. The estimation is based on the algorithm presented in Reference 9 to compute the most distinctive facial regions of an individual by comparing his/her face with a number of other faces. The concept of "facial differences" allows us to define the distinctiveness of faces [10] from a number of extracted photometric features. Each extracted facial feature is selected as distinctive if it is significantly different from any other feature in a given set (like a facial scar). In the context of short-time ageing, the same algorithm can be applied to a series of face images of the same person but sampled at different times. As a result, the facial regions which are more stable over time are located. These can be also considered as the most time-invariant regions in the face.

[2] The video streams are published as *Living my life faster*. Every subject acquired a picture of his/her face every day for a time period of 1–8 years.

14.2.1 Computing differences between faces

The computation of the changes in face appearance involves three steps: (1) extraction of face patches, (2) vectorization, and (3) *weighting*[3] of the patches.

Face analysis often requires processing of information embedded at several spatial resolutions, or multiple scales. Therefore, each face image patch is re-sampled using a log-polar mapping [11] to incorporate information at multiple spatial frequencies. Analytically, the log-polar scheme describes the mapping postulated to occur between the retina (retinal plane (r, q)) and the visual cortex log-polar or cortical plane (x, h). As illustrated in Figure 14.2(a) the size of the "receptive fields" follows a linear increment moving from the central region (fovea) outwards into the periphery. The resulting sampling process ensures that each patch contains both low-scale (fine resolution) and contextual (low resolution) information.

Even though any of the extracted face patches may be equally useful to quantitatively evaluate the ageing effects, a subset of the patches must be selected to bound the computational complexity of the algorithm. For this reason, as well as to ensure translation-independence and to side-step coarse registration issues, the patches are extracted at the loci centered on the edges extracted from the image. An example of a sampled face is presented in Figure 14.3. The white dots represent the positions where the log-polar re-sampling has been performed to extract the image patches. It is worth noting that extracting the patches at the edge points does not imply that the analysis occurs only at edges. The extracted points only represent the center (the *fovea*) of the patches, while the sampled area covers a wide neighborhood.

The set of log-polar image patches, sampled at each point, are vectorized, producing a set of feature vectors, each composed of 400 raw gray-level values. For each vector the similarity (the weight) between each patch in the reference face and all patches in the probe face is computed. This is performed by first computing the projection of all vectors into a higher dimensional space. The projected features belonging to the same face image will tend to cluster in the feature space.

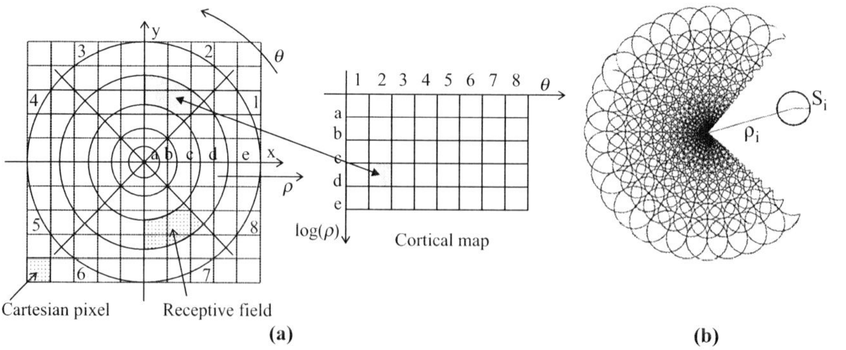

Figure 14.2 (a) Log-polar sampling strategy and (b) the adopted log-polar model

[3] The computed patch weight represents a measure of the difference of each feature in one face image from any other feature in another face image.

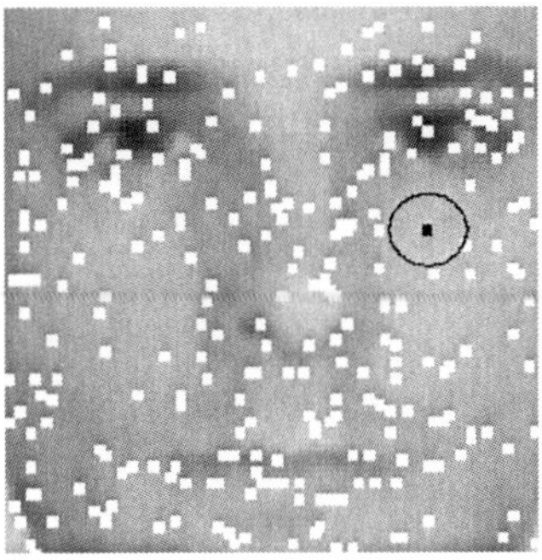

Figure 14.3 Points extracted for the face sampling. The white dots represent the center of the regions extracted to compute the differences. The circle shows the size of the log-polar sampling area

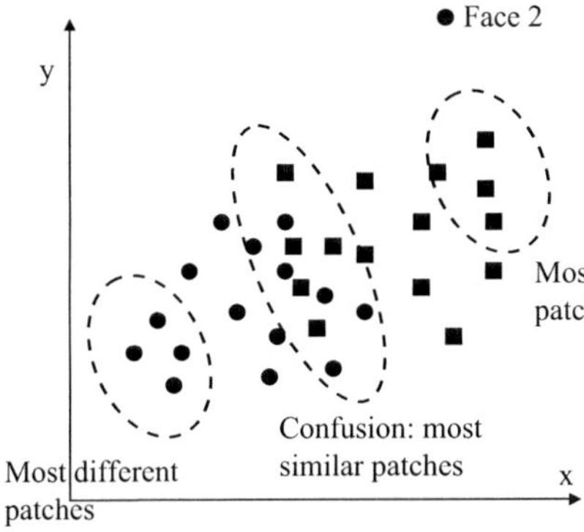

Figure 14.4 General framework of the algorithm applied to extract facial differences

The degree of similarity of two facial features extracted from two different face images is closely related to their mutual position in the higher dimensional space. This concept is exemplified by the diagram in Figure 14.4.

More formally, let P_1, P_2 be the two sets of patches extracted from the face images I_1 and I_2, respectively. The distinctiveness ω of a patch $p_{P_1}(x,y)$, belonging to set P_1 and centered at the pixel (x,y) of face I_1, is computed as:

$$\omega(p_{P_1}(x,y)) = d(p_{P_1}(x,y), P_2) \tag{14.1}$$

$$d(p_{P_1}(x,y), P_2) = \min_{(x',y')} d_M(p_{P_1}(x,y), p_{P_2}(x',y')) \tag{14.2}$$

where d_M is a distance metric between feature vectors, in this case the Euclidean distance. Other metrics could be employed as well.

14.2.2 How faces change over time

The algorithm defined to compute the face differences can be applied to obtain a quantitative measure of the differences of the face over time. To apply the algorithm in this context, we first define the time evolution for two samples. The time evolution between two image samples, I_1 and I_2, of the same face separated by a given time t is defined as the total difference between the two face appearance. Consequently, the difference between I_1 and I_2 is computed from the patch weights of both samples. The weights are averaged to obtain a single value of the difference.

The difference between I_1 and I_2 at a point (x, y) is computed considering only those patches extracted from I_1 and I_2 containing the point (x, y). Therefore, the difference $d_{I_1,I_2}(x, y)$, between two face samples I_1 and I_2 of a sequence of images from the same face, is defined by:

$$d_{I_1,I_2}(x,y) = \frac{1}{2} \left. \frac{1}{|P_{I_1}|} \sum_{p \in P_{I_1}} \omega(p) + \frac{1}{|P_{I_2}|} \sum_{q \in P_{I_2}} \omega(q) \right) \tag{14.3}$$

where P_{I_1} and P_{I_2} are, respectively, the sets of patches from I_1 and I_2 containing the point (x, y), $|P_{I_i}|$ represents the cardinality of the set P_{I_i} and ω is the weight determining the distinctiveness of a patch.

The computed difference value can be used as a measure of the time evolution between two samples. In a sequence of faces $S = I_1, I_2, \ldots, I_t, ..$, the evolution for a point (x, y), within the time span t, can be computed using $d_{I_1,I_{t+1}}$, that is, taking the initial sample as the reference point in time and computing the difference with the sample separated by the time t. A problem arising is that the results highly depend on the actual quality of both face samples being analyzed. Changes in illumination and image noise can introduce artifacts in the estimated time evolution. On the other hand, this approximation does not take advantage of the high temporal sampling of the sequences.

To overcome these issues, instead of a single difference value between two samples, the evolution within a time span is computed by considering a window of samples of size T and comparing each one of them with its respective sample separated by t. The average of the T differences returns a more robust evolution value, as noise and illumination changes are averaged out over time.

Following this concept, the evolution $E(x, y, t)$ in a sequence, for a point (x, y) and within the time span t, is defined by:

$$E(x,y,t) = \frac{1}{T} \sum_{i=1}^{T} d_{I_i,I_{i+t}}(x,y) \tag{14.4}$$

14.2.3 Temporal evolution of facial features

In order to assess the coherence of the ageing-related differences in different subjects, the quantitative results for standard facial features are compared in sequences from different individuals. The value of the time evolution for a *standard* feature (nose, eyes or mouth) of the same individual is expected to stay within a limited and small range. At the same time, when comparing samples from different individuals for the same feature, a substantial, consistent difference in the range of values is expected over time.

If the ranges of the time evolutions for different individuals can be separated, it would be possible to classify each individual on the basis of the face evolution as a signature. In order to verify these assumptions, six standard facial features are analyzed: the nose, eyes, mouth, chin, cheeks, and subnasal area. In this case, a feature in the face is defined as a window in the image. For every feature (or landmark) L the evolution over time t is defined by:

$$E(L,t) = \frac{1}{w_L h_L} \sum_{(x,y)\in L} E(x,y,t) \tag{14.5}$$

where w_L, h_L are, respectively, the width and the height of the region around the feature L and $E(x,y,t)$ is the evolution function for an image point as defined in (14.4).

14.2.4 Experimental evaluation

In order to analyze the effects of short-time ageing and its implications, two experiments have been carried out using sequences of faces over a time frame from one to four years. The data used in the reported experiments were extracted from a series of publicly available videos which can be downloaded from some Internet websites.[2] This dataset was chosen because the high temporal sampling in the video streams allows us to analyze the time evolution of facial features over a short time span.

Three video sequences (S_1, S_2, and S_3) from three different individuals have been used. The resolution of the images in all the sequences is 352×240 pixels. In Figure 14.5 some samples of the sequences are shown. S_1, S_2, and S_3 cover, respectively, a time span of 4, 3, and 1 year(s). As can be noted, the faces have been acquired approximately centered in the image and with the head always in the same position and orientation. For this reason the position of facial features does not vary over time, thus simplifying the localization process. The data are highly sampled in time at a rate of one frame per day. This allows us to process close in time images thus reducing the effects of noise and temporal aliasing. Only a very limited preprocessing was performed on the face images. In order to avoid face differences due to illumination changes, the samples are preprocessed with standard histogram equalization. To minimize the influence of outliers, the images with extreme variations such as eyeglasses (worn just once) or long beard were removed from the sequence.

A first series of experiments was devoted to the analysis of the global evolution of the face appearance. Toward this aim, the methodology described in section 14.2.2 has been applied to the sequence S_1. In Figure 14.6 the evolution for every face point over a time span of four years is shown. The brighter the value in the picture, the faster the temporal evolution of the point. These results are in line with the qualitative results reported in Reference 16.

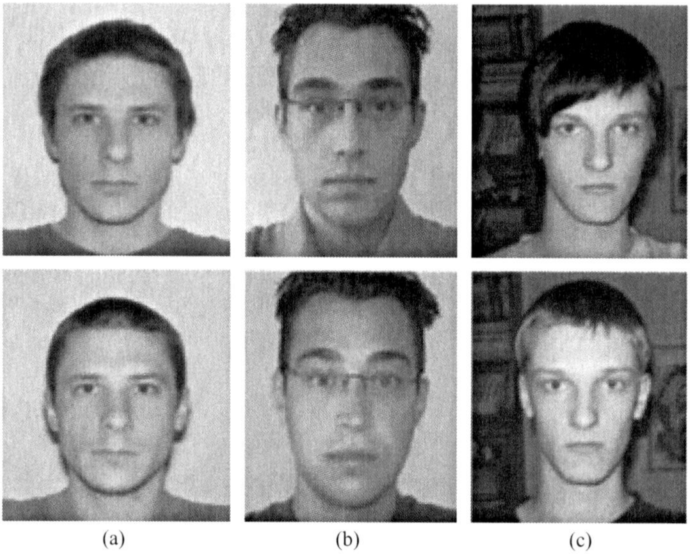

Figure 14.5 *Sample images extracted from the three image sequences used in the experiments. Images in (a), (b), and (c) correspond to images from sequence S_1, S_2, and S_3, respectively*

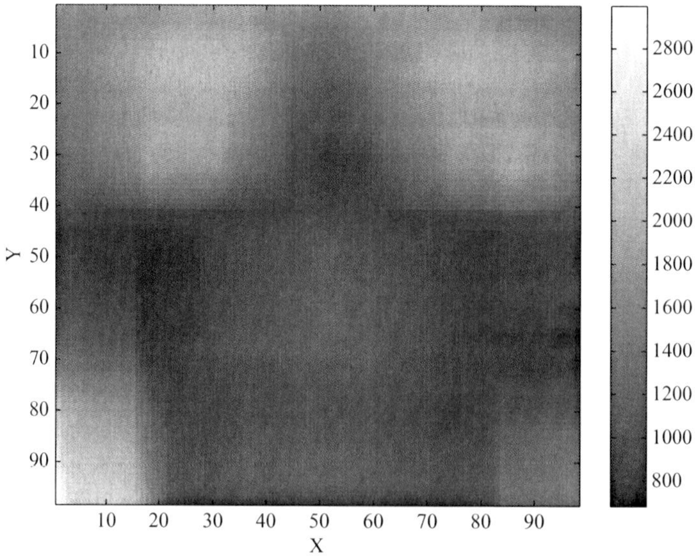

Figure 14.6 *Evolution for the subject in sequence 1 over four years. The area depicted corresponds to a 100 × 100 image window centered on the face. The brighter the pixel, the faster the temporal evolution of the corresponding face point. As expected, the most variable regions in the face appear to be those containing the eyes*

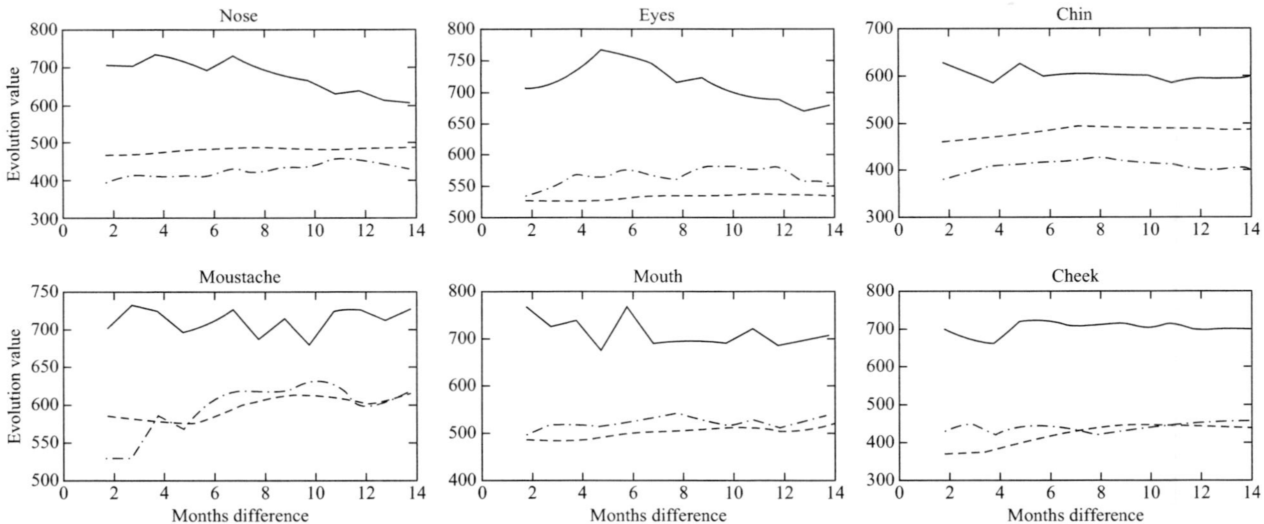

Figure 14.7 *Time evolution of facial differences obtained for six selected facial features: eyes, nose, mouth, chin, cheek, and subnasal area. The solid lines report, for each facial feature, the differences between samples from sequences of different subjects. The dashed lines represent the computed differences between samples of sequence S_2 and the dash-dotted lines correspond to the computed differences between samples of sequence S_3*

A second series of experiments was aimed at the analysis of the time evolution of specific, manually selected, facial regions. In particular, six regions were analyzed: the nose, eyes, mouth, chin, cheeks, and subnasal area. The evolution value $E(L, t)$ has been computed for sequences S_2 and S_3. Another temporal evolution has been synthesized by crossing samples of both sequences into a single image stream. This procedure allows a comparison of the range of variability of the differences related to samples of the same individual, and of different individuals as well.

In Figure 14.7 the results obtained analyzing different features over different time spans are shown. It is worth noting that the time evolution of every single feature, for the same subject, remains within a limited range. At the same time, the mean value of the temporal evolution of the same feature, but in different subjects, is always considerably different. The estimated difference in the range of values for different individuals is consistently in the order of 50–80%. This difference can be easily exploited to devise a robust template embedding the time evolution of each subject's feature for identification or verification purposes.

In order to make a consistent evaluation of the impact of the proposed algorithm on identity verification, an additional experiment has been performed. The MORPH [7] and FG-NET [8] databases have been used for the experiment. Both databases contain many heterogeneous face samples, captured over a long time span (Figure 14.8). The time frame between successive samples of the same subject is quite large, usually covering several decades.

The dataset used for the experiment is composed of 180 samples from the 2 databases. For each database 15 subjects are included, with 6 samples per subject. All samples have been captured from all subjects within the age interval of 25–50 years. To evaluate the identity verification capabilities of the algorithm, all samples are compared using the computation of photometric differences expressed in (14.3). Instead of using particular features, patches are extracted for the whole face region. The global difference value between two face images is computed as the average of all patche weights. In order to evaluate the identity verification capability of this approach, the false match rate (FMR) and false non match rate (FNMR) are computed as the matching threshold varies along the photometric difference space.

(a) (b) (c) (d)

Figure 14.8 Face samples from (a,b) MORPH database and (c,d) FG-NET database

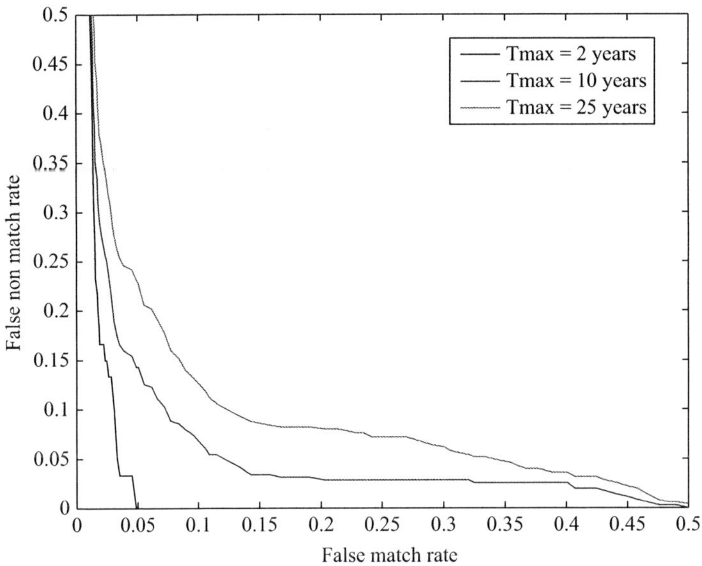

Figure 14.9 ROC curves considering maximum age comparison (T_{max}) of 2, 10, and 25 years. The equal error rates (EER) are respectively 0.04, 0.09, and 0.12

A parameter T_{max} is defined to tune the maximum age difference allowed between two samples to be matched. T_{max} allows evaluation of the performance of the algorithm depending on the time span of samples from same individuals. In Figure 14.9 the ROC curves obtained for $T_{max} = 2$, 10, and 25 years are shown. As expected, the error increases with the increase of the time difference between individual samples. The equal error rate (EER) when using $T_{max} = 2$ years is about 4% and increases up to 9% and 12% when T_{max} is raised, respectively, to 10 and 25 years. These results contribute to quantitatively determining adequate times to update an individual template in terms of expected error rate in an identity verification environment.

14.2.5 Discussion

Even the best learning algorithm cannot cope with every possible variation in the face appearance. This is especially true for natural developments and changes due to ageing. For this reason most identification algorithms fail to recognize individuals whenever the acquired face image is distant in time from the enrollment session. Sometimes even acquiring images of the same subject at different times in the day can result in noticeable changes in the face image. For these reasons, the analysis of long-time ageing effects must be coupled with a short-time quantitative estimation of the time variability of facial features.

The quantitative results obtained in the experimental analysis can be further exploited in a number of practical applications of biometric identification from face images. As an example we may consider the following:

- The time evolution can allow us to define the stability of facial features over time to devise more robust facial templates, which incorporate this measure in the face model.
- The analytical formulation of the temporal evolution of a subject's face may facilitate the update of a facial template. Both the global and local estimation of the temporal evolution can be applied as an indicator of the optimal time to update the template or part of it.
- The quantitative evaluation of the temporal evolution for a subject may allow a forecast of critical times when the template update or even reenrollment would be required in a verification environment.

14.3 Analysis of long-time ageing effects

The analysis of long-time ageing has been mainly studied within the field of human anatomy and visual perception [17, 18]. The research in this area helped in understanding and modeling craniofacial growth. The analysis of facial ageing involves two main issues:

- *Face appearance*: The way facial structure and appearance changes with age is not uniform during the lifetime. During formative years, the variations in the shape of the face are more prominent while in the later stages of adulthood texture variations such as wrinkles and pigmentation are more visible.
- *Computational processes*: The process of facial ageing is affected by several factors such as bone growth, facial muscle elasticity, and disease or environmental factors such as accidents and surgery. Further, ethnicity and gender may also influence the "ageing pattern" of the face.

The analysis of long-time ageing on human faces can be broadly divided into three categories: age estimation, computational models for age prediction, and face recognition across age progression.

14.3.1 Age estimation

For a given face image, Kwon and Lobo [5] used anthropometry of the face and wrinkle density to estimate the associated age group. They had three broad categories of ages: infants, young adults, and senior adults. For distinguishing infants from adults, they used six ratios of facial measurements such as eye-eye and eye-nose. On the other hand, for distinguishing between young and senior adults, they used facial wrinkle density. Ramanathan and Chellappa [6] proposed a method for estimating the age gap given a pair of face images as input. They had four age separations: 1–2, 3–4, 5–7 and 8–9 years. Their work was based on the fact that as age separation increases, facial appearance variations are more visible. They used a Bayesian framework for classifying pair of images.

Geng *et al.* [20] proposed the "Ageing pattern subspace: AGES" to characterize the cases when the ageing pattern is missing (having the images of an individual at each of the given age categories). In these cases, the algorithm tries to find the ageing pattern subspace from the available dataset which captures the appearance variations and predict the face for missing data points. Yang and Ai [21] used local binary patterns (LBPs) for texture classification. They identified a sequence of local features and combined into these a strong classifier to improve the performance.

14.3.2 Computational models for age progression

The process of building a model for estimating facial appearances across different ages requires developing a model which incorporates various factors such as gender, ethnicity, diet, and age group. Burt and Perrett [22] identified 7 age groups: 20–24 years, 25–29 years, up to 30–34 years, 35–39 years, 40–44 years, 45–49 years, and 50–54 years. They characterized the face shapes of each input face image by finding a set of fiducial features. They formed a composite image for each age group by averaging the shape and skin color from face images of the same age group. One limitation in their methodology is that the averaging process smoothes facial creases, hence the composite faces appeared younger than those from their own age groups. Tiddeman *et al.* [23] extended the idea proposed by Burt and Perrett and used wavelet-based methods for defining facial textures and creating "composite faces." They used edge strength and Markov Random Fields to retain edges (wrinkles) and improve texture characterization, respectively. Some of the results of facial ageing and rejuvenation approaches applied on some faces are shown in Figure 14.10.

In Reference 24, Ramanathan and Chellappa studied the facial ageing effects particularly for individuals of age 0–18 years. They observed that the primary reason for facial changes in formative years is the craniofacial growth. They combined cardioidal strain transformation along with age-based anthropometric data to develop a facial ageing model. In Reference 25, they introduced a shape and texture variational model for estimating facial ageing in adults. The shape variation model was developed by physical models that characterize the functionalities of different facial muscles, such as linear, sheet, and sphincter muscles. The texture variation model was applied to characterize facial wrinkles in predesignated facial regions such as the forehead and naso-labial region.

Suo *et al.* [26] proposed a technique based on a multi-resolution face model built upon "AND-OR" graphs. The "AND" nodes are used in finding the fine representation of faces, whereas the "OR" nodes are used in finding different possible configurations. Since the analysis is performed at multiple resolutions, the global appearance variations, such as skin color and hairstyle, are extracted at the lowest resolution level. At the next resolution some deformations in the facial components are extracted. Lastly, at the highest resolution, the finer details, such as skin pigments and wrinkles, are analyzed.

14.3.3 Face recognition across age progression

The developed techniques for face recognition across age progression can be further categorized as *generative* and *discriminative* methods [4]. As for *generative* methods, changes in facial appearance are first computed, and then applied to compensate for ageing variations in the input facial images. *Discriminative* methods do not involve any modification of the input face images. Rather, these techniques are based on the extraction of age-invariant signatures to be used for matching and identification.

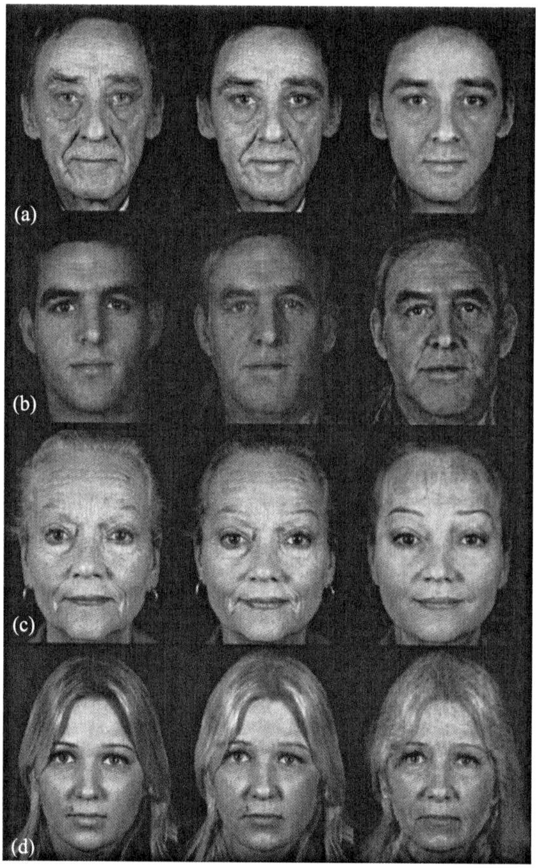

Figure 14.10 (a) Male rejuvenation, (b) male ageing, (c) female rejuvenation, and (d) female ageing. Image source: Reference 23

Ling *et al.* [3] proposed a discriminative method where the image gradient orientations are extracted as age-invariant features, which are proved to be stable across ages. They designed an ad hoc operator to extract the image gradient orientations from face images. A support vector machine (SVM) classifier was then applied to perform identity verification (genuine users vs. impostors) from faces across ages.

14.4 A novel face database to analyze long-time ageing effects

Several face databases have been collected over time, all with specific peculiarities related to the potential variabilities of human faces and in different acquisition scenarios. However, there are only a few datasets that include significant long-time ageing effects. Two notable examples are the FG-NET [8] and the MORPH [7] datasets. However, both datasets have been acquired under very controlled conditions. This makes it difficult to determine the performance of algorithms in real-life acquisition scenarios.

For this reason, a new dataset has been created by collecting images of celebrities publicly available on several websites. The collection procedure was similar to the one pursued to create the *Labeled Faces in the Wild* (LFW) dataset [31]. The face images in the dataset not only include a large time separation, but also variations in pose, facial expression, eye glasses, background, and indoor/outdoor environment.

14.4.1 Database collection

Around 300 images of 102 Indian celebrities were downloaded from the Internet using an image crawler. Several of these images contained faces with large pose variations, large occlusions, and with very small interocular distance (low resolution). Images with such extreme conditions were removed from the dataset. The sanitized database contains 2,618 images from 49 female and 53 male celebrities. The age span of the face images collected from the same subjects is between 4 and 88 years.

14.4.2 Database annotation

Since the images were downloaded from the Internet, these included many unexpected variations including makeup, grooming, pose, rotation, illumination, expression, and resolution. To provide a ground truth for researchers in processing the database, each face image has been accurately annotated according to the following features:

1. *Glasses*: Indicates whether the celebrity is wearing any goggles or spectacles in the image.
2. *Hair color*: Hair color of the celebrity in that image.
3. *Eye color*: Eye color of the celebrity.
4. *Surroundings*: Whether the image is taken in an inside or outside surrounding.
5. *Gender*: Gender of the celebrity, male or female.
6. *Beard*: Whether the celebrity has a beard or not in the image.
7. *Mustache*: Whether the celebrity has a mustache or not in that image.
8. *Age*: Approximate age of the individual in the image.
9. *Smiling or not*: If the celebrity is smiling or not in the image. This field would help in analyzing and testing accuracies of algorithms in different facial expressions.
10. *Pose*: Indicates if the pose of celebrity is frontal or side (right or left). This field would be helpful for finding out accuracies of algorithms on different poses.
11. *Full body or face image*: Shows if the image is a full body or just the face image of the celebrity.
12. *Ethnicity*: Shows whether the celebrity is of Indian or Western origin.
13. *Eyebrows*: If the celebrity has bushy eyebrows in the image or not.

During this process of manual annotation, the images which were small, redundant, irrelevant (e.g., not of the celebrity or unclear image) were also removed.

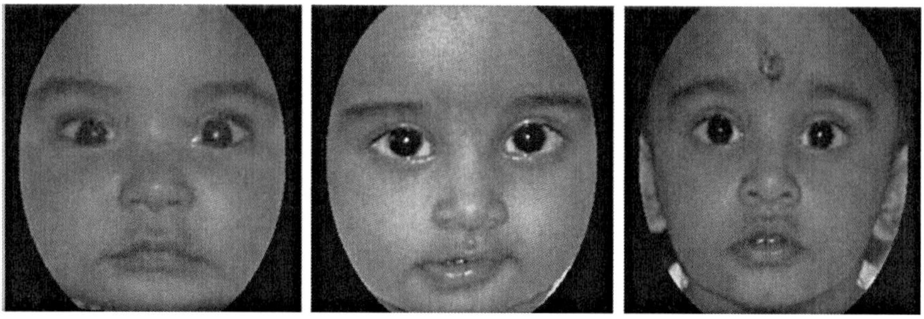

Figure 14.11 *Pre-processed images of subjects from FG-NET Facial Aging Database*

14.4.3 Image preprocessing

The images which were collected are mostly full body images. In many images, the faces are not aligned and due to different degree of close-up, face sizes are also not the same. So, the next step is to align all the images in our database based on the eye coordinates.

The eye coordinates of each image are manually annotated and the faces are aligned with a constant inter-eye distance of 90 pixels. After alignment, the Viola and Jones face detection algorithm [27] is applied to detect and extract the face from each image. Finally, the detected face images are resized to 128 × 128 pixel size. As there was variation in the background of the images, background removal was done by just keeping the elliptical face region and discarding the background. This has brought homogeneity in the database and helped in comparing the different face recognition algorithms. Examples of images after preprocessing are shown in Figure 14.11.

After extracting the face images from the crawled image database; binocular, left periocular, right periocular, and mouth regions from each of the face image are extracted.

14.5 Experimental validation and data statistics

In order to determine the recognition performance on the collected face ageing dataset, three different algorithms were tested corresponding to three different categories: matching of LBP representations, principal component analysis (PCA), and sparse representation. For each algorithm applied, the performance statistics were reported and analyzed with respect to the age variations embedded in the data.

14.5.1 LBP feature analysis

For the analysis of feature-based algorithm performance, LBPs [28] were computed from the whole face as well as from the right and left periocular regions, a rectangular region including both eyes and the mouth region. Both the three-patch and four-patch face descriptors [29] were evaluated. Every face image was compared against all the remaining face images in the database (full fold validation). For each image pair, the two extracted LBP descriptors were compared by computing the Chi-square distance between the histograms. Since each face image was divided into four different regions, also the fusion of the matching scores resulting from comparing each single region was computed. The scores computed from each region were first normalized using the min-max normalization technique and then fused using the sum rule. In particular the scores resulting from the following face sub-regions were fused together:

1. Binocular, left and right periocular regions.
2. Face and binocular regions.
3. Face, binocular, left and right periocular regions.
4. Face, left and right periocular regions.
5. Left and right periocular regions.
6. Face, binocular and mouth regions.
7. Mouth, left and right periocular regions.
8. Face and mouth regions.
9. Face, mouth, binocular, left and right periocular regions.

14.5.2 Baseline PCA

The PCA [30], or singular value decomposition, or *Karhunen-Love* decomposition, whatever terminology is used, is a well-known method for transforming multi-dimensional correlated data into a less than or at most equal linear uncorrelated orthonormal components. The linear uncorrelated principal components are ordered according to the associated eigenvalues. Consequently, the first principal component accounts for most of the variation in the data. The second principal component accounts for the variance which is not tapped under the first component. The third, fourth, and following components correspond to the information content with decreasing variance. This is a very efficient representation of the dataset if the data has a large number of dimensions and is very highly correlated. PCA helps in reducing the dimensions by finding out a linear combination of optimally weighted features (the principal components) which represent most of the data.

Two recognition experiments were performed using PCA to determine the recognition performances across age progression. In the former experiment, the probe set consisted of the latest (oldest) face image of every subject, while all remaining images formed the gallery or training set. In the second experiment, the probe set contained the earliest (youngest) face image of every subject, while all remaining images were used for training.

14.5.3 Sparse representation

Classification using sparse representation is being extensively studied for face recognition. The same experimental protocol described in V-A was applied. The Sparse coding has been applied first to the raw pixels and then also to the LBP features extracted from the face images.

14.5.4 Results and analysis

Figure 14.12 shows the cumulative matching curves (CMC) obtained from the scores computed matching the standard LBP as well as the three-patch and four-patch LBP features from the whole face and the fused scores. The highest accuracy, equal to *46.14%* at rank 1, is reported by the fusion of the matching scores from the whole face with the scores computed from all face sub-regions.

Table 14.1 reports, for each of four age groups, the number of probes which are correctly classified by matching the LBP histogram representations extracted from the face images.

Table 14.2 reports the same results as Table 14.1, but divided according to the gender of the probe subjects.

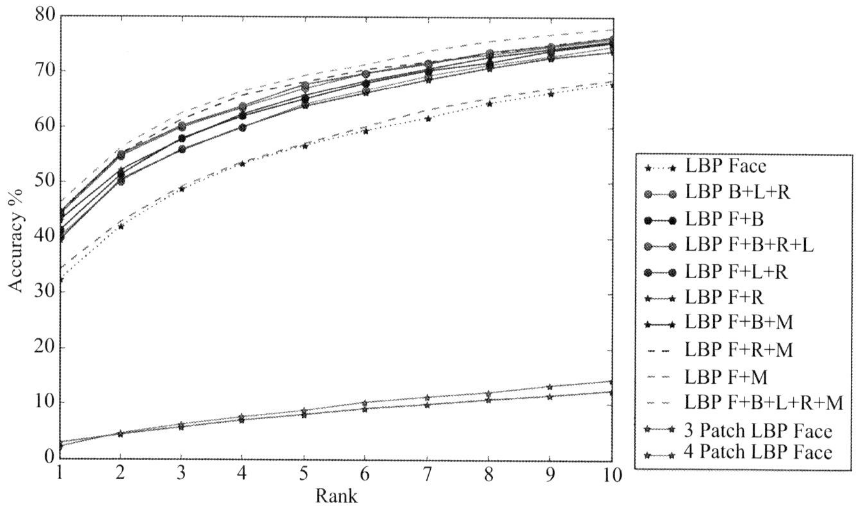

Figure 14.12 CMC curves obtained from the matching scores of the standard LBP and the three-patch and four-patch LBP representation of the whole face image and the LBP representations resulting from each facial sub-region. The results obtained from the fusion of the matching scores are also reported

Table 14.1 Number of correctly classified probes, across age progression

	0–18	**19–35**	**36–55**	**Above 55**
0–18	0	0	1 (0.038%)	0
19–35	0	302 (11.53%)	66 (2.52%)	20 (0.76%)
36–55	0	34 (1.3%)	254 (9.7%)	20 (0.76%)
Above 55	0	15 (0.57%)	31 (1.18%)	104 (3.97%)

Note: The scores were computed by matching the LBP histograms extracted from the face images.

Table 14.2 Number of correctly classified probes, according to the respective gender, across age progression

		Male		
	0–18	**19–35**	**36–55**	**Above 55**
0–18	0	0	1 (0.038%)	0
19–35	0	99 (3.78%)	38 (1.45%)	9 (0.34%)
36–55	0	18 (0.72%)	161 (6.14%)	8 (0.30%)
Above 55	0	7 (0.26%)	20 (0.76%)	39 (1.48%)
		Female		
	0–18	**19–35**	**36–55**	**Above 55**
0–18	0	0	0	0
19–35	0	203 (7.75%)	28 (1.07%)	11 (0.42%)
36–55	0	16 (0.61%)	93 (3.55%)	12 (0.45%)
Above 55	0	8 (0.30%)	11 (0.42%)	65 (2.48%)

As can be noticed from Table 14.2, male subjects are more accurately classified when the probe lies in the age range of 36–55 years, while female subjects are more accurately classified when the probe lies in the age range of 19–35 years.

Figure 14.13 shows the receiver operator characteristic (ROC) curves obtained from the scores computed matching the standard LBP as well as the three-patch and four-patch LBP features from the whole face and the fused scores. Table 14.3 reports the rank 1 accuracy results obtained by applying the PCA and the sparse representation-based classification to the dataset.

14.6 Discussion and conclusion

The effects of time elapsed on biological life forms are always evident and clearly perceivable. As for humans, changes in the face appearance can be categorized and purposively employed to infer the subject's age, as well as his/her tiredness, mood, and health state. The relatively young area of ageing analysis in biometrics has been often devoted to either define a computational model of long-time ageing or to compensate for ageing effects to improve recognition performance in the presence of time-separated face images of the same individuals.

The objective of this chapter has been to go beyond the classic approach of facial ageing and provide a more comprehensive analysis including both short-time and long-time ageing effects. The rationale for this approach is the intrinsic difference between these two processes. While long-time ageing can be relatively easy to record and model, short-time ageing is difficult to define and can produce a number of unexpected changes in the face appearance. The reason is that short-time face ageing can be produced by a number of either internal or external factors. Just as an example, a simple case of flu can dramatically change the face appearance and unexpectedly impair identification.

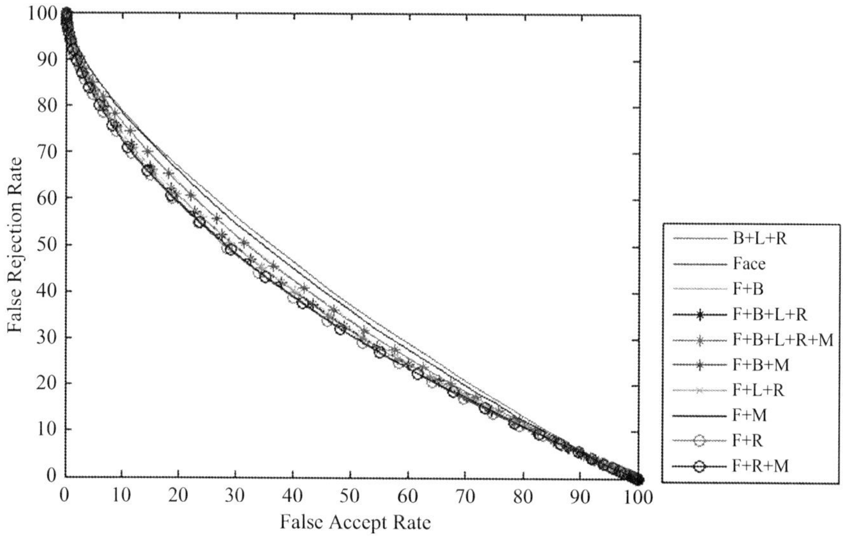

Figure 14.13 ROC curves obtained from the matching scores of the standard LBP and the three-patch and four-patch LBP representation of the whole face image and the LBP representations resulting from each facial sub-region. The results obtained from the fusion of the matching scores are also reported

Table 14.3 Rank 1 accuracy obtained from the PCA and the sparse representation-based classification applied to the dataset

Algorithm	**Probe Image**	**Accuracy**
PCA	Latest image	3.92%
PCA	Youngest image	4.9%
Sparse	Latest image	21.42%
Sparse	Youngest image	15.71%
Sparse on LBP images	Latest image	8.57%
Sparse on LBP images	Youngest image	10%

Even though it is beyond the scope of this chapter to take into account all possible sources of transient face changes developing within a short time span, a computational model has been proposed to account for and predict these changes. The reported experiments suggest that there are a number of specific areas within the face that are best candidates for transient changes in appearance. As such, these areas should be considered with care in component-based or feature-based face image matching. On the other hand, the specific transient variations of these areas can be modeled to design an *adaptive* template, where the adaptivity is related not only to the rate of change of the whole face, but also to the confidence to be associated to each topological area within the face. Thus, short-time ageing brings to the fore a new concept of template adaptation, not only as a mechanism to update the template according to the long-time changes in the biometric trait, but also as a method to produce flexible representations, capable of tolerating transient changes in the biometric trait due to short-time ageing. This concept of "flexible representation" could also find other potential applications, such as to cope with cosmetics or light surgery.

Long-time ageing has been extensively studied in the last decade and several algorithms have been proposed to model ageing effects in face images. This chapter did not propose a new approach or algorithm for template updating and age recovery or to compensate for ageing effects for recognition purposes. Rather, an analysis of the current state of the art has been performed and a novel face image database has been introduced, the *Delhi face ageing database*. The main purpose of this new database is to produce a data corpus which includes a large number of image variations which may well represent the acquisition variability experienced in real-life scenes. A number of comparative experiments were performed to validate the usefulness of the database. The reported results well demonstrate the need for better algorithms to cope with the superimposition of several processes, including changes in illumination, pose, and occlusions, with the long-time ageing. In particular, it seems that selectively processing local information may lead to a better identification despite the ageing effects. This also recalls the need to *stir in* a priori knowledge about the time variability of specific face areas. This additional information is expected to greatly improve the performance of any classification algorithm applied to face images with a large time separation. The reported experiments also show a different behavior in facial ageing for subjects belonging to different gender or age classes.

The research described in this chapter indicates the need to provide better models for both short-time and long-time ageing to either predict the age or to compensate for photometric changes in the images due to the time lapse. At the same time it is expected that research in template adaptation and flexible face representations will lead to the development of subject-specific models which take into account the peculiar variabilities of face areas due to transient or permanent changes over time.

References

[1] W. Zhao, R. Chellappa, P.J., Phillips, and A. Rosenfeld, 'Face recognition: A literature survey', *ACM Computing Surveys,* vol. 35, no. 4, pp. 399–458, 2003.

[2] M.L. Yui, D. Bolme, P.J. Phillips, J.R. Beveridge, and B.A. Draper, 'Preliminary studies on the Good, the Bad, and the Ugly face recognition challenge problem', *Proceedings of the IEEE Computer Vision and Pattern Recognition Workshops*, Providence, RI, pp. 9–16, 2012.

[3] H. Ling, S. Soatto, N. Ramanathan, and D. Jacobs, 'A study of face recognition as people age', *Proceedings of the IEEE International Conference Computer Vision*, pp. 1–8, 2007.

[4] N. Ramanathan, R. Chellappa, and S. Biswas, 'Age progression in human faces: A survey', *Journal of Visual Languages and Computing*, vol. 20, pp. 131–144, 2009.

[5] Y.H. Kwon and N. da Vitoria Lobo, 'Age classification from facial images', *Computer Vision and Image Understanding*, vol. 74, pp. 1–21, 1999.

[6] N. Ramanathan and R. Chellappa, 'Face verification across age progression', *IEEE Transactions on Image Processing*, vol. 15, no. 100, pp. 3349–3362, 2006.

[7] K. Ricanek Jr. and T. Tesafaye, 'MORPH: A longitudinal image database of normal adult age-progression', *Proceedings of the IEEE International Conference on Automatic Face and Gesture Recognition*, Washington DC, pp. 341–345, 2006.

[8] http://www-prima.inrialpes.fr/FGnet/

[9] M. Bicego, E. Grosso, and M. Tistarelli, 'On finding differences between faces', T.Kanade, A.Jain, N.Ratha, Eds. *Proceedings of Audio- and Video-based Biometric Person Authentication, LNCS*, New York, NY: Springer, vol. 3546, pp. 329–338, 2005.

[10] M. Bicego, G. Brelstaff, L. Brodo, E. Grosso, A. Lagorio, and M. Tistarelli, 'Distinctiveness of faces: A computational approach', *ACM Transactions on Applied Perception*, vol. 5, no. 2, pp. 11:1–11:18, 2008.

[11] E. Grosso and M. Tistarelli, 'Log-polar stereo for anthropomorphic robots', *Proceedings of European Conference on Computer Vision*, Dublin, Ireland, vol. 1, pp. 299–313, 2000.

[12] A. Lanitis, C.J. Taylor, and T.F. Cootes, 'Toward automatic simulation of aging effects on face images', *IEEE Transactions on Pattern Analysis and Machine Intelligence*, vol. 24, no. 4, pp. 442–455, 2002.

[13] T. Miyoshi and M. Hyodo, 'Aging effects on face images by varying vertical feature placement and transforming face shape', *Proceedings of IEEE International Conference on Systems, Man and Cybernetics*, Tainan, Taiwan, vol. 2, pp. 1548–1553, 2006.

[14] U. Park, Y. Tong, and A.K. Jain, 'Face recognition with temporal invariance: A 3D aging model', *Proceedings of IEEE Conference on Automatic Face and Gesture Recognition*, Amsterdam, The Netherlands, 2008.

[15] J. Wang, Y. Shang, G. Su, and X. Lin, 'Simulation of aging effects in face images', *LNCS*, vol. 345, pp. 517–527, 2006.

[16] E. Patterson, A. Sethuram, M. Albert, K. Ricanek, and M. King, 'Aspects of age variation in facial morphology affecting biometrics', *Proceedings of IEEE Conference on Biometrics: Theory, Applications, and Systems*, Washington, DC, 2007.

[17] M.S. Zimbler, M.S. Kokoska, and J.R. Thomas, 'Anatomy and pathophysiology of facial aging', *Facial Plastic Surgery Clinics of North America*, vol. 9, no. 2, pp. 179–187, 2001.

[18] K.A. Deffenbacher, T. Vetter, J. Johanson, and A. J. O'Toole, 'Facial aging, attractiveness, and distinctiveness', *Perception*, vol. 27, no. 10, pp. 1233–1243, 1998.

[19] N. Ramanathan and R. Chellappa, 'Face verification across age progression', IEEE *Conference on Computer Vision and Pattern Recognition*, Washington, DC, vol. 2, pp. 462–469, 2005, Washington DC, US.

[20] X. Geng, Z. H. Zhou, and K. Smith-Miles, 'Automatic age estimation based on facial aging patterns', *IEEE Pattern Analysis and Machine Intelligence*, vol. 29, no. 12, pp. 2234–2240, 2007.

[21] Z. Yang and H. Ai, 'Demographic classification with local binary patterns', *Proceedings of the Second International Conference on Biometrics,* Seoul, Korea, pp. 464–473, Springer, 2007.

[22] M. Burt and D. I. Perrett, 'Perception of age in adult caucasian male faces: Computer graphic manipulation of shape and colour information', *Journal of Royal Society*, vol. 259, pp. 137–143, 1995.

[23] B. Tiddeman, D. M. Burt, and D. Perrett, 'Prototyping and transforming facial texture for perception research', *IEEE Computer Graphics and Applications*, vol. 21, no. 5, pp. 42–50, 2001.

[24] N. Ramanathan and R. Chellappa, 'Modeling age progression in young faces', IEEE *Conference on Computer Vision and Pattern Recognition*, New York, NY, vol. 1, pp. 387–394, 2006.

[25] N. Ramanathan and R. Chellappa, 'Modeling shape and textural variations in aging adult faces', *IEEE International Conference on Automatic Face and Gesture*, 2008.

[26] J. Suo, F. Min, S. Zhu, S. Shan, and X. Chen, 'A multi-resolution dynamic model for face aging simulation', *IEEE Conference on Computer Vision and Pattern Recognition*, June 2007.

[27] P. Viola and M. Jones, 'Rapid object detection using a boosted cascade of simple features', *Proceedings of IEEE Conference on Computer Vision and Pattern Recognition*, Hawaii, pp. 511–518, 2001.

[28] T. Ojala, M. Pietikanien, and D. Harwood, 'A comparative study of texture measures with classification based on feature distributions', *Pattern Recognition*, vol. 29, pp. 51–59, 1996.

[29] L. Wolf, T. Hassner, and Y. Taigman, 'Descriptor based methods in the wild', *Proceedings of the ECCV Workshop on Faces in Real-Life Images: Detection, Alignment, and Recognition*, Marseille, France, 2008.

[30] K. Pearson, 'On lines and planes of closest fit to systems of points in space', *Philosophical Magazine*, vol. 2, no. 6, pp. 559–572, 1901.

[31] G.B. Huang, M. Ramesh, T. Berg, and E. Learned-Miller, *Labeled Faces in the Wild: A Database for Studying Face Recognition in Unconstrained Environments*, University of Massachusetts, Amherst, Technical Report 07-49, October 2007.

Section 4

Future trends and research challenges

Chapter 15

Analysis of ageing effects in biometric systems: difficulties and limitations

Meryem Erbilek and Michael Fairhurst

15.1 Introduction to physical ageing effects in biometric systems

It is, of course, well known that an individual's physiological and behavioural characteristics change with time, and that generally this natural 'age progression' poses some challenging issues in biometric systems. This is because such systems identify/verify individuals by using their physical and/or behavioural characteristics, and since many relevant personal characteristics can vary with time, key biometric data are thus also likely to change, sometimes significantly, with the passage of time. Age progression has an impact on both behavioural modalities (such as signature [1] and voice [2]) and physiological (such as iris [3] and fingerprint [4]) biometric modalities, even those which are regarded as generally more stable over time. As a result, developing applications, whether for security, for forensics or other purposes for long-term use can be a particularly challenging task.

Thus, the impact of age progression on biometric systems has become a very important topic in biometrics, and is currently generating a large volume of research. In general, the research studies published in the literature which address the impact of ageing on biometric systems can be classified into two broad categories, related respectively to *physical* and *template* ageing. In fact, both physical ageing (the effect in a general sense of the passage of time on the biometric measurements used in the identification process) and template ageing (changes occurring specifically across the time elapsed between enrolment – when the template is created – and use – when the current measurement values are matched against those stored on the template) are important in a practical context, and while these are obviously related phenomena, they are not the same thing, and it can be very useful to consider them separately.

This chapter will explore some of the limitations and difficulties which may be experienced specifically in the investigation and analysis of physical ageing issues in biometric systems. Considerable advances have been made in recent years by researchers seeking to investigate and explore the physical ageing process and its effects on most of the common biometric modalities such as iris [3], signature [1, 5–6],

University of Kent, UK

fingerprint [4], voice [2] and face [7–10]. However, there are several significant problems which can limit a comprehensive age progression analysis in studying ageing effects, and which indeed can also lead to apparently inconsistent conclusions about the effect of age progression on performance. Thus, in this chapter, some of the most important limitations which challenge research in this area will be investigated and explored, while some suggestions will also be presented about how to manage these issues in the analysis of physical ageing in biometric systems.

15.2 Limitations in the analysis of physical ageing effects in biometric systems

Ideally, any study exploring the impact of physical ageing would work with data corresponding to measurements on a population where the same individuals are monitored and measured at different points across their lifetime. This, of course, is not typically a practical possibility, especially in an area such as biometrics which has a relatively short history with respect to large scale practical application. In order to define and investigate the limitations and challenges arising in the analysis of physical ageing effects in biometric systems, it is therefore first necessary to understand how physical ageing effects are generally investigated with respect to any particular biometric modality. First, biometric data is captured on the basis of a snapshot of a population of individuals to form an experimental database. For example, the biometric data might be defined by the capture of an image of the eye or by dynamic/static measures generated by the production of a handwritten signature in the case of the iris and the signature modalities respectively. In addition to explicit biometric data, each individual's true identity and other information (for example, physical age is of specific interest in our present context) are stored for further analysis. Second, the target population is divided into *age-bands*, since age is a continuous variable, and changes associated with ageing generally occur within a longer time frame rather than in the form of step changes within shorter time periods. Finally, physical ageing analysis is carried out based on the designated age groupings for a chosen modality and with an experimental procedure matched to the research question of interest.

Since the first two steps are fundamental to any research study and necessary for any subsequent analysis of age-related issues in biometric systems, it is useful to define, explain and summarise their importance and limitations in greater detail as follows.

15.2.1 Age

Physical age in an individual is a continuous variable, with age-induced changes occurring progressively, generally relatively slowly and in a way which varies between one individual and the next. This means, of course, that we would not expect to see discernible changes in characteristics which depend on age actually occur from one day to the next, nor, in some cases, even over the timeframe of weeks or months. Thus, observations of an individual at, say, 30 years and 2 months of age are not likely to be significantly different from those made at an age of, say, 30 years and 3 months, and so on. This makes the division of a given target population into age-bands a necessity in any meaningful analysis of the effects of ageing in a broad population, as is evident in the studies cited in the previous section. For example, it has been shown that features extracted from face [11] and signature [1] biometric data exhibit particular differences as a function of age. In the case of the signature, age-bands of '26–40', '41–60' and '>60' were adopted in Reference 1 while in the case of face [11], age-bands of '13–24', '25–34' and '35–45' were used. A similar ageing scenario for the iris biometric modality was explored in Reference 3, but with respect to defined age-bands of '<25', '25–60' and '>60'. Looking more widely, the quality of fingerprints in different age populations has been investigated in Reference 12 on the basis of defined age-bands of '18–25' and '>62'. Thus, the age-bands adopted in ageing-related research are found to vary considerably across studies and, to an extent, with the biometric modality used, and the choices made – often with no specified rationale – make inter-study comparisons extremely difficult. Therefore, a better understanding of how to define appropriate age-bands is a challenging issue, yet it is necessary and important that this issue is considered if we are to be equipped to manage ageing effects most effectively.

15.2.2 Biometric data

The biometric database adopted for any empirical study of ageing issues may have an impact on the analysis in several different ways. For instance, the way in which the biometric data are collected plays a very important role in any ageing-related analysis. This includes the environment which is used for the data collection – for example, whether data are collected in a controlled or uncontrolled environment is a very important issue. If the data collection process is undertaken in a controlled environment, then a supervisor carrying out the data collection process becomes an important concern, especially when this is the person who decides, for example, the quality of the collected biometrics and whether this is acceptable.

Thus, in the further sections of this chapter, we will investigate these two issues more broadly, empirically explore the limitations that lie behind each and present suggestions about how to manage these issues in the analysis of physical ageing in biometric systems. Since we will illustrate our discussion with experimental results, in the next section we will present the basic experimental setup and important practical details to be used in all the experiments which we will later report.

15.3 Experimental infrastructure

In sections 15.4, 15.5 and 15.6 of this chapter, we will report an experimental study which we use to illustrate, and perhaps provide preliminary solutions for, some important practical questions about how to design a robust framework for research into the effects of ageing in biometric systems. For completeness, one behavioural and one physiological biometric modality (the signature and iris modalities respectively) are adopted to represent illustrative examples for the experimental analysis. We first briefly describe the tools and the experimental setup used for the experimentation as follows.

15.3.1 Biometric databases

Three different databases are used to form particular datasets for the experimental investigation which we will report in the following sections.

The first database which we draw upon is *Data Set 2 (DS2) of the BioSecure Multimodal Database (BMDB)* [13] which provides both iris and signature samples. The biometric samples in this database were collected as part of an extensive multimodal database by 11 European institutions participating in the BioSecure Network of Excellence [13], and is a commercially available database. This particular database contains biometric samples from 210 users, who are within the age range of 18–73.

Eye images in this database were acquired in a standard office environment, managed by a supervisor, and using the LG Iris Access EOU3000 system. During the acquisition, spectacles were not allowed, although contact lenses were allowed. Four eye images (two left and two right eye images) were acquired in two different sessions per user, each image with a resolution of 640 × 480 pixels [14]. Acquisition of signature samples in this database was carried out in a standard office environment using a Wacom Intuos 3 A6 graphics tablet with a sample rate of 100 Hz, under the guidance of a supervisor [14]. Thirty signature samples were acquired from each user, collection split across 2 different sessions.

In addition, we make use of two further databases. These are the *BioSecure Kent* database [15] and a further one which we designate as the *BTG* database [16], which we use to provide further samples for the iris and the signature modality respectively.

The biometric samples in the *BioSecure Kent* database [15] were collected as part of a more general and extensive multimodal database in the School of Engineering and Digital Arts at the University of Kent as part of a Europe-wide project undertaken by the BioSecure Network of Excellence [15]. This database is locally restricted to the School of Engineering and Digital Arts at the University of Kent. This particular database contains biometric samples from 79 users, who are within the age range of 18–73. Eye images and signature samples in this database were acquired by adopting the same protocol as described above for the *DS2 of BMDB* database. In fact, it is useful to report that part of the *BioSecure Kent* database (specifically, 36 individual's biometric data) is included as a part of *Data Set 2 (DS2) of BioSecure Multimodal Database (BMDB)* although the remaining part (specifically, 43 individual's biometric data) was discarded as far as a contribution to the larger database is concerned. The underlying criteria used to determine which samples to discard in this operation are not known.

The signature samples in the *BTG* database [16] were collected as part of an extensive collaborative study between the University of Kent and the British Technology Group (BTG) in the United Kingdom. This database is locally restricted to the School of Engineering and Digital Arts at the University of Kent. Acquisition of the signature samples was carried out using a graphics tablet with a sample rate of 100 Hz at the Hedge End Post Office and Newsagent, Southampton, UK and collected from members of the general public without the guidance of a supervisor [16]. This particular database contains 359 participants with a varying number of samples (between 2 and 79 per subject), but generating a total of 7,428 signatures. The subjects providing the samples contained in this database are divided into the age groupings of <26, 26–40, 41–60 and >60. These groupings were defined at the time of individual enrolment, and no information is available to change these bandings (i.e. the specific age of each subject is not recorded).

The iris samples of 10 subjects were found to be wrongly categorised (some of the left eye images labelled as right or vice versa) in the *DS2 of BMDB* database. Hence, this in practice decreases the available number of subjects for the iris modality to 200. For all iris datasets, in view of the established independence of left/right eye pairs within an individual, here we treat left and right eye sources as originating from different individuals. This effectively increases our population size by a factor of 2. For ease of subsequent reference, we designate the iris datasets from *DS2 of BMDB* and the *BioSecure Kent* datasets as I-Bio and I-BioKent respectively. The available number of samples per subject is different for signature databases, and we therefore randomly selected 15 samples per subject from each database for our experimental study. In the case of the *BTG* database, this required us to discard the 131 subjects who had donated fewer than 15 samples. For ease of subsequent reference, we designate the signature datasets from *DS2 of BMDB*, *BioSecure Kent* and the *BTG* datasets as S-Bio, S-BioKent and S-Btg respectively.

Relevant information about all of the iris and signature datasets is summarised in Table 15.1.

15.3.2 Biometric data processing

After the definition of our experimental datasets for the iris and signature modalities, each individual sample is processed to form a biometric template for the proposed experimental studies.

In the case of the iris modality, the eye images are first segmented by using a robust segmentation algorithm developed in-house [3, 17]. In the event of segmentation failure, we segment the irises manually and make sure that all eye images are correctly segmented, thus ensuring that subsequent analysis can focus on issues

Table 15.1 Dataset information

Modality	Database	Datasets	Number of Subjects	Number of Samples per Subject	Total Number of Samples
Iris	DS2 of BMDB	I-Bio	400	4	1600
	BioSecure Kent	I-BioKent	158	4	632
Signature	DS2 of BMDB	S-Bio	210	15	3150
	BioSecure Kent	S-BioKent	79	15	1185
	BTG	S-Btg	228	15	3420

related to age effects rather than any limitations on the reliability of the segmentation process. Subsequently, the correctly segmented irises are normalised and their features extracted using Masek's generally available and widely adopted software [18]. In a previous study [3], we have shown that this improved iris recognition software is able effectively to support the investigation of ageing effects in iris biometrics.

In the case of the signature modality, 35 features (see Table 15.2) which the literature shows are commonly used in signature processing experiments [1, 5, 19–23] are chosen to be extracted from each signature sample. The extracted features in each

Table 15.2 Extracted features

Feature	Feature Type	Feature Description
F1	Dynamic	Total distance of pen travel
F2	Dynamic	Total time taken to execute the signature
F3	Dynamic	Number of times the pen was removed from the tablet
F4	Static	Width of the signature
F5	Static	Height of the signature
F6	Static	Division of the signature height by the signature width
F7	Static	Standard deviation of x coordinate values
F8	Static	Standard deviation of y coordinate values
F9	Dynamic	Average pen velocity in x
F10	Dynamic	Average pen velocity in y
F11	Dynamic	Number of zero velocity sample points in x
F12	Dynamic	Number of zero velocity sample points in y
F13	Dynamic	Maximum pen velocity in x – Average pen velocity in x
F14	Dynamic	Maximum pen velocity in x – Minimum pen velocity in x
F15	Dynamic	Maximum pen velocity in y – Average pen velocity in y
F16	Dynamic	Maximum pen velocity in y – Minimum pen velocity in y
F17	Dynamic	Maximum pen velocity in x – Minimum pen velocity in y
F18	Dynamic	Average pen acceleration in x
F19	Dynamic	Average pen acceleration in y
F20	Dynamic	Number of zero acceleration sample points in x
F21	Dynamic	Number of zero acceleration sample points in y
F22	Dynamic	Maximum pen acceleration in x – Average pen acceleration in x
F23	Dynamic	Maximum pen acceleration in x – Minimum pen acceleration in x
F24	Dynamic	Maximum pen acceleration in y – Average pen acceleration in y
F25	Dynamic	Maximum pen acceleration in y – Minimum pen acceleration in y
F26	Dynamic	Maximum pen acceleration in x – Minimum pen acceleration in y
F27	Dynamic	The number of times pen passes through the midline of the signature
F28	Static	Number of points comprising the image
F29	Static	Sum of x coordinate values
F30	Static	Sum of y coordinate values
F31	Static	Horizontal centralness
F32	Static	Vertical centralness
F33	Dynamic	Azimuth
F34	Dynamic	Altitude
F35	Dynamic	Pressure

dataset are then mean and variance normalised (MVN). In order to use only unique and non-redundant features, for each dataset, Spearman's rank correlation [24] test is applied to the extracted and MVN features as in Reference 23.

15.3.3 Biometric data classification and performance evaluation

After forming a biometric template, an appropriate classification and evaluation methodology is established for the proposed experimental studies.

The available samples are divided into testing and training sets by using a *K*-fold cross-validation methodology. In *K*-fold cross-validation, the training set is divided equally into *K* different subsets. *K*-1 of the *K* subsets are used to train the classifier and the *K*th subset is used as a test set. This procedure is repeated *K* times and each time a different subset is used as the test set. For the experiments reported here, the value of *K* is chosen according to the available number of samples per subject involved in the specific experimental setup, in order to involve same number of samples per subject in each subset. For example, if there were 15 available samples per subject, then *K* was chosen to be 5, in order to include 3 samples per subject in each test subset and 12 samples per subject in each training subset. In general, the value of *K* is defined as 4 for the iris, and 5 for the signature modality-based experiments.

In using the defined testing and training sets, a simple *k-nearest neighbour (k-NN)* classifier [19] is used to perform matching between testing and training sets, and the predicted training label for each test subject is stored. This is a simple classifier that does not need any sophisticated training phase. Its only requirement is that labels of classes represented by the samples are available. Thus, to find the nearest neighbour, an appropriate distance metric according to the type of the extracted features (squared Euclidean distance and Hamming distance in the case of the signature and iris modalities are used respectively in this study) is calculated between each test sample and all training samples. The first *k* distances measured are then ranked in ascending order and the class label which most appears in the first *k* elements of the list is taken to determine the assigned class label (majority voting strategy).

Then, if the predicted training and test subjects are from the same class, the decision is designated a correct match; otherwise it is recorded as an incorrect match. The overall error rate is evaluated as the percentage of incorrect matches out of all matches.

15.4 Investigating ageing issues

As noted earlier, a significant issue which generally limits a comprehensive age progression analysis in biometric systems is that 'age' is a continuous variable, with age-induced changes occurring progressively and generally slowly. This makes the division of a given target population into age-bands a necessity in any type of ageing analysis. Thus, our experimentation and analysis is carried out with the aim of improving our understanding of the ageing process in relation to individual biometric-related data, which is an important issue in terms of ensuring a definition of age-bands which reflect relevant characteristics, and which helps to aid and unify our understanding of how to manage such effects. Hence, initially the effects of choice of age-bands will be investigated methodically with respect to an age-based analysis of the performance of practical biometric systems. Subsequently, we will use the results and observations to point towards the possibility of minimising the sensitivity of disparate investigations by a more considered choice of age-banding, in relation to both physiological and behavioural biometric modalities.

15.4.1 Effects of age-bands

Empirical investigation and analysis is carried out in order to help us better understand the effects of different choices of age-banding on how we can more reliably discern age-related trends, and we do this by considering both physiological and behavioural biometric modalities.

For these experiments, the error-based performance of both iris and signature modalities are therefore analysed with respect to four different possible age-band specifications based on three broad age groupings (which may be described as relating to the general categories 'young', 'middle' and 'older').

The I-BioKent and S-BioKent datasets are utilised for the iris and the signature modalities in this experiment. Each sample is processed to form a biometric template for our further experiments as explained in section 15.3.2. In the case of signature samples, 35 features are extracted (see Table 15.2), mean and variance normalised (MVN), and the correlation test applied. As a result, 19 uncorrelated features of the dataset are selected for our further experiments. These features are F1, F3, F6, F7, F9, F10, F13, F15, F18–F20, F23, F24, F27 and F31–F35 (see Table 15.2 for description of each feature).

In order to analyse physical ageing patterns, error-based performance in recognition is evaluated. All the available samples are classified to determine an estimate of the error rates incurred in each of the age-bands, as determined by using the tools described in section 15.3.3. The recognition error rate for all samples within each age-band is evaluated as the percentage of incorrect matches out of all test subjects of that age-band.

The evaluated error-based performances are shown in Figure 15.1(a) and (b) for the iris and the signature modality respectively. Considering the iris modality, the general ageing trend is the same for four of the age-band configurations, where error rate decreases as a function of age, suggesting that there might not be any significant sensitivity to age-band choice for this modality. However, it should be noted that there is a very little difference between the lower bound of '<25' and the middle bound of '25–54' in age-band 2, which suggests that caution should be exercised in drawing a firm conclusion here. Considering the signature modality, the broad ageing trend is the same for age-band 1 and age-band 3, where the minimum error rates are achieved for lower bounds ('<25' and '<31') and the maximum error rates achieved by the middle bound ('25–60' and '31–60'). However, the ageing trend between other age-bands is different, suggesting that there might be significant sensitivity to age-band choice for this modality.

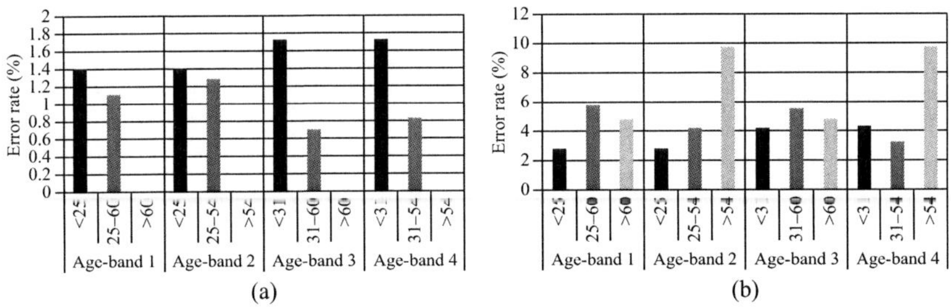

Figure 15.1 Error rates of (a) iris and (b) signature modalities with respect to different age-bands

These results illustrate that age-related performance trends are likely to change according to the adopted age-bands for both physiological and behavioural biometric modalities, illustrating the importance of understanding these processes in managing age-related issues in biometric systems. Possible approaches to address this limitation will be discussed in the next section.

15.4.2 Managing age-band effects

In the light of these initial observations, a more extensive error-based analysis was carried out with respect to several different defined age-bands on both the iris and the signature modalities in order to try to determine upper and lower bounds corresponding to age-bands which can most effectively reflect age-related performance patterns.

In this experiment, the same datasets (I-BioKent and S-BioKent) are utilised and each sample is processed to form a biometric template for the iris and the signature modalities as in the previous section. The recognition error-based performance is evaluated for both the iris and signature modalities with respect to a range of different age-bandings determined by the following inter-band boundaries:

- Lower bounds
  <25, <27, <29, <31, <33, <35
- Middle bounds
  25–50, 27–50, 29–50, 31–50, 33–50, 35–50
  25–52, 27–52, 29–52, 31–52, 33–52, 35–52
  25–54, 27–54, 29–54, 31–54, 33–54, 35–54
  25–56, 27–56, 29–56, 31–56, 33–56, 35–56
  25–58, 27–58, 29–58, 31–58, 33–58, 35–58
  25–60, 27–60, 29–60, 31–60, 33–60, 35–60
- Upper bounds
  >50, >52, >54, >56, >58, >60

In order to analyse the respective physical ageing patterns, error-based performance in a standard recognition task is evaluated. All the available samples are classified to determine an estimate of the error rates incurred in each of the age-bands considered by using the tools described in section 15.3.3. The error rate observed for each age-band is evaluated as the percentage of incorrect matches out of all test subjects of that age-band.

The measured error rates as a function of lower age bound chosen for the two modalities are shown in Figure 15.2(a) and (b) respectively. The '<25' lower age bound seems to reflect the ageing pattern better than the others (i.e. '<27', '<29', '<31', '<33', '<35') especially for the signature modality, since error rates start to increase and then remain stable after the age of 25 (i.e. these samples are more likely to belong to the same age group since there is no change in error rates).

The measured error rates as a function of the chosen upper age bound for the iris and signature modalities are shown in Figure 15.3(a) and (b) respectively. In this case, the '>60' upper age bound seems to reflect an ageing pattern better than the others (i.e. '>50', '>52', '>54', '>56', '>58') for the signature modality, since the maximum change in error rates occurs from the '>58' age bound to the '>60' age bound (error rate differences between upper age bounds with order: '>50'–'>52' = 0.56, '>52'–'>54' = 0.85, '>54'–'>56' = 1.86, '>56'–'>58' = 0.55, '>58'–'>60' = 2.57) (i.e. these samples are more likely to belong to the same age group since there is no change in error rates). In another words, except for the case of the '>60' upper age bound, the others are relatively stable in terms of error rates. However, in the case of the iris modality, there is no difference in terms of error rate for the upper age bounds.

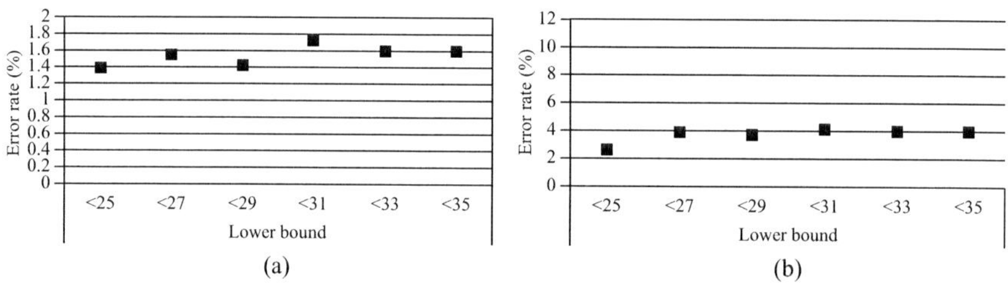

Figure 15.2 Error rates for lower age bounds for (a) iris and (b) signature modalities

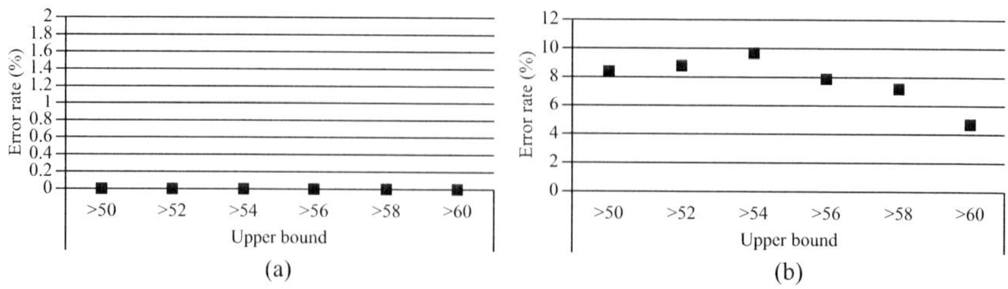

Figure 15.3 Error rates for upper age bounds for (a) iris and (b) signature modalities

In order to investigate the optimality of the selected middle age bounds, first we keep the upper bounds the same and increase the lower bound in a stepwise progression. The measured error rates are shown in Figure 15.4(a) and (b) for the iris and signature modality respectively. For both modalities, it is seen that whatever upper bound is adopted for the middle-aged bands, the best lower bound for the middle-aged banding is ‘25’, since error rates start to decrease and then remain stable for all other lower bounds for the middle-aged bands.

Subsequently, we keep the lower bound the same and increase the upper bound in a stepwise progression. In the case of the signature modality (see Figure 15.5(b)), it is observed that whatever the lower bound adopted, the best upper bound for the middle-aged banding which best reflects ageing pattern is ‘60’. As shown in Figure 15.5(a), it is harder to observe the difference between upper bounds to decide the best upper bound for the middle-aged banding. This is not surprising since we observed earlier that there is no difference in terms of error rates for the adopted upper age bounds for the iris modality.

Thus, the best age-bands to adopt in the sense of choosing those which most truly reflect the age-related performance variability pattern are seen to be ‘<25’, ‘25–60’ and ‘>60’ according to our experiments based on the iris and signature modalities. However, as can be seen from the results of our experimentation, the iris modality is more robust with respect to the chosen age-bands than is the case for the signature modality. This is perhaps not surprising, since iris is a physiological biometric and signature a behavioural biometric modality. Taking an overview of all individual results therefore leads us to an interesting conclusion, that we

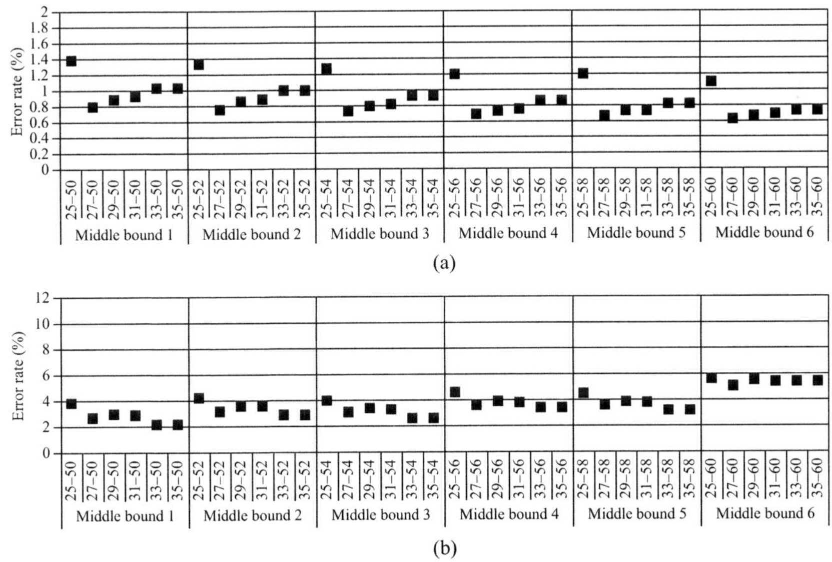

Figure 15.4 Error rates for middle age bounds for (a) iris and (b) signature modalities by keeping upper bound constant

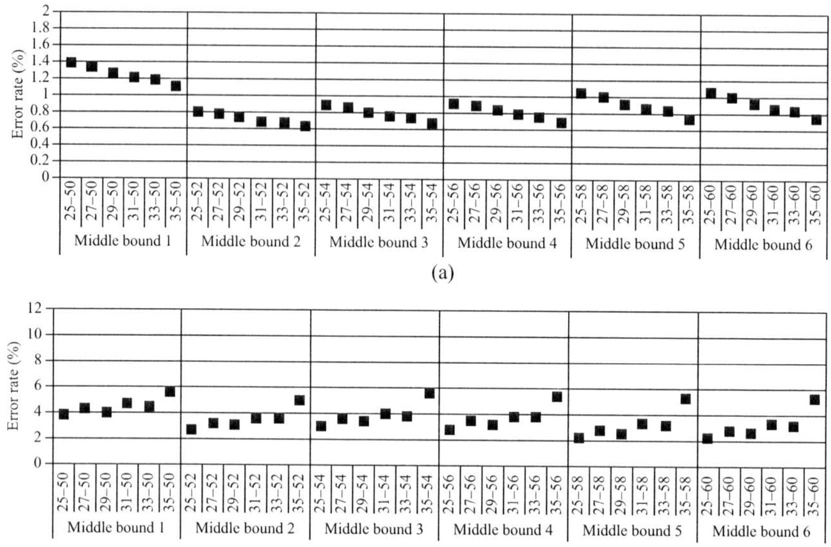

Figure 15.5 *Error rates for middle age bounds for (a) iris and (b) signature modalities by keeping lower bound constant*

can suggest that the choice of age-bands ('<25', '25–60' and '>60') is the one which seems to reflect a reliable understanding of ageing trends, and it is perhaps not unreasonable to consider that this may be a guide for other modalities as well.

We therefore suggest that this experimental study may be a helpful guide for further research both in terms of our suggested approach to determining optimal age-bands to adopt in specific circumstances, but also because our reported results should be especially valuable in providing some evidence that the age-bands determined here best reflect the way in which user age impacts on error-rate performance in a typical biometric system based on the iris modality or the signature modality.

Following the discussion set out above about how to define appropriate age-bands in order to manage ageing effects most effectively, we have next considered the other most important issue which is the implications of choice of biometric databases in the analysis of physical ageing issues. In order to investigate and further explore these issues, we have considered two different perspectives. Initially, we have investigated database implications with respect to the relation to biometric sample quality for the iris biometric modality. Subsequently, we analyse database implications in relation to biometric sample features with respect to the signature biometric modality. These will be discussed and presented in the next sections.

15.5 Database issues and biometric data quality

We have specifically investigated database implications in the context of biometric data quality. Since the question of how to define quality metrics in behavioural biometric modalities is a particularly difficult one, which could easily be a study in itself (although the work reported in Reference 25, for example, offers some useful pointers), we focus on the iris modality for this specific investigation, since there is a relative abundance of work to be found in the literature reporting studies related to the quality of iris images (References 26 to 29 provide some good examples), and it may be considered a simpler task to investigate quality issues in physiological biometrics.

However, we are particularly interested in our experimental study to investigate issues arising using different databases, but where their constituent biometric samples are collected under the same capture protocol, even where the collection environments are different. This allows us to understand more clearly how the quality of the biometric samples in a particular database can play a very important role in the analysis of ageing effects.

15.5.1 Database effects in terms of biometric data quality

Thus, we specifically investigate the effects of the database choice on the analysis of age-related effects on the performance of practical biometric systems. The recognition error-based performance for the iris modality is evaluated with respect to the defined age-bands as described below.

The I-Bio and I-BioKent datasets are utilised in this experiment. Each individual sample is processed to form a biometric template for our further experiments as explained in section 15.3.2.

After designation of our datasets and forming a biometric template for each iris sample, we evaluate the error-based recognition performance with respect to the '<25', '25–60' and '>60' age-bands which we have shown in section 15.4.2 to be best able to reflect the age-related performance variability pattern in the most stable way. In order to evaluate error rates, all the available samples are classified to determine an estimate of the error rates incurred in each of the age-bands considered, using the tools described in section 15.3.3. The error rate for each age-band is evaluated as the percentage of incorrect matches out of all test subjects of that age-band.

Experimental results are shown in Figure 15.6. It is seen that error rates decrease as a function of age for I-BioKent while minimum and maximum error rates are achieved by the older and middle-aged groups for I-Bio. Thus, a different ageing pattern is observed with respect to the different databases selected, and it is useful to

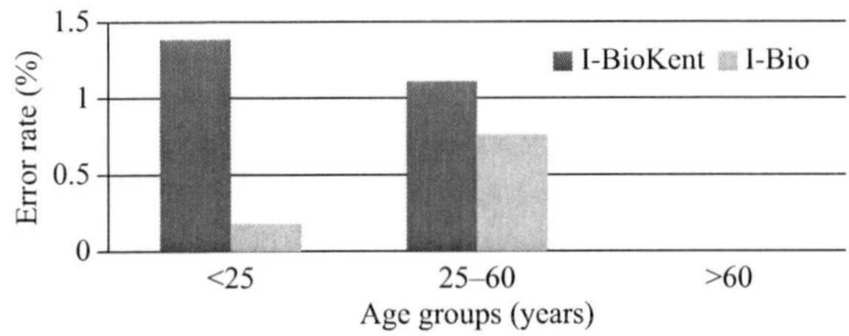

Figure 15.6 Error rates of iris modality with respect to two different datasets

determine the relationship between these observations and the quality of the samples in the databases. Even where the quality of the utilised databases is considered to be good (i.e. samples collected in a controlled environment), detailed visual inspection of the individual iris samples reveals that the only characteristic of any significance which impacts adversely on quality here is the problem of iris occlusion. In order to investigate this more rigorously, we measure the occlusion value (a value between 0 and 1) of each iris image in both databases using the formulation defined in (15.1).

$$Occlusion = \frac{Number\ of\ occluded\ pixels}{Total\ number\ of\ pixels} \tag{15.1}$$

For easier visualisation, a histogram showing the distribution of individual occlusion values within each iris database is illustrated in Figure 15.7. As this figure reveals, irises are more occluded in the I-Bio than in the I-BioKent database, which increases the plausibility of our assumption that biometric data quality differences between the databases have implications in the analysis of the effect of ageing. Thus, this issue needs to be investigated and explored in greater detail, as we shall see in the next section.

15.5.2 *Managing database effects in terms of biometric data quality*

As we have mentioned in the previous section, inconsistent ageing patterns are discovered when we evaluate the error rates of an iris recognition system on different databases, and we have shown that this is likely to be because the I-Bio database contains more occluded iris samples than the I-BioKent database.

In order to investigate further the impact of data sample quality of the analysis of the effects of ageing, we consider in more detail the effects of iris occlusion. Using the occlusion metric defined in (15.1), we first define an occlusion threshold value to work with and choose this empirically as 0.45 according to the data in Figure 15.7. We then remove those subjects in each database whose occlusion value is greater than 0.45. The number of removed subjects in each age-band for each database can be seen from Figure 15.8. It can be readily observed that in the

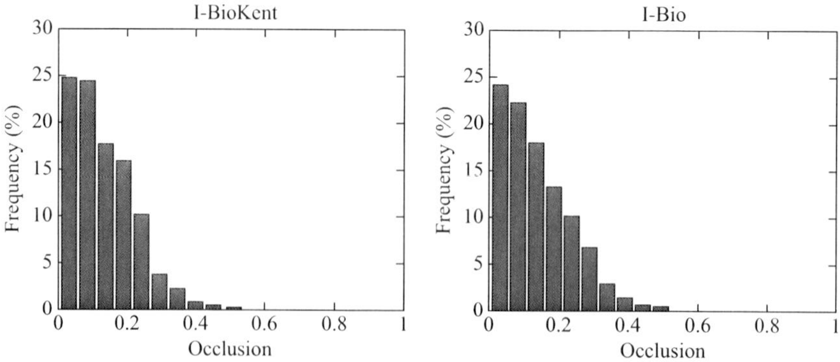

Figure 15.7 Occlusion histogram of each iris dataset

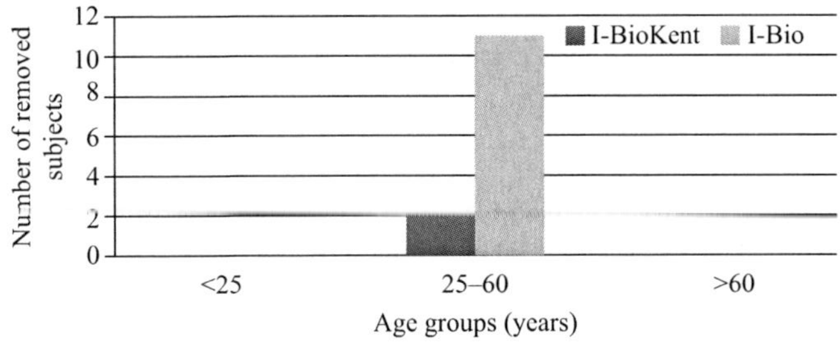

Figure 15.8 Number of subjects removed from database

Table 15.3 Error rates of iris modality without occluded images

Age-Bands	**I-BioKent**			**I-Bio**		
	Number of subjects	**Mean dilation ratio**	**Error (%)**	**Number of subjects**	**Mean dilation ratio**	**Error (%)**
<25	54	0.432	1.389	140	0.417	0.179
25–60	88	0.381	0.568	219	0.376	0.114
>60	14	0.324	0	30	0.343	0

I-Bio database, the middle of the chosen age groupings has the largest number of occluded images, suggesting that this issue of data quality is likely to be a principal reason for the inconsistent scenarios illustrated by the previous experiment.

We therefore measured the error rates with respect to the different age groups for each dataset with the occluded images removed (where ‘occluded’ is defined according to the threshold noted above), using the same setup as in section 15.5.1. The results obtained are shown in Table 15.3.

It can be seen, therefore, that if the variability in quality is removed as a factor which differs across the databases, the same trend with respect to ageing is now observed. It is known that a crucial factor in determining the effects of ageing in iris biometrics is pupil dilation, which does not directly change the inherent features of the iris but increases access to the unique iris features differentially with age progression. This is because the ability of the pupil to fully dilate in response to incident light decreases with age, resulting in a tendency to increasing occlusion of the inherent iris patterning at a given incident light level as age increases [3]. This small study has shown how, when comparing like with like across databases (in this example, looking at a broad ‘quality’ measure based on occlusion), the true effect of ageing can more readily be determined. In this case, for example, we can draw the conclusion (Table 15.3) that error rates decrease as a function of age since the dilation decreases with age.

Here, it is very important to understand the difference between the occlusion and the dilation factors. As explained in Reference 6, a likely reason for performance degradation when pupil dilation increases is the fact that in these circumstances only a small area of the iris region is available from which to compute the required subject-specific defining features. As described earlier, the recognition software adopted [7] transforms the iris region into a fixed rectangular block by using an appropriate sampling approach. In highly dilated images, since there is a smaller available iris region, there are fewer distinct pixels to resample, and this results in poorly characterized irises. However, in the case of occlusion, occluded parts of the iris are stored directly in a fixed rectangular block and will not be taken into account for the later stages of processing. Hence, an increasing degree of occlusion directly decreases the available area of iris which can be used to support successful processing in the later stages of the iris recognition system.

15.6 Database issues and biometric features

For our final experiment we specifically investigate and explore database issues in the context of biometric data at the feature level. The signature modality, being a behavioural biometric and thus generating features which can be expected to exhibit considerable variability, provides an excellent opportunity to illustrate some of the important issues which arise when addressing database effects at the feature level.

The principal aim of this experimental study is to investigate the implications, in terms of database characteristics, of the analysis of physical ageing effects by using databases comprising biometric samples collected under different protocols and different environments and in different acquisition conditions (for example, with or without the guidance of a supervisor). This allows us to show how differences in terms of the features extracted from biometric samples play a very important role in the analysis of ageing effects.

15.6.1 Database effects in terms of biometric features

The S-Bio and S-Btg datasets are adopted in this experiment. Each signature sample is processed to form a biometric template for our further experiments as explained in section 15.3.2. The first 32 features (defined in Table 15.2) are extracted, normalised (using the MVN approach described) and the correlation test applied.

First, we examine the results of the feature correlation study. Our investigation showed that features F1, F3, F6–F10, F13, F15, F18–F20, F23, F27, F31 and F32 were found not to be correlated in S-Bio while in S-Btg, features F1, F3, F6, F7, F9, F10, F13, F15, F18–F21, F23, F27, F31 and F32 were found to be uncorrelated.

Feature F8 is found to be uncorrelated in S-Bio and correlated with F1 and F5 in S-Btg. Feature F21 is found to be uncorrelated in S-Btg and correlated with F2, F11, F12, F20 and F28 in S-Bio. These minor differences occurring across the datasets are not surprising because signature samples from S-Bio were collected in a standard office environment with the guidance of a supervisor, while signature samples from S-Btg were collected in an unregulated public retailing environment without the guidance of a supervisor. Also, another important issue here concerns the nature of the target box drawn on the signature capture paper, specifying the space within which users are requested to sign. This space was smaller in the S-Btg collection environment than in the case of S-Bio. This difference is likely to limit the height and the velocity of an individual's signature, which can result in a different correlation pattern in S-Btg. Thus it is apparent that different individual scenarios of signature data collection can result in a different pattern of correlation between the features.

Following these findings, in order to ensure like-for-like analysis as much as possible across the 2 different datasets, 15 common uncorrelated features of the 2 datasets are selected for our further experiments. These features are F1, F3, F6, F7, F8, F10, F13, F15, F18, F19, F20, F23, F27, F31 and F32.

After determining the uncorrelated common features, a feature selection method based on distance discriminant (FSDD) [30] is adopted for the selection and analysis of features from samples from each of the 3 age-bands independently, and for each dataset independently. For this experimentation, the age-bands are defined to be '<26', '26–60' and '>60'. In section 15.4.2, we have shown that assignment of the '<25', '25–60' and '>60' age-bands are found to reflect most stably the age-related performance variability pattern. However, the lower boundaries are chosen to be 26 for this particular experiment, since the subjects providing the samples contained in the S-Btg database are divided into the age groupings of '<26', '26–40', '41-60' and '>60' and the data required to re-band this population are not available.

The results of this feature ranking process within each dataset are shown in Table 15.4. It is observed that different feature rankings are obtained for different age groups, and for the same age groups in different datasets. An important point which this illustrates here is that feature ranking depends on both the dataset used (i.e. issues around the data collection process) and on the age groups considered. This also confirms previous findings, namely that the distributions of feature values within age groups are indeed different [1, 5, 31]. However, it is clear that there is not any observable general trend here in terms of feature ranking and age groups, as Table 15.4 illustrates. These issues will be investigated and possible ways to manage them in a practical and sensible way will be briefly introduced in the next section.

Table 15.4 Feature ranking by using samples from age groups

Feature Number	Rank<26	Rank 26–60		Rank >60	
	S-Bio	S-Bio	S-Btg	S-Bio	S-Btg
F1	2	1	2	1	2
F3	1	2	11	6	3
F6	4	4	3	5	11
F7	13	6	13	4	1
F9	3	13	5	2	5
F10	5	3	1	3	13
F13	11	11	9	11	6
F15	6	5	10	13	4
F18	7	7	4	8	7
F19	15	8	6	15	12
F20	8	15	7	7	8
F23	14	10	12	9	9
F27	12	12	8	10	10
F31	10	14	15	12	15
F32	9	9	14	14	14

15.6.2 Managing database effects in terms of biometric features

In the light of the above findings, in order to explore how these feature rankings affect the error rate of a signature recognition system with respect to each dataset and across the chosen age groups, the error rate for each age group for each dataset is evaluated using an incremental process of adding one ranked feature at a time to the current operational feature vector.

In order to evaluate error rates, all the available samples are classified to determine an estimate of the error rates incurred in each of the age-bands defined in section 15.6.1, using the tools described in section 15.3.3. The error rate of each age group is defined as the percentage of incorrectly matched test subjects belonging to that age group out of all test subjects of that age group.

The error rates generated from samples from the three different age groups using age-specific ranked features are presented in Figures 15.9 and 15.10 for the S-Bio and S-Btg databases respectively. For both databases, if we consider the error rates

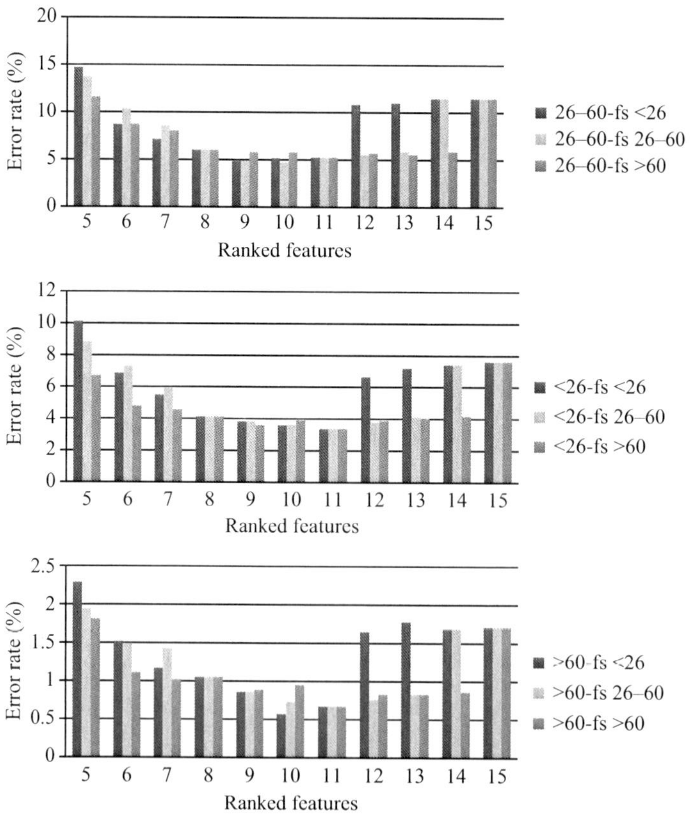

Figure 15.9 Error rate of each age group with the feature selection (fs) based on samples from different age groups for S-Bio

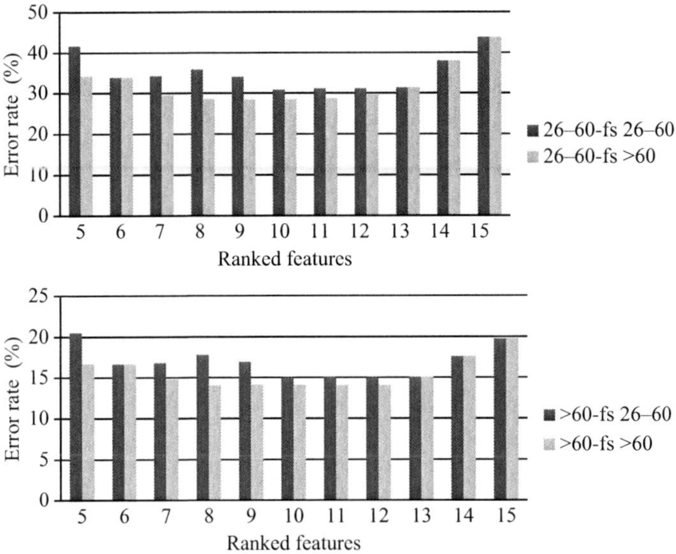

Figure 15.10 Error rate of each age group with the feature selection (fs) based on samples from different age groups for S-Btg

generated based on feature sets comprising respectively the first 5, 6 and 7 ranked features, there is no consistent trend with respect to the different age groups for all age-specific ranked features. In other words, selecting features from different age groups may decrease or increase the error rate for samples within the corresponding age-band. This further emphasises how difficult it is to discern absolute trends which relate attainable error rate to age group.

However, for all datasets, whatever sample population (i.e. regardless of age group) is used for feature selection, if the optimum number (around 11) of features is used in executing the recognition process for each age group, approximately the same error rates are achieved. This suggests that a powerful strategy for optimising performance reliability is to use feature selection without regard to age group and then optimise the feature set size in order to remove the potential variability in individual feature impact across age grouping and databases.

In order to emphasise further and more clearly demonstrate the value of our experiments, the error rate with respect to the whole population for each dataset is evaluated by using age-specific ranked features, adopting exactly the procedure as explained above, based on an incremental process of adding one ranked feature at a time to the operational feature vector. While testing on the whole population, we show the results when the training patterns are chosen respectively from each of the three age groups.

The results are shown in Figure 15.11, where it can be seen that for all datasets and for all age-specific feature selections forming the training set, a minimum error rate is achieved when the number of ranked features included in the active feature

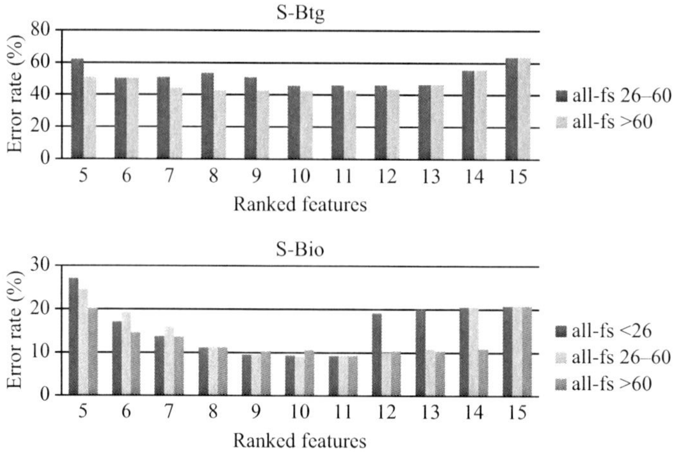

Figure 15.11 Error rate of all samples with the feature selection (fs) based on samples from different age groups

vector is around 10 or 11. This is a further illustration of the value of this type of investigation, since it shows how a practical population which is heterogeneous with respect to the age profile of its members can be processed effectively despite the perturbations in performance which age factors can otherwise introduce.

This investigation also provides some further insight into issues concerning the practical management of template ageing in signature-based biometrics. Our results suggest, for example, that the need for frequent template updating with physical ageing might be a less sensitive issue than might have been imagined, although further work is necessary in this respect, especially with regard to tracking changes across age for individual signers.

15.7 Conclusions

In this chapter, we have presented and discussed a detailed investigation of a methodical approach to exploring the impacts, limitations and challenges involved in the analysis of ageing issues with respect to the performance of biometric systems. We have tried to identify several of the most important factors which need to be considered in practical studies of ageing effects and provided some pointers to how this might help us to manage the ageing issue, especially as we look to the longer term future.

Recognising the difficulty of ageing as a continuous and incremental process, which is by no means a uniform process across populations, we have investigated and analysed the effects of choosing different age-bands to adopt in trying to characterize changes which occur as a result of the physical ageing process. Such a choice is sometimes made randomly, but is almost always, it seems, a relatively arbitrary decision, or one based on the desire for comparisons, and not generally considered on the basis of how well it reflects system sensitivity to age thresholds. Our results have illustrated methodically and quantitatively, for both a physiological and a behavioural modality, how different choices can lead to very different performance levels and thus different conclusions, limiting our understanding of age-related performance trends and how to manage these. Additionally, we have proposed a way to determine an age-banding structure which most accurately reflects age-related trends. Our proposed approach can be used as a guide in the determination of age-bands for any kind of age-based analysis of biometric data processing.

Although, of course, other modalities may require further analysis, our experimental results based on the iris and signature modalities suggest an age-band structure (i.e. '<25', '25–60' and' >60') which appears best to reflect age-related trends in performance for these 2 specific modalities, providing useful information to support both the analysis and practical management of age-related factors in biometric systems.

Furthermore, we have investigated how certain important characteristics of an adopted biometric database for any empirical study of ageing issues may have an impact on the analysis in terms of biometric data quality or the nature of the features extracted from the biometric samples. This analysis and discussion also shows that it is necessary to analyse ageing issues for different biometric modalities by considering their differences and adopting an appropriate experimental approach.

References

[1] R. Guest, 'Age dependency in handwritten dynamic signature verification systems', *Pattern Recognition Letters,* vol. 27, pp. 1098–1104, 2006.

[2] H. Beigi, 'Effects of time lapse on speaker recognition results', *16th International Conference on Digital Signal Processing 2009*, Santorini-Hellas; pp. 1–6, 2009.

[3] M. Fairhurst and M. Erbilek, 'Analysis of physical ageing effects in iris biometrics', *Computer Vision, IET*, vol. 5, pp. 358–366, 2011.

[4] S. K. Modi, S. J. Elliott, J. Whetsone and K. Hakil, 'Impact of age groups on fingerprint recognition performance', *IEEE Workshop on Automatic Identification Advanced Technologies 2007,* Alghero; pp. 19–23, 2007.

[5] J. Chapran, M. C. Fairhurst, R. M. Guest and C. Ujam, 'Task-related population characteristics in handwriting analysis', *Computer Vision, IET*, vol. 2, pp. 75–87, 2008.

[6] M. Abreu and M. Fairhurst, 'Analyzing the impact of non-biometric information on multiclassifier processing for signature recognition applications', *2nd IEEE International Conference on Biometrics: Theory, Applications and Systems 2008. BTAS 2008*, Arlington, VA; pp. 1–6, 2008.

[7] K. Ricanek Jr. and T. Tesafaye, 'MORPH: A longitudinal image database of normal adult age-progression', presented at the *Proceedings of the 7th International Conference on Automatic Face and Gesture Recognition*, Southampton, 2006.

[8] R. Singh, M. Vatsa, A. Noore and S. Singh, 'Age transformation for improving face recognition performance', *Pattern Recognition and Machine Intelligence*, in A. Ghosh, R. De and S. Pal (Eds.). vol. 4815, Springer, Berlin/Heidelberg, pp. 576–583, 2007.

[9] W. Ho, P. Watters and D. Verity, 'Are younger people more difficult to identify or just a peer-to-peer effect', in *Computer Analysis of Images and Patterns*. W. Kropatsch, *et al.* (Eds.). vol. 4815, Springer, Berlin/Heidelberg, pp. 351–359, 2007.

[10] L. Yui Man, D. Bolme, B. A. Draper, J. R. Beveridge, G. Givens and P. J. Phillips, 'A meta-analysis of face recognition covariates', *IEEE 3rd International Conference on Biometrics: Theory, Applications, and Systems, 2009*, Washington, DC; *BTAS '09*, pp. 1–8, 2009.

[11] Y. Sun, J.-M. Zhang, W. Liang-Min, Z. Yong-Zhao and S. L. Song, 'A novel method of recognizing ageing face based on EHMM', *Proceedings of 2005 International Conference on Machine Learning and Cybernetics*. Guangzhou, China; pp. 4599–4604, 2005.

[12] N. C. Sickler and S. J. Elliott, 'An evaluation of fingerprint image quality across an elderly population vis-a-vis an 18–25 year old population', *39th Annual International Carnahan Conference on Security Technology 2005. CCST '05*, Las Palmas de Gran Canaria, Spain; pp. 68–73, 2005.

[13] BioSecure, 'Biometrics for secure authentication, FP6 NoE, IST-2002-507634', http:\\www.biosecure.info, 2004.

[14] J. Ortega-Garcia, J. Fierrez, F. Alonso-Fernandez, J. Galbally, M. R. Freire and J. Gonzalez-Rodriguez, 'The multiscenario multienvironment BioSecure Multimodal Database (BMDB)', *IEEE Transactions on Pattern Analysis and Machine Intelligence*, vol. 32, pp. 1097–1111, 2010.

[15] J. Ortega-Garcia, F. Alonso-Fernandez, J. Fierrez-Aguilar, C. Garcia-Mateo, S. Salicetti and L. Allano, 'Software tool and acquisition equipment recommendations for the three scenarios considered', Report No.: D6.2.1. Contract No.: IST-2002-507634, 2006.

[16] R. M. Guest, 'The repeatability of signatures', *Ninth International Workshop on Frontiers in Handwriting Recognition 2004. IWFHR-9 2004*, Washington, DC; pp. 492–497, 2004.

[17] M. Erbilek and M. Fairhurst, 'Evaluating iris segmentation for scenario optimisation', *4th International Conference on Imaging for Crime Detection and Prevention*, Kingston, London, UK; 2011.

[18] L. Masek and P. Kovesi, 'MATLAB source code for a biometric identification system based on iris patterns', *School of Computer Science and Software Engineering*, University of Western Australia, 2003.

[19] M. Abreu and M. Fairhurst, 'Improving identity prediction in signature-based unimodal systems using soft biometrics', In *Biometric ID Management and Multimodal Communication*. J. Fierrez, *et al.* (Eds.), vol. 5707, Springer, Berlin/Heidelberg, pp. 348–356, 2009.

[20] M. Fairhurst and M. Abreu, 'An investigation of predictive profiling from handwritten signature data', *10th International Conference on Document*

Analysis and Recognition 2009. Barcelona; *ICDAR '09*, pp. 1305–1309, 2009.

[21] L. L. Lee, T. Berger and E. Aviczer, 'Reliable online human signature verification systems', *IEEE Transactions on Pattern Analysis and Machine Intelligence*, vol. 18, pp. 643–647, 1996.

[22] H. Lei and V. Govindaraju, 'A comparative study on the consistency of features in on-line signature verification', *Pattern Recognition Letters*, vol. 26, pp. 2483–2489, 2005.

[23] M. Erbilek and M. Fairhurst, 'A framework for managing ageing effects in signature biometrics', *Biometrics, IET*, vol. 1, pp. 136–147, 2012.

[24] C. Spearman, 'The proof and measurement of association between two things', *American Journal of Psychology*, vol. 15, pp. 72–101, 1904.

[25] J. Galbally, J. Fierrez, M. Martinez-Diaz and R. Plamondon, 'Quality analysis of dynamic signature based on the sigma-lognormal model', *2011 International Conference on Document Analysis and Recognition (ICDAR)*, pp. 633–637, 2011.

[26] N. D. Kalka, Jinyu Zuo, N. A. Schmid and B. Cukic, 'Estimating and fusing quality factors for iris biometric images', *IEEE Transactions on Systems, Man and Cybernetics, Part A: Systems and Humans*, vol. 40, pp. 509–524, 2010.

[27] N. D. Kalka, J. Zuo, A. N. Schmid and B. Cukic, *Image Quality Assessment for Iris Biometric*. vol. 6202. ETATS-UNIS: Society of Photo-Optical Instrumentation Engineers, Bellingham, WA, 2006.

[28] E. Krichen, S. Garcia-Salicetti and B. Dorizzi, 'A new probabilistic iris quality measure for comprehensive noise detection', *First IEEE International Conference on Biometrics: Theory, Applications, and Systems, 2007. BTAS 2007*, Crystal City, VA; pp. 1–6, 2007.

[29] L. Guangmmg, Q. Jiayin and L. Qingmin, 'A new scheme of iris image quality assessment', *Third International Conference on Intelligent Information Hiding and Multimedia Signal Processing, 2007. IIHMSP 2007*, pp. 147–150, 2007.

[30] J. Liang, S. Yang and A. Winstanley, 'Invariant optimal feature selection: A distance discriminant and feature ranking based solution', *Pattern Recognition*, vol. 41, pp. 1429–1439, 2008.

[31] R. Mergl, P. Tigges, A. Schröter, H.-J. Möller and U. Hegerl, 'Digitized analysis of handwriting and drawing movements in healthy subjects: Methods, results and perspectives', *Journal of Neuroscience Methods*, vol. 90, pp. 157–169, 1999.

Chapter 16

Usable biometrics for an ageing population

M. Angela Sasse and Kat Krol

16.1 Introduction

In this chapter, we examine how ageing affects human interaction with biometric systems. We focus on the ability and willingness of older users to engage with advanced technologies, and use them successfully. Ageing can change physical and behavioural characteristics, and this can impact performance – see Chapter 2 by Lanitis *et al.* in this volume; in this chapter, we focus on the change in general physical and cognitive capabilities, and how this impacts the ability to interact with the systems in which biometric recognition is embedded. If older users find it hard to present their biometrics correctly, or find the behaviour required to present unacceptable, this can lead to exclusion – this and other broader ethical and social implications of the issues we observe are examined in detail in Chapter 3 (Rebera and Mordini, this volume).

We conclude that the current usability challenges that biometrics presents for elderly users are not insurmountable – they can be overcome by systematic application of human–computer interaction knowledge and principles, and a determined empirical effort to improve performance through testing and adaption. Once this is achieved, biometric technology has the potential to be a key enabler of access to new technology and services for elderly users, who struggle with current security mechanisms such as passwords and CAPTCHAs. Biometric technology that can be adapted to meet requirements of older users and performs reliably would transform the interaction of older users with technology.

16.2 What is usability?

Usability is defined by the International Organization for Standardization (ISO) as 'The extent to which a product can be used by specified users to achieve specified goals with effectiveness, efficiency, and satisfaction in a specified context of use' (ISO, 2010).

UCL, UK

So to design a usable product or system, the design needs to fit with the characteristics of the target users, their goals and tasks, and the physical and social context in which they interact with that technology. We examine what each of these characteristics means for the usability of biometrics.

16.2.1 User characteristics

From ergonomics, usability has inherited the notion of general and specific user characteristics. *General user characteristics* are physical and mental characteristics shared by all – or rather, nearly all – humans, such as having two arms and ten fingers, being able to see, hear and speak. But there are three reasons why the idea of general physical and mental characteristics is problematic.

1. Within the general characteristics, there are ranges. Taking body height as an example, the difference between the tallest (2.51 metres) and shortest (0.54 metre) living person is nearly 2 metres. These are of course extremes, and traditional ergonomic design practice was to design the physical aspects of systems to accommodate that 95% of the population. Variations that can pose particular challenges for biometric systems include those in skin and eye colour. There are also ranges in perceptual capabilities, such as visual acuity and hearing, and mental ones – speed of processing, number of items a person can recall.
2. Even though the characteristics are supposed to be general, not all humans fit the template: some are born without certain limbs or digits, others may lose them during their lifetime because of accidents or illness. While a few decades ago, humans thus affected were labelled as 'disabled' and not expected to participate in many work or leisure activities (such as travel) in many parts of the world, we now regard them as differently abled, and adapt the physical environment to allow them to participate. This notion extends to technology – the principle of *universal access* states that technologies must be designed so they do not exclude users with disabilities. This is enshrined in both the ISO usability standard, and many countries compel public and commercial sector organisations to design facilities and technologies that allow access.
3. Some characteristics change during the lifetime of a human – humans grow from 'Fetus into Man' (Tanner, 1989) – in terms of height, physical maturity is reached around age 20, but different body tissues mature at different rates. Body height declines from the age of 50 onwards due to bone shrinkage, and the acuity of sight and vision can start to decline even earlier. As we shall see later, the age-related changes can affect both biometric characteristics themselves, and make it more difficult to operate the systems through which users interact with them. Lanitis *et al.* (in this volume) conclude that ageing affects the biometric templates of all biometric traits in different ways.

Given this level of variation, the notion of *general human characteristics* that a biometric solution could build on is starting to look problematic – at least for systems that need to accommodate a large, heterogeneous user population – for example, in a national context, biometrics are added to passports.

The more a system can be targeted to a specific user group and its capabilities, existing experiences and expectations, the more usable it will be. With homogeneous user populations, solutions can be developed to match the *specific user characteristics* of that group – if we design a system for, say, Formula 1 drivers, we would have to accommodate few of the variations above – those resulting ethnic differences being a notable exception.

Sometimes, making a system accessible for differently abled or elderly users can result in improved convenience that appeals to users beyond the target population. The television remote control was originally developed for users with reduced mobility, who had difficulty getting up to change the television channel. But the convenience it offered was so attractive to able-bodied users that Robert Adler from Zenith developed the prophetically named 'lazy bones' version for the wider consumer market, creating several generations of couch potatoes in the process.

Making a system senior-friendly and improving navigation performance among younger users as a side-effect has also been documented in web usability. In their study, Chadwick-Dias, McNulty and Tullis (2003) show that after a redesign of a website in line with the feedback received from older users, user performance improved for both younger and older participants in a subsequent study.

Other adaptions made for those with reduced capabilities do not appeal beyond the target user group. A good example is when consumer technologies are made simpler or bigger. Many older users find modern smartphones too complex to understand, and too small to operate effectively, so mobile phone companies have started to offer phones with larger displays and buttons that offer only voice calls. These 'simple' phones with reduced functionality will not appeal to a wider user group of younger users who want to access email and Facebook while on the move.

Whenever possible, designers must fit the system to the capabilities of the user, and not vice versa. In usability, fitting the user to the system, through the use of education and training for example, is always a secondary option, which is only deployed when technology cannot be adapted to human users, or the cost of doing so would outstrip the utility of the system.

16.2.2 User goals

Human behaviour is largely goal-driven. While there are some users for whom playing with technology is an enjoyable pastime, most use it in pursuit of work or private goals. Most older people who become silver surfers do so because they want to stay in touch with their grandchildren, clear out their garages, or find health information or cheaper medication online. Users carry out *tasks* they see as steps towards achieving those goals. Goals can be composed into *primary tasks* (chat to my grandchild via Skype, sell my collection of motorbike tools on eBay); but in the course of these, users are confronted with a number of *secondary* or *enabling* tasks (e.g., creating a Skype or eBay account). Users can understand that these are necessary, but try to get them done as quickly as possible, so they can return to the primary tasks on the path towards achieving their goals – meaning they will not take pride in doing it well, or see it as something worth investing time and effort in.

Security is a secondary task – users don't want to give it much time or attention. If its secondary tasks become a noticeable drain on resources or disrupt the path to the primary goal, users will resent it. If users are not able to complete a security task at all (something that happens to many confronted with a CAPTCHA, see section 16.4.2) they are barred from the service altogether – a Lose-Lose result for customers and service providers alike.

Whether a security mechanism 'fits' with primary tasks depends on a number of factors:

1. The frequency with which the mechanism is used. Adams and Sasse (1999) found that many people can manage relative complex passwords if they use them several times a day – with frequent usage, retrieval from memory becomes automatic.
2. The number of similar credentials or actions. Even frequently used passwords are likely to be confused by users who have many similar memory items, or in the first couple of days after changing it.
3. Time taken to complete primary and security task. The time and effort associated with a secondary task has to be proportionate to the primary task – see also the Compliance Budget below. Brostoff and Sasse (2000) found that the introduction of a service graphical authentication mechanism reduced access to the system by 60% compared to passwords (which users previously said they did not like), because it took too long to authenticate.
4. The disruptiveness of the secondary task. The way in which the security mechanism fits with the primary task is crucial – if it interrupts the user at a critical time of executing the primary task, or competes for the same physical and mental resources, it will affect performance and perception.

People make implicit cost-benefit assessments as they go through life. If the perceived effort involved in a secondary task exceeds the perceived benefits, they experience what Beautement, Sasse and Wonham (2008) call *friction*. Most users recognise that security mechanisms have some benefit for them or their organisation, but since they usually rate the risk differently from security experts, their perception of whether it is worth making that effort is different. Herley (2009) argues that users' assessment of risk makes more sense from an economic point of view than that of security experts – who 'value users' time at zero' and ignore the workload and disruption that cumbersome and complex security mechanisms create.

Most employees and consumers now have to comply with a multitude of security mechanisms – especially around authentication. Beautement *et al.* describe how, if there are several such tasks, friction accumulates and users approach their *compliance threshold*. When close to their compliance threshold, users will try to shortcut the security tasks (e.g., by choosing *c@t1c@2c@3c@t4* when told they have to have a 12 character cryptographically strong password). Such actions often reduce or cancel the effectiveness of the security mechanism. As described in section 16.4.1, the load of knowledge-based authentication has become such that many users now suffer from authentication fatigue, and this presents an opportunity for well-designed biometric solutions.

16.2.3 Context of interaction

Users interact with a system in a physical and social context, which can create constraints in terms of usability. For instance, as we discussed above, many users succeed in recalling and entering a long and complex, but frequently used password in a traditional office environment. Entering the same password on a touchscreen will take significantly longer, and fail more often, even with frequent use. The same voice recognition can work well in quiet environments, but fail when the user is in a place with high background noise. Physical and social environments constraints also affect how secure a mechanism is: while receiving entry feedback from your iPhone may improve accuracy, it becomes a threat when you have to enter it many times with colleagues looking over your shoulder.

16.2.4 Implications for design

The implications of the usability framework present a challenge to the designers of security systems today: they are used to choosing from a relatively small set of standard mechanisms. For authentication purposes, for instance, passwords continue to be widely used, even though the usability problems and impact on personal and organisational productivity have been documented (Adams & Sasse, 1999; Inglesant & Sasse, 2010). A 'usable security' authentication solution would mean selecting a mechanism that creates minimal physical or mental workload for users, does not disrupt them in the execution of their primary tasks, and can successfully operate and be operated in the physical and social context. Biometrics have the potential to offer reliable authentication with low physical and mental workload – especially if the biometric sample for recognition can be captured as a by-product of the activity users perform in pursuit of their primary goal (see section 16.4.1). If designers think along these lines, it would particularly benefit elderly users because it would also minimise the need to learn or remember anything about the system (see next section).

16.3 The impact of cognitive ageing on the use of technology

Akatsu and Miki (2004) stress that for a long time usability for elderly focussed on assisting with visual and physical impairments by increasing contrast, font size or making buttons easier to press. However, they postulate that in order to increase usability, it is necessary to also include cognitive considerations. They suggest a three-layer model addressing cognitive ageing and how it influences elderly people's use of technology (Figure 16.1).

16.3.1 Deterioration of cognitive capabilities

With age people's ability to acquire new skills declines; they struggle when they have to master a new skill, particularly if it has a high level of complexity. In what follows, we describe the five main signs of age-related cognitive decline and their impact on interaction with technology.

(c) Factors associated with cultural and social values

- Values, knowledge and framework for each generation.
- Not wanting others to find out about the deterioration in capabilities.

(b) Factors associated with a lack of knowledge and mental models

- Knowledge and mental models concerning particular modes of operation of equipment.
- Knowledge relative to the concept of the information itself.

(a) Factors associated with decline of cognitive capabilities of the elderly

- Deterioration of inhibition functions.
- Decrease in short-term memory capacity.
- Delays in comprehension.

Figure 16.1 Ease of use and cognitive ageing: a three-layered factor model by Akatsu and Miki (2004)

16.3.1.1 Longer reaction times

As we age our reaction times become slower. It is a well-researched phenomenon and its impact has been studied in many areas, most notably in driving (e.g., Makishita & Matsunaga, 2008; Cantin *et al.*, 2009).

Related to longer response times is the elderly's difficulty in collecting information quickly. This can be due to the fact that the space dimension of perception shrinks with age. Space dimension of perception is the amount of information that can occupy a person's attention at any one time (Worcester, Loustau & O'Connor, 1990). This quantity of information is subject to age-related changes and young people are able to absorb more information than older ones. Studies in the field have shown that younger people scored better when presented with unfamiliar information accompanied by distractions than the elderly. Nevertheless, elderly participants performed better when they were given less but familiar material with no distractions present (Botwinick, 1984).

For that reason, it is crucial that the elderly are given enough time when interacting with a system. In a study by Akatsu and Miki (2004), some elderly participants experienced timeouts when they took too long to complete the experimental tasks on an ATM. A timeout meant that they needed to start the task from the beginning. In comparison with university students, the elderly in the study required twice as long for the procedure of a fund withdrawal and three times as long for a fund transfer.

16.3.1.2 Attention

There are also significant changes in attention, especially in selective attention. Younger people have been shown to perform better in selective attention tasks than younger. Rogers (2000) showed that three factors reduce this difference: experience, cues and the heterogeneity of target-distracting objects.

16.3.1.3 Memory decline

Cross-sectional and longitudinal studies have shown a decline in working and episodic memory (Hedden & Gabrieli, 2004). With age, there are significant declines in abilities such as the encoding of new memories and the processing speed of new information. There is also a decline in working memory which is the simultaneous short-term maintenance and manipulation of information involving executive processes.

Usability studies involving cross-generational comparisons also showed that older participants received lower scores for verbal memory than younger participants (Ziefle & Bay, 2006). In a learning and memory test, younger participants received an average score of 13.3 out of 15 while older participants 11.4. This difference was found to be statistically significant ($p = .005$). Memory ability and age were found to be negatively correlated ($r = -.46$, $p = .008$).

Research by Johnson, Reeder, Raye and Mitchell (2002) has shown that the elderly have a deficit in information refreshing, that is, in the simple reflective operation of thinking of a just-presented item. The researchers stress that this can have an impact on other age-related cognitive phenomena. Activated but irrelevant items persist more for older than younger people. Older adults might lack the ability to suppress the activation of task-irrelevant information (Hasher, Zacks & May, 1999). Johnson *et al.* (2002) suggest that the failure to suppress irrelevant information may be related to the failure to activate relevant information. This is a possible explanation for why older participants sometimes become lost in system menus (Akatsu & Miki, 2004) or struggle when confronted with detailed information (Chadwick-Dias *et al.*, 2003).

16.3.1.4 Spatial abilities

Spatial ability consists of spatial memory and spatial scanning. Spatial memory is responsible for the mental visualising of the spatial location of some elements. Spatial scanning is the action of visually scanning through available information. With age, spatial abilities decrease in precision and quickness. In order to measure spatial abilities of different age groups, Ziefle and Bay (2006) used the paper-folding test in their study on the usability of mobile phone menus. While the younger participants had a mean performance of 13.4 (SD = 3.5) out of 20, the older ones had 9.6 (SD = 3.6). This difference was found to be statistically significant ($p = .004$). A correlation between spatial abilities and age showed that spatial abilities decrease as a function of age ($r = -.53$, $p = .002$).

In a study on navigation through a menu of interactive TV, Obrist, Bernhaupt, Beck and Tscheligi (2007) found that elderly participants explored the interface as fast as younger ones but they lacked accuracy. Elderly participants struggled with the use of navigational keys. They were always slower in viewing, searching and finding of information than younger participants. The authors attributed it to elderly people not being familiar with current conventions of information presentation and navigation. They also struggled due to visual impairments when the colour contrast was not high enough or the font size too small.

16.3.2 Lack of knowledge and mental models

Mental models are the user's mental representations of how a particular system works and how it is structured (Gentner & Stevens, 1983). Mental models are formed through increasing expertise and combining prior experience, cognitive schemata and problem-solving strategies (Johnson-Laird, 1983). Age and lack of experience make it less likely for users to develop a correct and useful model of a system (Calero-Valdez, Ziefle, Alagöz & Holzinger, 2010). Research has shown that a good mental model of a system improves user performance. In a study on a small-screen diabetes assistant, Calero-Valdez *et al.* (2010) found that having a simplified spatial representation of a device menu improves user performance. Even those having an inaccurate mental model performed better in the experimental tasks than those who did not have a mental representation of the system at all. The study also showed that previous experience with similar devices matters. Experience with mobile phones was significantly correlated with the quality of the arranged model of the diabetes assistant ($r = -.43$).

16.3.3 Cultural and social values

Within the realm of the cultural and the social, we identified three major factors that influence elderly people's interaction with biometric systems. These are (i) the tension between generations, (ii) habits and experience and (iii) beliefs and convictions.

16.3.3.1 Tensions between generations

Akatsu and Miki (2004) reported that some of their participants mentioned that they did not want to be a nuisance to others and did not want people around to become aware of their deteriorated cognitive capacities. Participants in a study by Sayago (2006) reported that the younger generation discouraged them from using computers. Some were told they might break them or delete important information.

16.3.3.2 Habits and experience

Elderly people often struggle with cash machines, thus they decide to go to a teller. Studies show that the older the customer, the more likely they are to go to a teller instead of using a cash machine (Akatsu & Miki, 2004).

When it comes to experience, studies have shown that elderly people do not realise the shortcomings of computer systems as younger people do. For example, in the ATM study by Akatsu and Miki (2004), a voice message prompting the user to enter their name was given at a time after they already did so. Nevertheless the elderly often started entering their name again, although it was already entered in the field there. A younger user would have realised that there is a glitch in the software and the machine did not register that they had already entered their name, but an elderly person did not question the system and complied with the request.

The elderly are used to more personal service, to speaking to a cashier, getting things explained and their questions answered. Nowadays, the industry is pushing towards automation of interactions in banking and retail. Manned tills gradually disappear and customers are encouraged to use automated check-outs. One might think it is becoming more anonymous but this is not entirely the case. Customers are anonymous in the sense that they do not have face-to-face interactions with the cashier, but they are not anonymous in the sense that the shop or the bank does not know who has just used their services. Because of identity verification in a bank or loyalty cards in a supermarket, these institutions have more information about their customers than they did in the past. Thus, elderly users are more used to face-to-face interaction, and they are likely to find the cold automated way of doing things unfriendly and invasive in the way that gathered information about their habits is stored and processed by machines rather than remembered by the shopkeeper who knew them by their first name.

Another factor here is the fact that elderly people crave human contact. If they have the option to choose between using a system that is difficult to understand and has too much information or a friendly helpful cashier providing them with all necessary information, they will always decide for a cashier. And this is true for young users too. Our recent study showed that in a work context employees are more likely to call up their helpdesk if they feel that looking for information online is not going to give them the answers quickly. If a service provider's aim is to automate their services they have to make these systems highly usable and make them have a clear advantage over manned counterparts.

While younger users tend to be more confident and would blame the system for hold-ups, older users would tend to blame themselves for failing to accomplish a task. Studies have shown that older users take longer to complete a set task because unlike digital natives they read all the information on the screen (e.g., Groff, Liao, Chaparro & Chaparro, 1999; Chadwick-Dias *et al.*, 2003). For that reason, instructions given should be concise to be comprehensible and quick to read for both younger and older users.

16.3.3.3 Beliefs and convictions

Older adults are more likely to exhibit computer anxiety (Ellis & Allaire, 1999). A range of studies has shown that the elderly do not understand the benefits of the use of computers (Czaja & Sharit, 1998; Melenhorst, Rogers & Caylor, 2001) and think that the benefits from using technology outweigh the effort that has to be put into learning how to use it (Holzinger, Searle, Kleinberger, Seffah & Javahery, 2008). In the case of biometric systems not only the unfamiliarity with technology could be a reason for resenting the use of biometric systems but also the association elderly users could have of fingerprints with criminality (Davies, 1994).

16.3.4 Behavioural implications

Studies into web usability and age have come up with lists of what is different about the elderly in terms of their interaction with devices. Chadwick-Dias *et al.* (2003) show that elderly users

- are more cautious to click on links,
- sometimes repeatedly click widgets that are not links,
- access highly detailed information and never move beyond it,

- struggle with web terminology and jargon,
- spend more time to read text and instructions,
- struggle to understand their location and path,
- have difficulty with window management, scrolling and tabbed navigation.

16.3.5 Improving elderly users' interaction with biometric systems

The cognitive challenges in the use of biometrics that we have highlighted can be attributed to two main factors, *ability* and *attitude*. Although not much can be done about the deteriorating cognitive capacities that impact ability in elderly users, the interfaces can be designed in order to make the interaction more intuitive and pleasant. Cognitive decline is a well-researched topic and we briefly highlighted its impact on the interaction with technology in section 16.3.1. There is an increasing body of research on how the elderly use technology and there are a number of guidelines for making interfaces senior user-friendly (Hodes & Lindberg, 2002). At the same time, attitudes towards technology have to do with beliefs and convictions which are shaped by preferences and experience. We suggest that the following four measures could help diminish the impact of cognitive decline and have a positive impact on changing attitudes towards biometric systems.

16.3.5.1 Training and feedback improve interaction

Research has shown that exposure, training and practice can make the elderly more successful in interacting with biometric systems. Theofanos, Micheals, Scholtz, Morse and May (2006) showed that when receiving feedback over the course of the study, the fingerprint quality of older participants improved significantly and at the end, prints of older women were of the best quality and also fewer attempts were needed to collect them.

Theofanos, Stanton, Orandi, Micheals, and Zhang (2007) conducted a study on how different instruction modes and an interaction with an operator influence the ease of collecting 10 fingerprints from users. They found that with operator assistance 98% of the participants were able to successfully provide their fingerprints. At the same time, only 56% of the participants who received poster instructions were successful. They also took longer than those receiving video or verbal instructions. The high success rate for participants receiving instructions from an operator shows that at least initially users do require face-to-face guidance and it can be highly effective.

Nevertheless, both studies mentioned above were conducted on a subpopulation of knowledge workers who were highly computer literate and motivated. The reality of where such a biometric reader would be deployed would be very different. As a usability person one has to also keep the worst case scenario in mind. In this situation, let us imagine an elderly person is waiting in a queue at an airport, the atmosphere is tense, they are under pressure to be quick to avoid embarrassment, they are unfamiliar with the technology, tired from the trip and overwhelmed. In this case, if something goes wrong, the user is 'rejected by the system' or they need elaborate instruction to present their biometric feature they will feel *they failed* and not *the system failed*.

Although as the studies mentioned above show, training can be very successful, elaborate instruction and explanation of how to use biometric systems should remain a secondary option. Striving for high system usability should be the solution. Being rejected by a piece of technology, ordered around or instructed, is simply bad user experience and can be embarrassing or even humiliating. It gives users, elderly or not, the impression that there is something wrong with them and they do not conform to a 'norm'. The truth is the technology has its shortcomings, either the back-end has problems with identity verification or the user interface is unusable.

16.3.5.2 Careful design makes seniors more successful

In general, devices should be designed to optimise some capabilities while compensating for others (Hart, Chaparro & Halcomb, 2008). Elderly users are unlikely to develop a new mental model for that reason, newly created devices and systems need to fit into existing requirements, preferences and thought patterns (Holzinger *et al.*, 2008).

Research has shown that pure compliance with guidelines is not sufficient. A study by Hart, Chaparro and Halcomb (2004, 2008) shows that sites designed with senior friendly guidelines in mind did result in higher task success but did not mean participants navigated through them quickly and efficiently. Compliance with these guidelines also did not result in higher user satisfaction or preference. For that reason, the researchers suggest using guidelines as a starting point and then combining them with usability testing to achieve a better result.

16.3.5.3 Practice improves performance

As mentioned earlier, some participants in the ATM study by Akatsu and Miki (2004) experienced a timeout and needed to start the operations from the beginning. Although this might have led to frustration, the time measurements show that the participants learned from their first unsuccessful attempt and were quicker at the second attempt. Also Ziefle and Bay (2006) found that older participants in their study performed distinctly better when completing the set tasks a second time. In comparison with the first run, their performance improved by an average of 6%. Their speed improved by 28%, it went from 14:26 in the first attempt to 10:17 in the second. The researchers also measured disorientation through returns in menu hierarchy and returns to the top. For hierarchical returns, there was a decrease from 33.1 to 28 (16%) in the 2 runs. For returns to the top, there was an improvement from 4.9 to 2.9 (41%) returns.

16.3.5.4 Exposure increases acceptance

Ideally, the more exposure and experience people have with computers, the more positive attitude they have. Coventry, De Angeli and Johnson (2003) describe that in a focus group prior to interaction with an ATM using iris verification, participants showed negative attitudes towards this technology. After interacting with the system, attitudes changed positively provided that the interaction was successful. The researchers suggest that there are several factors that could alleviate initial negative attitudes towards biometric technology, such as a marketing campaign, exposure to and experience with the technology. They suggest that enrolment could be a good opportunity to educate users.

However, Sasse (2007) stresses that there has to be consistency between the enrolment procedure and the subsequent use of the system. For example using the IRIS system at London Heathrow turned out to be confusing in this respect since the enrolment and subsequent use happen in two different environments. During the enrolment the individual sits down and the camera is adjusted to them, whereas during the actual use one is standing in front of a panel with three cameras and has to adjust to one of them. The adjustment is first downwards to bring one's face into the field of view of the camera and second sideways to position one of the eyes in the target circles. Finally, one has to move forward or backward to be at the prescribed distance.

This "bendy shuffle" to make it possible for the limited technology to capture one's biometric features illustrates a wider problem with biometric technologies. Namely, they assume users would adjust themselves to the piece of equipment rather than the equipment would be adjusted to the user. And in the specific case of iris scanners, the pieces of technology that adjust themselves to the user, by being height adjustable and having autofocus, have been proven to perform significantly better (Secunet, 2005). Nevertheless, Coventry and Johnson (1999) stress that more adaptive biometric scanners come at a higher price and tend to take up more space. For that reason, purchasers are generally more reluctant to buy them.

16.4 Opportunities for biometrics

16.4.1 0-effort, 1-step, 2-factor authentication?

In a recent study funded by the US National Institute of Science and Technology (NIST), the authors detailed how the ever-increasing burden of knowledge-based authentication was damaging productivity and undermining security (Steves, Chisnell, Sasse, Krol & Wald, 2013). The authors found many examples of increasing 'friction' first identified by Beautement *et al.* (2008) because of increasing mental and physical workload associated with authentication. They were particularly incensed at the frequency with which they had to authenticate to the same systems and services over and over again – 'If I authenticated and did not move away from the computer, it should know that and not bother me again'. Steves *et al.* also discovered that the disruptiveness of authentication on primary tasks meant several of their participants were frustrated by the 'insidious escalation of disruption over time on any one account or application'. Participants reported that they had opted out of using a number of technologies and services because of the disruptive nature of authentication. Considering the particular requirements of older users, disruption compounds the memory and concentration problems many users suffer from (see section 16.3.1).

Steves *et al.* conclude that the time has come for a massive shift from these disruptive, explicit forms of authentication to *implicit* forms of authentication, which recognise users without disrupting or bothering them, leveraging user interaction carried out as part of the primary task and the mode of interaction instead. Biometrics are the natural method of achieving this: users interacting with keyboards could be identified from their typing patterns (e.g., Maxion & Killourhy, 2010). Users of touchscreens on smartphones and tablets can be identified through their gestures or touch, while those using audio or video can be identified through voice or facial recognition. If combined with a knowledge-based credential (a simple password the user types or speaks, or a shape entered on the screen, biometrics can provide a zero (or very low) effort, one-step, two-factor authentication). With this type of authentication, we can remove the barrier of secondary task authentication that older users otherwise struggle with. But realising this vision requires increasing accuracy of performance of these biometrics, and predicting and futureproofing these biometrics against predictable ageing effects. Biometrics requiring constant re-enrolments would be as disruptive as passwords are today.

Briggs & Olivier (2009) have presented an intriguing idea of overcoming this problem: the *biometric daemon*. Inspired by the demons that every human character in Philip Pullman's Northern Lights is accompanied by, they propose a biometric device which is initially imprinted with the fixed biometric properties of its owner, and is then regularly updated with the fluid biometric properties of its owner. This would overcome the usability problems caused by current high false rejection rates. But what is particularly intriguing for older users is that they propose the demon could also act as an electronic pet 'which (i) part-shares identity with its owner, (ii) needs nurturing and (iii) effectively dies when separated from its owner for any length of time'. Since the benefits of pet ownership and interaction on the physical and mental well-being of older adults are well-documented, the idea of a Tamagochi-type token being looked after by this user group, and offering friction-free authentication in return, is one worth exploring.

16.4.2 Replacing CAPTCHAs

In section 16.2.2, we discussed that security mechanisms that users cannot cope with can bar them from access to important services. One security mechanism that falls into this category are CAPTCHAs – 'Completely Automated Public Turing Test to Tell Computers and Humans Apart' (CAPTCHA, 2013). Service providers use this test to stop botnets from creating accounts which are then used for malicious purposes – such as sending spam, or in the case of a large European low-cost airline – to prevent screen-scraping of its prices by web comparison sites. While the security needs of the websites are understandable, it makes all legitimate human users do extra work to access the sites. In a column in *Scientific American* published in February 2012, the technology writer David Pogue calculated that every day 17 man-years of effort are spent on decoding them (Pogue, 2012). In addition to increasing workload for some users – many of whom need 2–3 attempts to access a site – CAPTCHAs prevent a significant number of users, who struggle to decipher the squiggly warped letters from accessing the site altogether (Bursztein, Bethard, Fabry, Mitchell & Jurafsky, 2010) – and many of these are older users, among whom vision impairments are common (see section 16.2.1). To improve accessibility for users who have problems with sight, audio CAPTCHAs like SoundsRight (Lazar *et al.*, 2012) have been developed, but these do not necessarily offer an improvement for older users, and have also been shown to be easier to attack, providing a 'back door' for attackers as Bursztein and Bethard (2009) showed when they were able to overcome 75% of eBay's audio CAPTCHAs.

The usability problem with current CAPTCHAs presents an enormous opportunity for biometrics, proving that you are human using biometrics could be achieved in a low-effort, natural interaction. Since older users in particular struggle with CAPTCHAs, they would be chief beneficiaries – but as with the TV remote control (see section 16.2.1) the increased convenience is likely to be welcome by all users.

16.5 Conclusions

As this chapter has shown, usability and security have not rubbed along well for quite some time, and users have paid the cost. If target users are not able to use the system correctly, it is not accessible – for the ageing population, the increasing workload and complexity of security solutions create the risk of exclusion.

In the context of security, accessibility is particularly important to consider for at least two reasons:

1. Many security mechanisms act as 'gatekeepers', providing access to systems or services. Thus, if the mechanism is not usable, it can undermine the accessibility not only of the gatekeeper but also of a wider range of systems.
2. When users find a system difficult or impossible to use, many of them will recruit help from others: friends, family, or employees of a service provider, for instance. The vast majority of helpers will be trustworthy, well-intended, and act in the users' best interest, but the suggested assistance of others often undermines the assumptions made by the designers of the security mechanisms, and security itself.

As discussed in the previous section, the increasing use of devices and services requires authentication by all segments of the population, including the ageing population. All services – including government ones – now expect users to trust them online. The low usability of existing knowledge-based authentication mechanisms has led to authentication fatigue and habitual casual circumvention of security mechanisms. There is now widespread recognition that purely knowledge-based authentication is not secure: with cloud computing, passwords can be cracked in increasingly short time – and length and complexity of passwords required to withstand such attacks makes them unusable (Arnell *et al.*, 2012). Add to this the proliferation of consumer-owned devices which are impossible to secure (unless Trusted Computing suddenly is adopted widely), and the need for two-factor authentication become obvious.

Most current two-factor solutions are token-based, and anything but usable: most involve at least three steps as well as two factors; the user has to remember to carry the device – something that older people may find problematic.

This current state of affairs presents a significant opportunity for biometrics to come to the rescue, and offer convenient solutions that minimise user workload and create minimal disruption to their primary tasks. But in addition to meeting the specific characteristics of the ageing user population, solutions must also fit into the tasks and contexts in which these users operate. That fit means we want biometrics to 'disappear' into users' tasks, which suggests a range of different biometrics embedded in devices and services.

This suggests a large infrastructure project, which needs to be tackled sooner rather than later. The UK government's approach to joint public and private sector using federated identity (Government Digital Service, 2013) mentions that authentication must be accessible as well as secure, but it has not specified specific usability criteria, and makes no specific mention of biometrics. In our view, this is a mistake, since a significant infrastructure project will be required to deliver usable and secure authentication for all sections of the population, and it requires policy, not just technology.

As Challenger and Clegg (2001) pointed out:

> "When designing and operating any new system it is critical to focus on and optimise both technical and social factors (e.g. Cherns, 1976, 1987). It is inevitable that changes to one part of a system will necessitate subsequent changes to other parts; thereby, to optimise success, the system should be considered holistically (e.g. Hendrick, 1997; Clegg & Shepherd, 2007). Thus, people, processes and procedures, goals, culture, technology, and buildings and infrastructure should all be viewed as interdependent and given joint consideration."

It is hard to imagine that individual service providers will take this perspective. There must be a concrete set of information about the users and their goals and tasks that the design can be fitted to – defining a usable mechanism without first defining the system it is to be used in can only result in wasted effort.

References

A. Adams & M.A. Sasse (1999): 'Users are not the enemy: Why users compromise security mechanisms and how to take remedial measures'. *Communications of the ACM*, 42(12), 40–46.

H. Akatsu & H. Miki (2004): 'Usability research for the elderly people'. *Oki Technical Review* (Special Issue on Human Friendly Technologies); 71(3), 54–57.

S. Arnell, A. Beautement, P. Inglesant, B. Monahan, D. Pym & M.A. Sasse (2012): 'Systematic decision making in security management modelling password usage and support'. *Proceedings of the International Workshop on Quantitative Aspects in Security Assurance (QASA 2012)*. Pisa, Italy.

A. Beautement, M.A. Sasse & M. Wonham (2008): 'The compliance budget: Managing security behaviour in organisations'. *Proceedings of the 2008 Workshop on New Security Paradigms*. ACM, Lake Tahoe, CA. pp. 47–58.

J. Botwinick (1984): *Aging and Behavior: A Comprehensive Integration of Research Findings (2nd edn.)*. New York, NY: Springer.

P. Briggs & P.L. Olivier (2008): 'Biometric daemons: authentication via electronic pets'. *CHI'08 extended abstracts on Human Factors in Computing Systems*. Florence, Italy. ACM, pp. 2423–2432.

S. Brostoff & M.A. Sasse (2000): 'Are passfaces more usable than passwords? A field trial investigation'. *People and Computers XIV-Usability or Else!*, 405–424, Springer London.

E. Bursztein, S. Bethard, C. Fabry, J.C. Mitchell & D. Jurafsky (2010): 'How good are humans at solving CAPTCHAs? A large scale evaluation'. *2010 IEEE Symposium on Security and Privacy (SP)*. IEEE. pp. 399–413.

E. Bursztein & S. Bethard (2009): Decaptcha: Breaking 75% of eBay audio CAPTCHAs. *Proceedings of the 3rd USENIX Conference on Offensive Technologies*. USENIX Association, p. 8.

V. Cantin, M. Lavallière, M. Simoneau & N. Teasdale (2009): 'Mental workload when driving in a simulator: Effects of age and driving complexity'. *Accident Analysis & Prevention*, 41(4), 763–771.

A. Calero-Valdez, M. Ziefle, F. Alagöz & A. Holzinger (2010): 'Mental models of menu structures in diabetes assistants'. In: K. Miesenberger, J. Klaus,W. Zagler & A. Karshmer (Eds.), *ICCHP 2010*. LNCS, vol. 6180, pp. 584–591.

CAPTCHA (2013): http://www.captcha.net/

R. Challenger & C.W. Clegg (2011): 'Crowd disasters: A socio-technical systems perspective'. *Contemporary Social Science*, 6(3), 343–360.

L. Coventry & G.I. Johnson (1999): 'More than meets the eye! Usability and iris verification at the ATM interface'. In: S. Brewster, A. Cawsey & G. Cockton (Eds.), *Human-Computer Interaction: Interact 99*, IOS Press/IFIP, pp. 151–156.

L.M. Coventry, A. De Angeli & G. Johnson (2003): 'Usability and biometric verification at the ATM interface'. *Proceedings of the SIGCHI Conference on Human Factors in Computing Systems*. ACM, pp. 153–160.

S.J. Czaja & J. Sharit (1998): 'Age differences in attitudes toward computers'. *Journal of Gerontology*, 53B(5), 329–340.

S.G. Davies (1994): 'Touching Big Brother: How biometric technology will fuse flesh and machine', *Information Technology & People*, 7(4), 38–47.

R.D. Ellis & J.C. Allaire (1999): 'Modeling computer interest in older adults: The role of age, education, computer knowledge, and computer anxiety'. *Human Factors*, 41(3), 345–356.

D. Gentner & A.L. Stevens (1983): *Mental Models*. Hillsdale: L. Erlbaum.

Government Digital Service (2013): 'ID assurance', http://digital.cabinetoffice.gov.uk/category/id-assurance/

L. Groff, C. Liao, B. Chaparro & A. Chaparro (1999): 'Exploring how the elderly use the web'. *Usability News*, 1(2), 1–2.

T.A. Hart, B.S. Chaparro & C.G. Halcomb (2004): 'Designing websites for older adults: The relationship between guideline compliance and usability'. *Proceedings of the Human Factors and Ergonomics Society Annual Meeting* (vol. 48, no. 2, pp. 271–274). Sage New Orleans, Louisiana.

T.A. Hart, B.S. Chaparro & C.G. Halcomb (2008): 'Evaluating websites for older adults: Adherence to 'senior-friendly' guidelines and end-user performance'. *Behaviour & Information Technology*, 27(3), 191–199.

L. Hasher, R.T. Zacks & C.P. May (1999): 'Inhibitory control, circadian arousal, and age'. In: D. Gopher & A. Koriat (Eds.), *Attention and Performance XVII: Cognitive Regulation of Performance. Interaction of Theory and Application* (pp. 653–675). Cambridge, MA: MIT Press.

T. Hedden & J.D.E. Gabrieli (2004): 'Insights into the ageing mind: A view from cognitive neuroscience'. *Nature Reviews Neuroscience*, 5, 7–96.

C. Herley (2009): 'So long, and no thanks for the externalities: The rational rejection of security advice by users. *New Security Paradigms Workshop (NSPW).* Oxford, UK.

R.J. Hodes & D.A.B. Lindberg (2002): 'Making your website senior friendly'. http://www.nlm.nih.gov/pubs/checklist.pdf

A. Holzinger, G. Searle, T. Kleinberger, A. Seffah & H. Javahery (2008): 'Investigating usability metrics for the design and development of applications for the elderly, In: K. Miesenberger, J. Klaus, W.L. Zagler, A.I. Karshmer (Eds.), *ICCHP 2008*, Springer, Berlin Heidelberg.

P.G. Inglesant & M.A. Sasse (2010): 'The true cost of unusable password policies: Password use in the wild, *Proceedings of the 28th International Conference on Human Factors in Computing Systems (CHI'10)*. Atlanta, USA. pp. 383–392.

International Organization for Standardization (ISO) (2010): ISO 9241-210: 'Ergonomics of human-system interaction—part 210: Human-centred design for interactive systems, http://www.iso.org/iso/iso_catalogue/catalogue_tc/catalogue_detail.htm?csnumber=52075

P.M. Johnson-Laird (1983): *Mental Models: Towards a Cognitive Science of Language, Inference, and Consciousness*. Harvard University Press, Harvard.

M.K. Johnson, J.A. Reeder, C.L. Raye & K.J. Mitchell (2002): 'Second thoughts versus second looks: An age-related deficit in reflectively refreshing just-active information'. *Psychological Science*, 13, 64–67.

J. Lazar, J. Feng, T. Brooks, G. Melamed, B. Wentz, J. Holman, . . . & N. Ekedebe (2012): 'The SoundsRight CAPTCHA: An improved approach to audio human interaction proofs for blind users'. *Proceedings of the 2012 ACM Annual Conference on Human Factors in Computing Systems.* Austin, TX. ACM, pp. 2267–2276.

H. Makishita, & K. Matsunaga (2008): 'Differences of drivers' reaction times according to age and mental workload'. *Accident Analysis & Prevention*, 40(2), 567–575.

R.A. Maxion & K.S. Killourhy (2010): 'Keystroke biometrics with number-pad input'. In *IEEE/IFIP International Conference on Dependable Systems & Networks* (DSN-10, Chicago, IL. IEEE Computer Society Press, Los Alamitos, CA. pp. 201–210).

A.S. Melenhorst, W.A. Rogers & E.C. Caylor (2001): 'The use of communication technologies by older adults: exploring the benefits from the user's perspective'. *Proceedings of the Human Factors and Ergonomics Society 45th Annual Meeting*. Human Factors and Ergonomics Society, Santa Monica, CA.

M. Obrist, R. Bernhaupt, E. Beck & M. Tscheligi (2007): 'Focusing on elderly: An iTV usability evaluation study with eye-tracking. In: P. Cesar, K. Chorianopoulos, and J.F. Jensen (Eds.), 5th European Conference on

Interactive TV – EuroITV 2007 May 24, 25, 2007, Amsterdam, The Netherlands, pp. 66–75.

D. Pogue (2012): 'Time to kill off CAPTCHAs', *Scientific American*, http://www.scientificamerican.com/article.cfm?id=time-to-kill-off-captchas

W.A. Rogers (2000): 'Attention and ageing', In: D.C. Park and N. Schwarz (Eds.), *Cognitive Aging: A Primer* (pp. 57–73). New York, NY: Psychology Press.

M.A. Sasse (2007). 'Red-eye blink, bendy shuffle, and the yuck factor: A user experience of biometric airport systems'. *IEEE Security & Privacy*, 5(3), 78–81.

D.E.A. Sayago (2006): 'Some aspects of ICT accessibility, usability and design methods with the young elderly', http://www.dtic.upf.edu/~ssayag/dea_sayago_July06.pdf

Secunet (2005): 'Untersuchung der Leistungsfähigkeit von Biometrischen Informationssystemen—BioPII' (A study of the performance of biometric systems), Öffentlicher Abschlussbericht, https://www.bsi.bund.de/ContentBSI/Publikationen/Studien/biop/BioPII.html

M. Steves, D. Chisnell, M.A. Sasse, K. Krol & H. Wald (2013): *Report: Authentication Diary Study*, National Institute of Standards and Technology, Gaithersburg, MD, USA.

J.M. Tanner (1989): *Foetus into Man: Physical Growth from Conception to Maturity*. 2nd edn. Castlemead, Ware.

M. Theofanos, R. Micheals, J. Scholtz, E. Morse & P. May (2006): 'Does habituation affect fingerprint quality?' *Proceedings of the 24th International Conference on Human Factors in Computing Systems (CHI'06)*. Montréal, Canada.

M. Theofanos, B. Stanton, S. Orandi, R. Micheals & N.F. Zhang (2007): 'Ten-print fingerprint capture: Effect of instructional modes on user performance'. *Proceedings of the Human Factors and Ergonomics Society Annual Meeting* (vol. 51, no. 10, pp. 597–601). Sage. New Orleans, Louisiana.

M.I. Worcester, A. Loustau & K. O'Connor (1990): 'Tailoring teaching to the elderly in home care', *Home Health Care Services Quarterly*, 11(1–2), 69–120.

M.I. Worcester (1999): 'Tailoring teaching to the elderly in home care', *Home Health Care Services Quarterly*, 11(1–2), 69–120.

M. Ziefle & S. Bay (2006): 'How to overcome disorientation in mobile phone menus: A comparison of two different types of navigation aids'. *Human Computer Interaction*, 21(4), 393–432.

Chapter 17

Ageing effects and implications for biometric template protection

Christian Rathgeb[1], Andreas Uhl[1] and Christoph Busch[2]

17.1 Introduction

Ageing causes modifications of biometric features which affect the performance of a biometric system [25]. Long time intervals between the initial registration (enrolment) of a subject and authentication attempts might yield degrading biometric performance, referred to as the template ageing effect. Biological ageing, subject behaviour, sensor ageing, changes in place of acquisition, etc. represent the main reasons for biometric template ageing. In recent years biometric template ageing has received significant research attention and several investigations on diverse biometric characteristics such as face [24] or iris [11] have been conducted. It has been found that, in practice, biometric template ageing causes larger intra-class variabilities, due to biometric-dependent reasons, e.g. reduced skin elasticity and increased possibility of injuries in the case of fingerprints. As a consequence, FRRs are expected to increase for a constant decision threshold (see Figure 17.1(a) and (b)).

Most concerns about the use of biometrics arise from the storage and misuse of biometric data, e.g. tracking persons without consent. Biometric cryptosystems and cancellable biometrics represent emerging technologies of biometric template protection addressing these concerns and improving public confidence and acceptance of biometric systems [36]. Low intra-class variability at high inter-class variability is considered a fundamental premise of biometric template protection, i.e. biometric traits need to be captured under favourable conditions in order to provide practical recognition rates. Significant performance degradations have been reported on less constrained datasets [4]. It has been demonstrated that distinct template protection schemes appear robust in the presence of a moderate amount of noise [37]. So far, studies on the effects of ageing on template protection have remained evasive.

[1] University of Salzburg, Austria
[2] Center of Applied Security Research Darmstadt (CASED), Germany

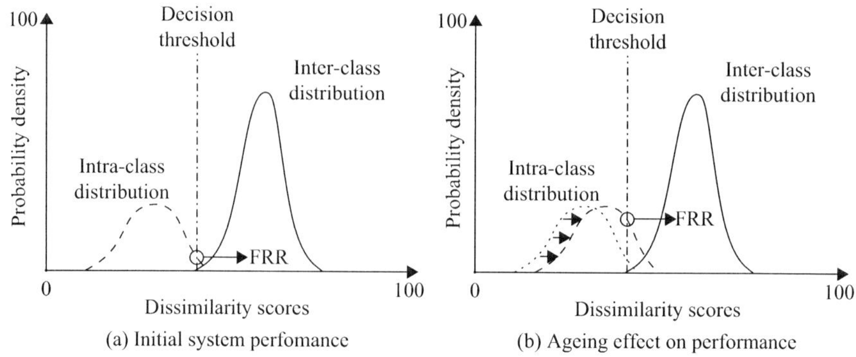

Figure 17.1 Biometric template ageing effect: intra-class variability increases over time

In this work the impact of template ageing on iris biometric template protection is investigated. Based on detailed theoretical analysis of possible effects and implications for template protection schemes, experiments are presented in the form of a case study on a biometric cryptosystem, in particular the fuzzy commitment scheme [20], in which chosen keys prepared with error correction information are bound to binary biometric feature vectors. Stability over a subject's lifetime has long been claimed as a key advantage of the iris [29], which represents a protected internal organ formed in the third month of gestation [26]. Recent studies have shown that for a state-of-the-art iris comparator, and a three-year time lapse, a 153% increase in the FRR was observed at a decision threshold corresponding to a one in two million false acceptance rate (FAR) [11], which might be considered a quite rigorous bound causing high FRRs.

The remainder of this work is organised as follows: Section 17.2 introduces biometric template protection and summarises the current state-of-the-art. Possible implications of ageing effects on biometric template protection schemes are discussed in section 17.3. Experiments based on iris-biometric cryptosystems are presented in section 17.4. A discussion is given in section 17.5. Conclusions are drawn in section 17.6.

17.2 Biometric template protection

The industry has long claimed that one of the primary benefits of biometric templates is that original biometric signals acquired to enrol a data subject cannot be reconstructed from stored templates. Several techniques, e.g. for fingerprints [5] or iris [45], have proven this claim wrong. Since most biometric characteristics are largely immutable, a compromise of raw biometric data or biometric templates results in permanent loss of a subject's biometrics. Biometric template protection technologies offer significant advantages to enhance the privacy and security of biometric systems, providing reliable biometric authentication at a high security level.

17.2.1 Categorisation

Biometric template protection schemes are commonly categorised as (1) biometric cryptosystems (also referred to as helper data-based schemes) and (2) cancellable biometrics (also referred to as feature transformation). Biometric cryptosystems are designed to securely bind a digital key to a biometric or generate a digital key from a biometric [6], offering solutions to biometric-dependent key-release and biometric template protection [7, 18]. Replacing password-based key-release, biometric cryptosystems bring about substantial security benefits. Due to biometric variance, conventional biometric systems perform ‘fuzzy comparisons’ by applying decision thresholds which are set up based on score distributions between genuine and non-genuine subjects. In contrast, biometric cryptosystems are designed to output stable keys which are required to match a hundred percent at authentication. Original biometric templates are replaced by biometric-dependent public information (auxiliary data – AD) which assists the key-release process.

Cancellable biometrics consist of intentional, repeatable distortions of biometric signals based on transforms which provide a comparison of biometric templates in the transformed domain [33]. The inversion of such transformed biometric templates must not be feasible for potential impostors. In contrast to templates protected by standard encryption algorithms, transformed templates are never decrypted since the comparison of biometric templates is performed in transformed space which is the very essence of cancellable biometrics.

Compared to existing biometric systems, biometric cryptosystems tend to reveal noticeably inferior performance [44]. This is because within a biometric cryptosystem the enrolled template is not seen and, therefore, cannot be adjusted for the direct comparison with a given biometric sample. In addition, biometric recognition systems are capable of setting more precise thresholds to adjust the tolerance of the system. Biometric cryptosystems and cancellable biometrics are designed to meet two major requirements of biometric information protection (ISO/IEC 24745): (1) irreversibility, i.e. it should be computationally hard to reconstruct the original biometric template from the stored reference data (protected template), while it should be easy to generate the protected biometric template; (2) unlinkability, i.e. different versions of protected biometric templates can be generated based on the same biometric data (renewability), while protected templates should not allow cross-matching (diversity). Illustrations of both properties are shown in Figure 17.2(a) and (b).

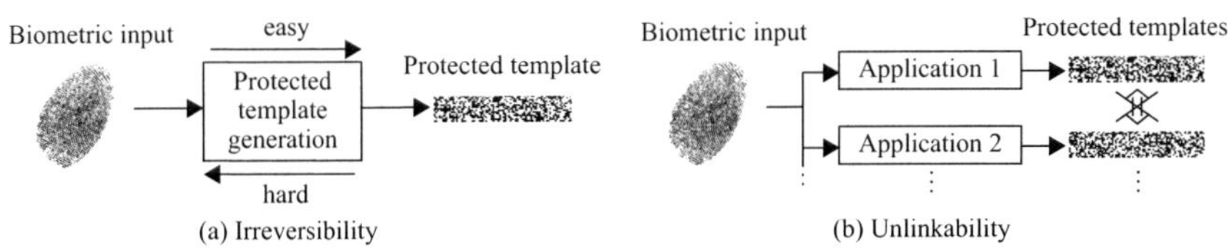

Figure 17.2 Biometric template protection: properties of irreversibility and unlinkability

17.2.2 The state-of-the-art

Focusing on the current state-of-the-art in biometric template protection, key approaches to biometric cryptosystems and cancellable biometrics are summarised in Tables 17.1 and 17.2 respectively. Representing one of the simplest key-binding approaches the fuzzy commitment scheme [20] has been successfully applied to iris [16] (and other biometrics). The fuzzy vault scheme [19] which represents one of the most popular biometric cryptosystem has frequently been applied to fingerprints. Early approaches (e.g. Reference 8), which required a pre-alignment of biometric templates, have demonstrated the potential of this concept. Several techniques (e.g. References 43 and 32) to overcome the shortcoming of pre-alignment have been proposed. Quantisation schemes (e.g. References 46 and 41) have been applied to several physiological and behavioural biometrics, and while focusing on reported performance rates, these schemes require further studies. Besides, approaches which aim at 'salting' existing passwords with biometric features have been proposed [30]. In the BioHashing approach [12] biometric features are projected onto secret domains applying user-specific tokens prior to a key-binding process. Variants of this approach have been exposed to reveal

Table 17.1 Experimental results of key approaches to biometric template protection

Reference	Applied Technique	Characteristic	FRR/FAR (%)	Remarks
[16]	Fuzzy commitment	Iris	0.42/0.0	Small test set
[3]			5.62/0.0	Short key
[8]	Fuzzy vault	Fingerprints	20–30/0.0	Pre-alignment,
[32]				>1 enrol sample
[48]			4.0/0.004	>1 enrol sample
		Iris	5.5/0.0	–
[10]	Quantisation	Online signature	28.0/1.2	>1 enrol sample
[46]			7.05/0.0	Short key
[30]	Password-hardening	Voice	>2.0/2.0	Short key
[42]	BioHashing	Face	0.0/0.0	Non-stolen token

Table 17.2 Experimental results of key approaches to biometric cancelable biometrics

Reference	Applied Technique	Characteristic	FRR/FAR (%)	Remarks
[34]	Block permutation,	Fingerprints	~35/10^{-4}	–
	surface folding		~15/10^{-4}	–
[2]	Revocable BioTokens		~0.08 EER	
[28]	BioConvolving	Online signature	10.81 EER	–
[13]	BioHashing	Face	0.0002 EER	Non-stolen token

impractical performance rates under the non-stolen-token scenario [23]. With respect to recognition rates, the vast majority of biometric cryptosystems are by no means comparable to conventional biometric systems. While numerous approaches to biometric cryptosystems generate rather short keys at unacceptable performance rates, several enrolment samples may be required as well (e.g. 4 samples in Reference 8). Approaches which report practical recognition rates are tested on rather small datasets (e.g. 70 persons in Reference 16) which must not be interpreted as significant. In addition, the introduction of additional tokens, whether random numbers or secret PINs, often clouds the picture of reported results.

First approaches to non-invertible transforms [33] (representing an instance of cancellable biometrics), which have been applied to face and fingerprints, include block-permutation and surface-folding. Diverse proposals (e.g. References 49 and 15) have shown that recognition performance decreases noticeably compared to original biometric systems while sample images of transformed biometric images render non-invertibility doubtable. BioHashing [12] (without key-binding) represents the most popular instance of biometric salting yielding a two-factor authentication scheme. Since additional tokens have to be kept secret (e.g. References 39 and 47), result reporting turns out to be problematic. Perfect recognition rates have been reported (e.g. in Reference 13), while the opposite has also been found to be true [23].

17.3 Age factors and template protection

As previously mentioned, the effects of template ageing do not coincide with the requirements of biometric template protection schemes. While template ageing causes increased intra-class variability [25], template protection schemes require low intra-class variability [36]. In this section possible effects of template ageing on the technologies of biometric cryptosystems and cancellable biometrics are discussed and consequential issues and challenges are summarised.

17.3.1 Ageing effects on biometric template protection

Biometric cryptosystems deal with the natural variability of biometric samples either by the application of error correction (e.g. in References 20 and 19) or by means of quantisation (e.g. in References 46 and 41). This coarse categorisation of biometric cryptosystems involves different types of implications of ageing effects:

- *Schemes based on error correction*: In order to handle increased intra-class variabilities error correction capacities could be expanded. However, the fact that FARs are lower bounded by error correction capacities [21] represents a great challenge since an unbounded use of error correction (if applicable) makes the system even more vulnerable [40].
- *Schemes based on quantisation*: Stronger quantisation (e.g. construction of larger intervals in Reference 46) could be applied in order to adapt the system to larger intra-class distances. Again, more quantisation causes security vulnerabilities since inter-class distances must not be expected to vary, i.e. while FRRs may be maintained FARs will increase at the same time.

Within the vast majority of cancellable biometric systems performance evaluations do not differ from that of underlying biometric recognition systems (e.g. in Reference 33). While transforms are applied to biometric signals, in most cases comparators (and feature extractors) remain unaltered (e.g. Hamming distance-based comparison for cancellable iris biometrics in References 49 and 15). Hence, effects of biometric ageing on generic biometric recognition systems can be directly transferred to cancellable biometrics, where additional challenges, e.g. template alignment, are expected to cause more drastic effects in terms of accuracy degradation.

17.3.2 Issues and challenges

Ageing factors raise several issues and challenges to be solved. In the case of a stronger overlap of genuine and impostor score distributions capabilities are quite limited. Depending on the type of template protection scheme an increased usage of error correction as well as stronger quantisation imply security vulnerabilities [40].

In order to tackle the problem at source, underlying recognition algorithms have to provide a certain resilience to ageing effects. Ageing-invariant feature extractors, e.g. age-independent face features [24], could serve as a stable source for template protection schemes. In order to extract or predict age-invariant biometric features (e.g. for fingerprints [17]) further modality-dependent investigations are required, since existing research is still *in statu nascendi*. Reliable component selection which represents a related approach is frequently applied within template protection schemes. Shortened feature vectors, composed only of the most reliable and discriminative features, are expected to maintain small intra-class variabilities. Subject-specific reliable component selection requires several enrolment samples or a periodic update of corresponding auxiliary information.

Another way to stem ageing effects is the incorporation of multiple biometric characteristics. Conventional multi-biometric systems have improved the accuracy and reliability of biometric systems [38]. Increased performance provided by multi-biometric recognition systems forms a suitable basis for long-term biometric recognition. Several approaches to multi-biometric template protection, which mostly utilise feature level fusion, have been proposed (e.g. in References 22 and 31). Possible countermeasures against ageing effects on biometric template protection are summarised in Table 17.3.

17.4 Effects on iris biometric cryptosystems: a case study

In order to investigate the effects of ageing and resulting implications on template protection on an empirical basis a case study based on iris biometrics is presented. Different types of iris-based fuzzy commitment schemes are evaluated on an iris database acquired within a time period of two years.

17.4.1 Experimental setup

Experiments are carried out on the ND-Iris-Template-Aging-2008–2010 database [11], which comprises good quality NIR-illuminated iris images captured within a

Table 17.3 Possible countermeasures of template protection schemes to ageing effects

Countermeasure	Description
Improved error correction	A more effective use of error correction can handle increased intra-class variability
Improved quantisation	Improved quantisation increases the flexibility of the system
Age-invariant feature extraction	Feature extractors generate age-invariant features (e.g. by predicting biometric data)
Reliable component selection	A pre-selection of most stable and discriminative (subject-specific) features
Multimodal template protection	In general a fusion of multiple biometrics increases recognition accuracy allowing more tolerant decision thresholds

Table 17.4 Applied datasets of the ND-Iris-Template-Aging-2008–2010 database

Subset	Subjects	Images	Short-term Comparisons	Long-term Comparisons
2008–2009	88	~4,500	~12,000	~30,500
2008–2010	40	~2,100	~5,800	~14,300

time period of two years, more precisely from spring 2008 to spring 2010. For time lapses of one (2008–2009 and 2009–2010) and two years (2008–2010) baselines are provided by according short-term comparisons. Short-term comparisons represent comparisons based on images of identical subjects captured after a short time period (a few months), i.e. each dataset compromises images captured within three acquisition sessions (initial, after short-term, and after long-term). Experimental evaluations are performed for a time lapse of one year (2008–2009) and two years (2008–2010). Table 17.4 summarises the corresponding number of subjects, images, and short- and long-term comparisons in these datasets (datasets have been slightly reduced due to some segmentation errors in the preprocessing stage). Figure 17.3 depicts sample images of two subjects captured in 2008, 2009, and 2010 respectively. Changes in iris texture patterns may not be recognisable via visual inspection.

17.4.2 Iris recognition system

In the preprocessing step the pupil and the iris of a given sample image are located applying Canny edge detection and Hough circle detection. More advanced iris detection techniques are not considered; however, as the same detection is applied for both datasets the obtained results retain their significance. Once the pupil and

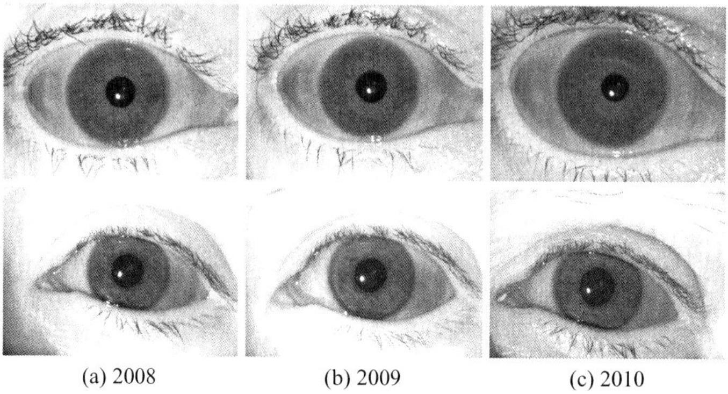

Figure 17.3 Sample images of the ND-Iris-Template-Aging-2008–2010 database captured in (a) 2008, (b) 2009 and (c) 2010

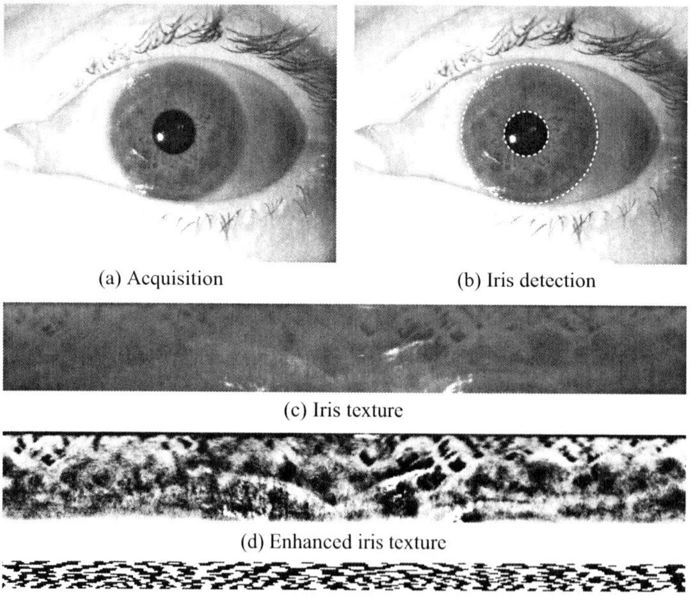

Figure 17.4 Processing chain: (a) image of eye, (b) detection of pupil and iris, (c) unrolled iris texture, (d) preprocessed iris texture and (e) sample iris code

iris circles are localised, the area between them is transformed to a normalised rectangular texture of 512×64 pixel, according to the 'rubbersheet' approach by Daugman [9]. As a final step, lighting across the texture is normalised using block-wise brightness estimation. For a sample image the entire preprocessing step is illustrated in Figure 17.4(a)–(d).

In the feature extraction stage a custom implementation[1] of the algorithm of Ma *et al.* [27] is employed. Within this approach the texture is divided into 10 stripes to obtain 5 one-dimensional signals, each one averaged from the pixels of 5 adjacent rows, hence, the upper 512×50 pixels of the preprocessed iris textures are analysed. A dyadic wavelet transform is then performed on each of the resulting 10 signals, and 2 fixed subbands are selected from each transform resulting in a total number of 20 subbands. In each subband all local minima and maxima above an adequate threshold are located, and a bit-code alternating between 0 and 1 at each extreme point is extracted. Utilising 512 bits per signal, the final binary feature vector comprises a total number of $512 \times 20 = 10240$ bits. Comparisons between pairs of images are performed by calculating the fractional Hamming distance of the appropriate iris codes. In order to compensate against small head tilts iris codes are circularly shifted in each direction (± 8 bits), seeking for an optimal alignment. For each dataset a balanced number of impostor comparisons is performed where pairs of iris codes of different subjects are chosen randomly.

For both time lapses, one and two years, increased FRRs are observed, plotted in Figure 17.5. While FRRs resulting from short-term comparisons remain almost identical a time period of one year causes a notable increase (see Figure 17.5(a)). An even more drastic increase is observed for a time lapse of two years (see Figure 17.5(b)). At the same time inter-class distributions remain almost unaltered. Table 17.5 summarises mean Hamming distances and corresponding standard deviations of inter-class distributions of each database. For long-term comparisons a slight decrease of means and a slight increase of standard deviations can be observed; changes are in the order of 10^{-4}.

Receiver operation characteristic (ROC) curves for both time lapses are plotted in Figure 17.6. For a FAR of 0.01% short-term comparisons reveal similar genuine acceptance rates (GARs) of 97.53% and 97.51%, resulting in EERs of 1.24% and 1.18%, respectively. In the case of a one-year time lapse a GAR of 96.28% is achieved at a FAR of 0.01% (see Figure 17.6(a)), resulting in an EER of 1.65%. While performance drops within the one-year period may still be acceptable, within the two-year period the GAR at a FAR of 0.01% drops to 89.39% and the EER increases to 3.47% (see Figure 17.6(b)). Obtained performance rates and observed degradations caused by ageing effects are summarised in Table 17.6. In contrast to the general thought that the iris pattern remains stable throughout a subject's lifetime, a drastic performance decrease is observed for a relatively short time period of two years.

As previously mentioned, sensor ageing or changes in the acquisition scenario (e.g. altered illumination) represent possible reasons for template ageing. It is not clear whether the observed biometric performance degradation is caused by biological ageing of iris textures (which may not appear obvious in the case of visual inspection) or other reasons such as sensor ageing or changes in the acquisition scenario.

[1] The University of Salzburg Iris Toolkit © USIT, http://www.wavelab.at/sources/

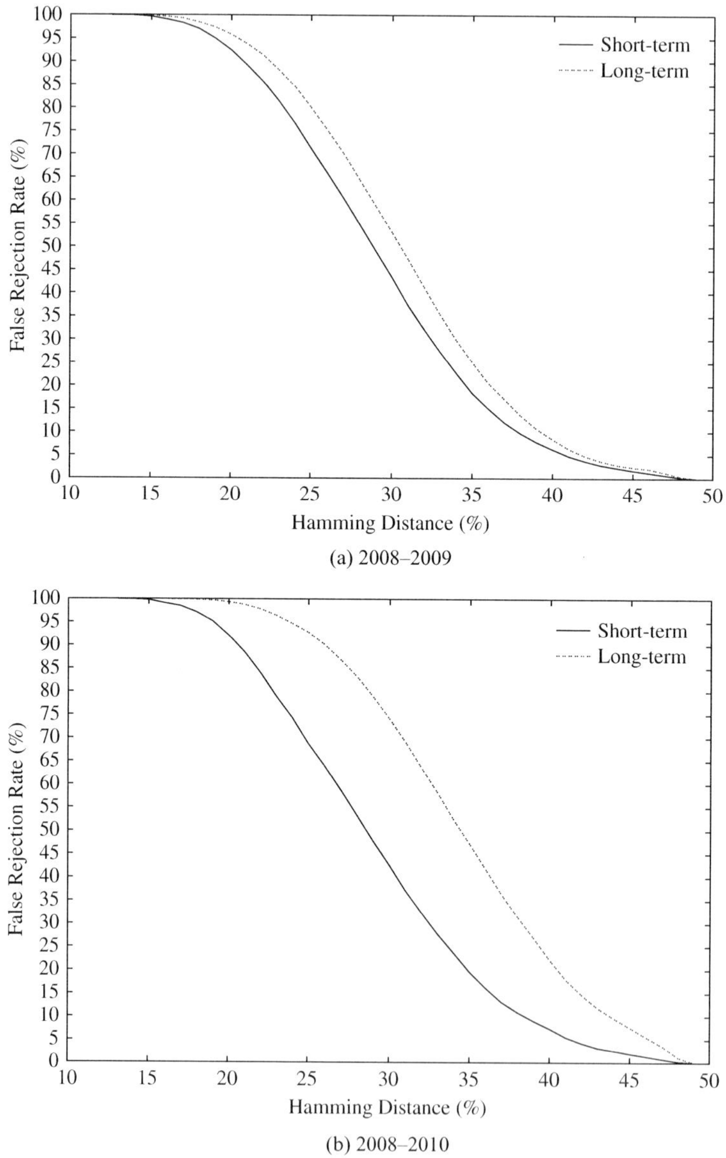

Figure 17.5 Increasing false rejection rates (FRRs) for time lapses of one and two years

17.4.3 Iris fuzzy commitments

The best performing iris biometric template protection schemes are based on the fuzzy commitment scheme [20] (see Table 17.1). Based on the previously described feature extractor two different fuzzy commitment schemes are applied. The

Table 17.5 Mean Hamming distances (HDs) and the corresponding standard deviations of inter-class distributions of each subset of the applied iris biometric ageing database

Dataset	Mean HD	Standard Deviation
2008–2009 short-term	0.48349	0.00964
2008–2009 long-term	0.48314	0.00980
2008–2010 short-term	0.48372	0.00946
2008–2010 long-term	0.48345	0.00985

first fuzzy commitment scheme follows the approach of Hao *et al.* [16]. In the original proposal a 140-bit cryptographic key is encoded with Hadamard and Reed–Solomon codes. While Hadamard codes are applied to correct natural variance between iris codes, Reed–Solomon codes handle remaining burst errors (resulting from distortions such as eyelids or eyelashes). For the applied algorithm of Ma *et al.* the application of Hadamard codewords of 128-bit and a Reed–Solomon code $RS(16, 80)$ reveals the best experimental results for the binding of 128-bit cryptographic keys [35]. At key-binding, a $16 \cdot 8 = 128$ bit cryptographic key R is first prepared with a $RS(16, 80)$ Reed–Solomon code. The Reed–Solomon error correction code operates on block level and is capable of correcting $(80-16)/2 = 32$ block errors. Then the 80 8-bit blocks are Hadamard encoded. In a Hadamard code, codewords of length n are mapped to codewords of length 2^{n-1} in which up to 25% of bit errors can be corrected. Hence, 80 8-bit codewords are mapped to 80 128-bit codewords resulting in a 10240-bit stream which is bound with the iris code by XORing both. Additionally, a hash of the original key $h(R)$ is stored as second part of the commitment. At authentication, key retrieval is performed by XORing an extracted iris code with the first part of the commitment. The resulting bit stream is decoded applying Hadamard decoding and Reed–Solomon decoding afterwards. The resulting key R' is then hashed and if $h(R') = h(R)$ the correct key R is released. Otherwise an error message is returned.

Again, template alignment is achieved by decommiting at various shifting positions. Since decision thresholds within a fuzzy commitment schemes are set at discrete intervals defined by error correction configurations, performance rates are plotted in terms of FRR and FAR at a distinct number of decoded block errors after bit-level error correction. FRRs and FARs for the fuzzy commitment scheme of Hao *et al.* and the related number of corrected block errors are plotted in Figure 17.7. Obtained FRRs for each dataset and according performance degradations caused by biometric ageing are summarised in Table 17.7. For each dataset the maximum block-level error correction capacities are exploited, i.e. 32 block errors. Compared to the original recognition system inferior performance rates are obtained, due to a non-ideal use of error correction and the non-ability of setting precise decision thresholds. At a FAR $\leq$0.01% a slight increase of $\sim$1% FRR is observed, 5.97–6.95%, for a time period of one year (see Figure 17.7(a) and(b)). In

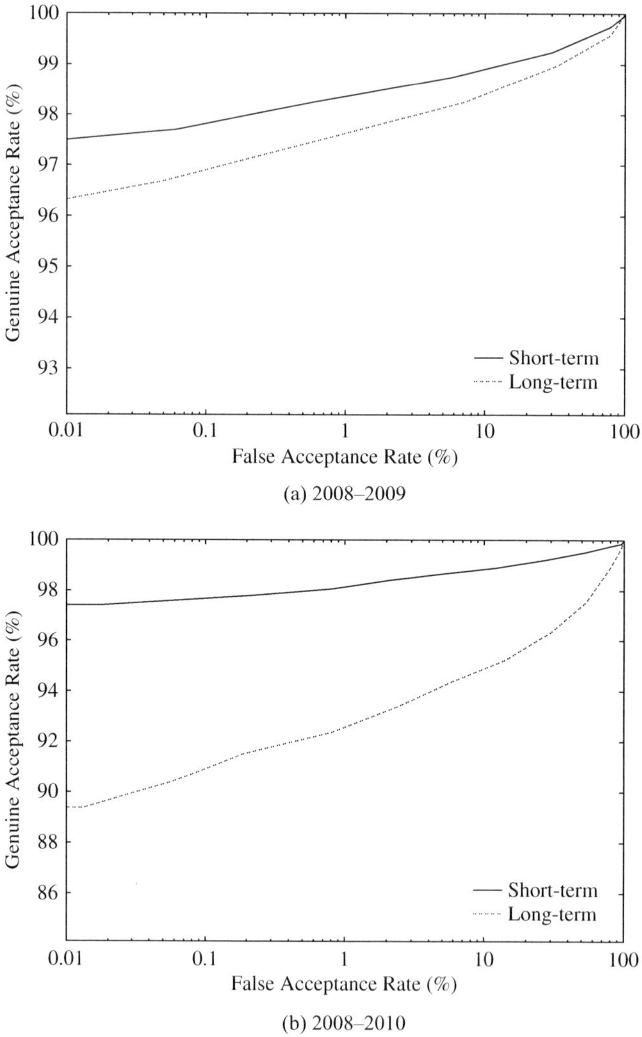

Figure 17.6 Receiver operation characteristic curves for time lapses of one and two years

the case of the two-year time lapse a more drastic effect of biometric ageing is observed. Compared to the original system, which exhibits a GAR of 89.39%, a FRR of 20.97% is obtained at a FAR $\leq$0.01% (see Figure 17.7(c) and (d)).

The second fuzzy commitment scheme was proposed by Bringer *et al.* [4]. Motivated by their observation that the system in Reference 16 does not provide the reported performance rates on datasets captured under unfavourable conditions, a more effective error correction decoding is suggested. The proposed technique which is referred to as Min-Sum decoding presumes that iris codes of 2,048 bits are

Table 17.6 Obtained performance rates for different time lapses and resulting degradations caused by iris biometric ageing (FRRs are measured at a FAR of 0.01%)

Time Lapse	Short-term		Long-term		Degradation	
	EER	GAR	EER	GAR	EER	GAR
2008–2009	1.24%	97.53%	1.65%	96.28%	0.41%	1.25%
2008–2010	1.18%	97.51%	3.47%	89.39%	2.29%	8.12%

arranged in a two-dimensional manner. In the original system a 40-bit key R is encoded with a two-dimensional Reed–Muller code such that each 64-bit line represents a codeword and each 32-bit column represents a codeword too. To obtain the helper data P the iris code is XORed with the two-dimensional Reed–Muller code. It is shown that by applying a row-wise and column-wise Min-Sum decoding the recognition performance comes near to practical boundaries. In order to adopt the system to the applied feature extractions, 8,192 bits of iris codes are arranged in 64 lines of 128 bits (best experimental results are achieved for this configuration). To generate the commitment, a 56-bit cryptographic key R is used to generate the error correction matrix. Since Reed–Muller codes are generated using Hadamard matrices and each line and each column of the resulting two-dimensional code has to be a codeword, $2^n + 1$ codewords define a total number of 2^{n+1} codewords. Due to the structure of the error correction code $2^{7 \cdot 8} = 2^{56}$ possible configurations of the $128 \times 64 = 8,192$-bit error correction code exist. At authentication a given iris code is XORed with the commitment and the iterative Min-Sum decoding is applied until the correct key R is retrieved or a predefined threshold is reached.

Performance rates, FRRs and FARs, and the related number of decoding iterations are plotted in Figure 17.8. Obtained rates, which are summarised in Table 17.8, confirm the claim that the fuzzy commitment scheme based on Min-Sum decoding reveals better results for challenging databases. For a time lapse of one year a FRR of 5.41% is obtained at a FAR $\leq$0.01% (see Figure 17.8(b)), outperforming the fuzzy commitment scheme based on two-stage error correction even in case of the short-term comparisons. While the size of the secret R is reduced, the second fuzzy commitment scheme almost maintains the accuracy of the original system for the short-term comparisons. The improved error correction based on iterative decoding allows a more precise adjustment of error correction capacities. Still, in the case of a two-year time lapse the FRR increases to 16.11% at a FAR $\leq$0.01% (see Figure 17.8(d)). As previously mentioned, the performance of the first fuzzy commitment scheme is lower bounded by the ability of correcting 32 block errors. For instance, if the first system were capable of correcting up to 39 block errors, FARs $\leq$0.01% would be maintained and FRRs would decrease to 3.35% and 3.91% for short-term and long-term time lapses of the one-year time period respectively. Again, for both fuzzy commitment schemes FARs remain rather

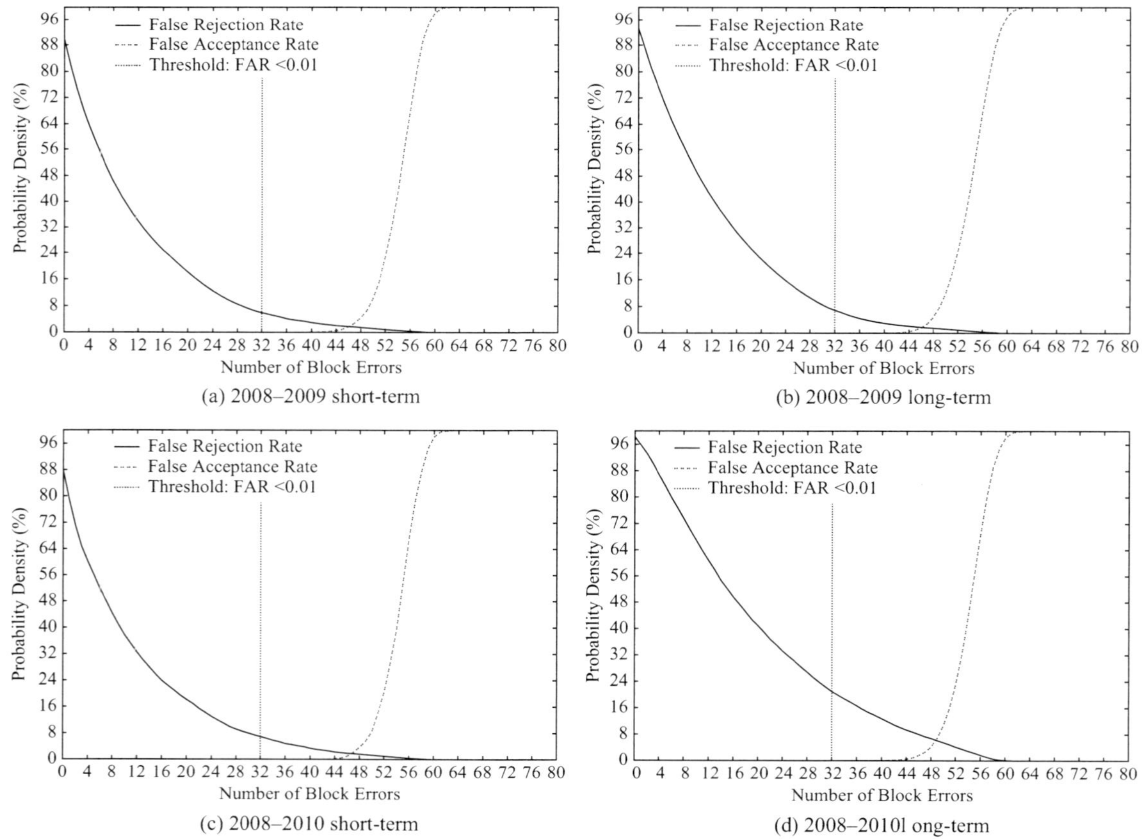

(a) 2008–2009 short-term

(b) 2008–2009 long-term

(c) 2008–2010 short-term

(d) 2008–2010l ong-term

Figure 17.7 Performance rates and the number of corrected block errors for the fuzzy commitment scheme of Hao et al. at time lapses of one and two years

Table 17.7 Obtained performance rates for the fuzzy commitment scheme of Hao et al. and resulting degradations (FRRs are measured at a FAR ≤0.01%)

Time Lapse	Short-term		Long-term		Degradation FRR
	FRR	Correlation blocks	FRR	Correlation blocks	
2008–2009	5.97%	32	6.95%	32	0.98%
2008–2010	6.95%	32	20.97%	32	14.02%

stable over time periods. Obviously, ageing factors reveal adverse consequences for template protection schemes, which appear more sensitive to increased intra-class variabilities.

17.5 Discussion

Ageing factors, which undoubtedly affect the recognition accuracy of the vast majority of existing biometric systems [24], require further investigations in order to establish proper modality-dependent countermeasures. So far, research has mainly been focused on face biometrics (e.g. in Reference 14), where ageing prediction appears feasible. First commercial systems, e.g. NEC Facial Recognition,[2] incorporating ageing prediction have been introduced. As opposed to face biometrics where ageing tell-signs are obvious, e.g. wrinkles and sagging [1], for other biometric modalities, e.g. fingerprints or iris, the present knowledge on detailed effects of ageing is rather limited.

Template protection technologies represent a rather recent field of research (commercial deployments are still rare). In the presented experimental study it is confirmed that, as expected, template ageing significantly affects the performance of template protection techniques, in particular iris biometric fuzzy commitment schemes. It is demonstrated that ageing effects cause a drastic decrease of accuracy, even if the applied template protection scheme provides a flexible means of handling increased intra-class variabilities, which is considered a premise to assist an underlying biometric system. In order to fully cope with ageing effects within template protection schemes, ageing issues have to be solved for conventional biometric systems for the first time. In cases where biometric recognition systems are capable of handling biometric ageing effects, e.g. by predicting ageing effects or extracting age-invariant features, template protection can be incorporated. Meanwhile short-term re-enrolment of registered subjects appears mandatory, especially for biometric systems incorporating template protection schemes.

[2] NEC Facial Recognition, © NEC Corporation, Japan, http://www.nec.com/, retrieved July 2012.

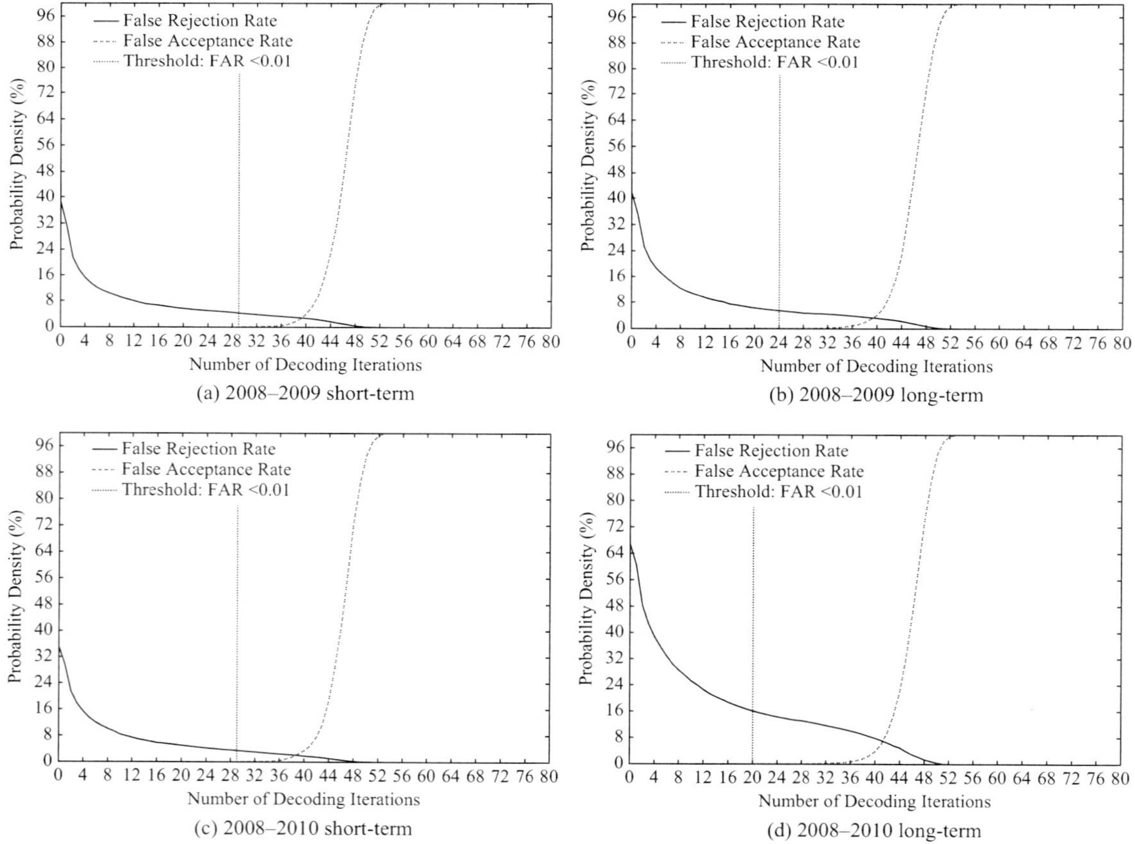

Figure 17.8 Performance rates and the number of corrected block errors for the fuzzy commitment scheme of Bringer et al. at time lapses of one and two years

Table 17.8 Obtained performance rates for the fuzzy commitment scheme of Bringer et al. and resulting degradations (FRRs are measured at a FAR ≤0.01%)

Time Lapse	Short-term		Long-term		Degradation FRR
	FRR	Iterations	FRR	Iterations	
2008–2009	4.22%	29	5.41%	24	1.19%
2008–2010	3.81%	29	16.11%	20	12.30%

17.6 Conclusion

In practice, biometric template protection schemes require favourable conditions, i.e. low intra-class variability [36]. Biometric template ageing, which has been most studied in the context of face recognition, causes increased intra-class variability [24], i.e. requirements on template protection do not coincide with ageing effects. In this work implications of ageing effects on template protection are discussed for various types of technologies and possible countermeasures are proposed. In a case study on iris biometric fuzzy commitment schemes the effects of ageing are investigated over a time period of two years. Experimental studies confirm recent investigations on iris template ageing effects [11], observing a significant decrease in the performance of fuzzy commitment schemes.

Acknowledgements

The authors would like to thank K. W. Bowyer and his group of the Computer Vision Research Laboratory (CVRL) at the Department of Computer Science and Engineering, University of Notre Dame, for providing the ND-Iris-Template-Aging-2008–2010 database [11].

References

[1] Albert, M., Sethuram, A., Ricanek, K.: 'Implications of adult facial aging on biometrics'. *Biometrics – Unique and Diverse Applications in Nature, Science, and Technology*, Southampton, UK, pp. 89–106. InTech (2011), Rijeka, Croatia

[2] Boult, T.: 'Robust distance measures for face-recognition supporting revocable biometric tokens'. *FGR '06: Proceedings of the 7th International Conference on Automatic Face and Gesture Recognition*, pp. 560–566 (2006)

[3] Bringer, J., Chabanne, H., Cohen, G., Kindarji, B., Zémor, G.: 'Optimal iris fuzzy sketches'. *Proceedings 1st IEEE International Conference on Biometrics: Theory, Applications, and Systems*, Washington, DC, pp. 1–6 (2007)

[4] Bringer, J., Chabanne, H., Cohen, G., Kindarji, B., Zémor, G.: 'Theoretical and practical boundaries of binary secure sketches'. *IEEE Transactions on Information Forensics and Security* **3**, 673–683 (2008)

[5] Cappelli, R., Lumini, A., Maio, D., Maltoni, D.: 'Fingerprint image reconstruction from standard templates. *IEEE Transactions on Pattern Analysis and Machine Intelligence* **29**(9), 1489–1503 (2007)

[6] Cavoukian, A., Stoianov, A.: 'Biometric encryption'. In: Stan Z. Li (Ed.), *Encyclopedia of Biometrics*. New York, NY, Springer-Verlag (2009), New York, NY, USA

[7] Cavoukian, A., Stoianov, A.: Biometric encryption: The new breed of untraceable biometrics. In: Biometrics: fundamentals, theory, and systems. Wiley (2009)

[8] Clancy, T.C., Kiyavash, N., Lin, D.J.: 'Secure smartcard-based fingerprint authentication'. *Proceedings ACM SIGMM 2003 Multimedia, Biometrics Methods and Applications Workshop*, Berkley, CA, pp. 45–52 (2003)

[9] Daugman, J.: 'How iris recognition works'. *IEEE Transctions on Circiuts and Systems for Video Technology* **14**(1), 21–30 (2004)

[10] Feng, H., Wah, C.C.: 'Private key generation from on-line handwritten signatures'. *Information Management and Computer Security* **10**(18), 159–164 (2002)

[11] Fenker, S.P., Bowyer, K.W.: 'Analysis of template ageing in iris biometrics'. *Proceedings of the IEEE Computer Society and IEEE Biometrics Council Workshop on Biometrics (CVPRW'12)* (2012). pp. 45–51

[12] Goh, A., Ngo, D.C.L.: 'Computation of cryptographic keys from face biometrics'. In: A. Lioy & D. Mazzocchi (Eds.), *Communications and Multimedia Security* (LNCS: 2828), Springer, New York, NY, pp. 1–13 (2003)

[13] Goh, A., Teoh, A.B.J., Ngo, D.C.L.: 'Random multispace quantization as an analytic mechanism for biohashing of biometric and random identity inputs'. *IEEE Transactions on Pattern Analysis and. Maching Intelligence* **28**(12), 1892–1901 (2006)

[14] Grother, P.J., Quinn, G.W., Phillips, P.J.: 'Report on the evaluation of 2d still-image face recognition algorithms'. NIST Interagency/Internal Report (NISTIR) – 7709, **56** pp. 2010

[15] Hämmerle-Uhl, J., Pschernig, E., Uhl, A.: 'Cancelable iris biometrics using block re-mapping and image warping'. *Proceedings of the Information Security Conference* 2009 *(ISC'09)* (LNCS: 5735), Pisa, Italy, pp. 135–142 (2009)

[16] Hao, F., Anderson, R., Daugman, J.: 'Combining cryptography with biometrics effectively'. *IEEE Transactions on Computers* **55**(9), 1081–1088 (2006)

[17] Hotz, T., Gottschlich, C., Lorenz, R., Bernhardt, S., Hantschel, M., Munk, A.: 'Statistical analyses of fingerprint growth'. *Proceedings of the International Conference of the Biometrics Special Interest Group (BIOSIG '11)*, pp. 11–19 (2011)

[18] Jain, A.K., Ross, A., Uludag, U.: 'Biometric template security: Challenges and solutions'. *Proceedings of the European Signal Processing Conference (EUSIPCO)*, Istanbul, Turkey (2005)

[19] Juels, A., Sudan, M.: 'A fuzzy vault scheme'. *Proceedings of the 2002 IEEE International Symposium on Information Theory*, Lausanne, Switzerland, p. 408 (2002)

[20] Juels, A., Wattenberg, M.: 'A fuzzy commitment scheme'. *6th ACM Conference on Computer and Communications Security*, Singapore, pp. 28–36 (1999)

[21] Kelkboom, E.J.C., Molina, G.G., Breebaart, J., Veldhuis, R.N.J., Kevenaar, T.A.M., Jonker, W.: 'Binary biometrics: An analytic framework to estimate the performance curves under Gaussian assumption'. *Transactions on System, Man, and Cybernetics – Part A: Systems and Humans* **40**(3), 555–571 (2010)

[22] Kelkboom, E.J.C., Zhou, X., Breebaart, J., Veldhuis, R.N.S., Busch, C.: 'Multi-algorithm fusion with template protection'. *Proceedings of the 3rd IEEE International Conference on Biometrics: Theory, Applications and Systems (BTAS'09)*, Washington, DC, pp. 1–7 (2009)

[23] Kong, A., Cheunga, K.H., Zhanga, D., Kamelb, M., Youa, J.: 'An analysis of BioHashing and its variants'. *Pattern Recognition* **39**, 1359–1368 (2006)

[24] Lanitis, A.: 'A survey of the effects of ageing on biometric identity verification'. *International Journal of Biometrics* **2**(1), 34–52 (2010)

[25] Lanitis, A., Tsapatsoulis, N.: 'Quantitative evaluation of the effects of aging on biometric templates'. *IET Computer Vision (Special Issue on Future Trends in Biometric Processing)* **5**, 338–347 (2011)

[26] Li, Y.H., Savvides, M.: 'Iris recognition'. In: N. V. Boulgouris, Konstantinos N. Plataniotis and Evangelia Micheli-Tzanakou (Eds.), *Biometrics: Theory, Methods, and Applications*, Hoboken, NJ, pp. 315–339. Wiley (2009)

[27] Ma, L., Tan, T., Wang, Y., Zhang, D.: 'Efficient iris recognition by characterizing key local variations'. *IEEE Transctions on Image Processing* **13**(6), 739–750 (2004)

[28] Maiorana, E., Campisi, P., Fierrez, J., Ortega-Garcia, J., Neri, A.: 'Cancelable templates for sequence-based biometrics with application to on-line signature recognition'. *Transactions on System, Man, and Cybernetics – Part A: Systems and Humans* **40**(3), 525–538 (2010)

[29] Monro, D.M., Rakshit, S., Zhang, D.: 'DCT-based iris recognition'. *IEEE Transactions on Pattern Analysis and Machine Intelligence* **29**(4), 586–595 (2007)

[30] Monrose, F., Reiter, M.K., Li, Q., Wetzel, S.: 'Using voice to generate cryptographic keys'. *Proceedings of the 2001: A Speaker Odyssey, The Speech Recognition Workshop*, **6**, pp. (2001)

[31] Nagar, A., Nandakumar, K., Jain, A.K.: 'Multibiometric cryptosystems based on feature level fusion'. *IEEE Transactions on Information Forensics and Security* **7**(1), 255–268 (2012)

[32] Nandakumar, K., Jain, A.K., Pankanti, S.: 'Fingerprint-based fuzzy vault: Implementation and performance'. *IEEE Transactions on Information Forensics And Security* **2**, 744–757 (2007)

[33] Ratha, N.K., Connell, J.H., Bolle, R.M.: 'Enhancing security and privacy in biometrics-based authentication systems'. *IBM Systems Journal* **40**, 614–634 (2001)

[34] Ratha, N.K., Connell, J.H., Chikkerur, S.: 'Generating cancelable fingerprint templates'. *IEEE Transactions on Pattern Analysis and Machine Intelligence* **29**(4), 561–572 (2007)

[35] Rathgeb, C., Uhl, A.: 'Systematic construction of iris-based fuzzy commitment schemes'. *Proceedings of the 3rd International Conference on Biometrics 2009 (ICB'09)* (LNCS: 5558), Alghero, Italy, pp. 947–956 (2009)

[36] Rathgeb, C., Uhl, A.: 'A survey on biometric cryptosystems and cancelable biometrics'. *EURASIP Journal on Information Security* **2011**(3), pp. 1–25 (2011)

[37] Rathgeb, C., Uhl, A.: 'Iris-biometric fuzzy commitment schemes under signal degradation'. *Proceedings of the 5th International Conference on Image and Signal Processing (ICISP'12)*, Springer (LNCS: 7340), New York, NY, pp. 217–225 (2012)

[38] Ross, A., Nandakumar, K., Jain, A.K.: *Handbook of Multibiometrics (International Series on Biometrics)*. Springer-Verlag, New York (2006)

[39] Savvides, M., Kumar, B., Khosla, P.: 'Cancelable biometric filters for face recognition'. *ICPR '04: Proceedings of the Pattern Recognition, 17th International Conference on (ICPR'04)* **3**, 922–925 (2004)

[40] Stoianov, A., Kevenaar, T., van der Veen, M.: 'Security issues of biometric encryption'. *Proceedings of the Toronto International Conference Science and Technology for Humanity (TIC-STH)*, Toronto, Canada, pp. 34–39 (2009)

[41] Sutcu, Y., Sencar, H.T., Memon, N.: 'A secure biometric authentication scheme based on robust hashing'. *MMSec '05: Proceedings of the 7th Workshop on Multimedia and Security*, New York, NY, pp. 111–116 (2005)

[42] Teoh, A.B.J., Ngo, D.C.L., Goh, A.: 'Personalised cryptographic key generation based on FaceHashing'. *Computers and Security* **2004**(23), 606–614 (2004)

[43] Uludag, U., Jain, A.K.: 'Fuzzy fingerprint vault'. *Proceedings of the Workshop: Biometrics: Challenges Arising from Theory to Practice*, Cambridge, UK, pp. 13–16 (2004)

[44] Uludag, U., Pankanti, S., Prabhakar, S., Jain, A.K.: 'Biometric cryptosystems: Issues and challenges'. *Proceedings of the IEEE* **92**(6), 948–960 (2004)

[45] Venugopalan, S., Savvides, M.: 'How to generate spoofed irises from an iris code template'. *Transactions on Information Forensics and Security* **6**, 385–395 (2011)

[46] Vielhauer, C., Steinmetz, R., Mayerhöfer, A.: 'Biometric hash based on statistical features of online signatures'. *ICPR '02: Proceedings of the 16th International Conference on Pattern Recognition (ICPR'02)*, Quebec, Canada, vol. 1, p. 10123 (2002)

[47] Wang, Y., Plataniotis, K.: 'Face based biometric authentication with changeable and privacy preservable templates'. *Proceedings of the IEEE Biometrics Symposium 2007*, Baltimore, MD, pp. 11–13 (2007)

[48] Wu, X., Qi, N., Wang, K., Zhang, D.: 'A novel cryptosystem based on iris key generation'. *Fourth International Conference on Natural Computation (ICNC'08)*, pp. 53–56 (2008)

[49] Zuo, J., Ratha, N.K., Connel, J.H.: 'Cancelable iris biometric'. *Proceedings of the 19th Internationl Conference on Pattern Recognition 2008 (ICPR'08)*, Tampa, FL, pp. 1–4 (2008)

Chapter 18

Future challenges for systems delivery using ageing tolerant biometrics

Peter Hawkes

A biometrically enabled transaction token was devised at the UK National Physical Laboratory as long ago as the late 1980s [1]. It was then envisaged that the use of this electronic token in, e.g. payment transactions, would be enabled by automatic personal identification methods. The methods suggested included the measurement of biometric characteristics such as 'written signatures, fingerprint and vein scanning'. It was suggested that the PIN be retained as a supplementary check on identity. Particular emphasis was given to the prospects for vein scanning as the basis of a new biometric [2].

These ideas were not followed up until the early 2000s. Then various Japanese companies including Hitachi and Fujitsu introduced vein scanning (vascular) biometric products for customer ID at cash dispensers [3]. In these systems the vein template is held in the chip card and not in a central database.

In contrast Toshiba has recently devised secure communications protocols which allow the template matching step to be carried out centrally. Known as ACBio these protocols have been standardised by ISO as ISO 2476.

More widespread adoption of vascular and other biometric methods of automatic personal identification now needs to overcome the apparent lack of a credible business case in payment and other applications where PINs are well established. 'Chip & PIN' payment card payment systems are a typical example.

The widespread use of Chip & PIN and the increase in B to C eCommerce have led to the need for consumers to remember large numbers of different PINs and passwords. Several possible early applications suitable for initial small-scale local deployment of biometric tokens are mentioned. If these take off commercially the prospects for the early introduction of multi-application biometric tokens should improve.

A possible new taxonomy for integrating biometric ID verification into applications is suggested. The proposed 'Ageing Adaptation Engine' (AAE) could be accommodated in a central server. It would communicate with client tokens via the ACBio protocols. However, the commercial introduction of ageing adapted

Consultant, previously with British Technology Group, UK

biometrics on these lines may well be stifled by the introduction of new privacy protection legislation.

18.1 Introduction

As shown in previous chapters the reference templates for many biometric methods need to be adjusted to compensate for effects arising from the ageing of the subject person and infrequent use. The challenge is to compensate for entropy increases as the subject person ages. This leads to increased recognition errors such as false non-match rates (FNMR).

As is evident from earlier chapters several compensation techniques are now available. More such techniques can be expected to appear in the near future.

Ideally the subject person needs to revisit the trusted enrolment centre for manually supervised re-enrolment. This is tolerable for the compulsory renewal of biometric passports once every ten years. However, it is unpopular with the public and expensive. The national government can compel the traveller to re-enrol and to pay the bill.

A possible alternative that would cost less and be more convenient to travellers is remote re-enrolment. Preferably the re-enrolment would generally be computer controlled with light manual supervision rather than 100% manually supervised. The use of such an automatic teleservice would be popular with both citizens and the government. Given adequate security it could eventually be used for the renewal of government identity documents such as passports and visas. Depending on the particular biometric and its sensitivity to ageing the use of the remote re-enrolment would be scheduled as frequently as is needed for optimum results such as minimum FNMR without excessive failure to recognise wanted persons.

For most private sector identity management services re-enrolment by teleservices or some other form of automatic in-service adaptation will generally be preferred to visiting a manually operated re-enrolment centre.

These template 'tuning' measures could be carried out by a generic AAE. To function an adaptation engine has to keep some records of past verification bids, successful as well as unsuccessful. So an extensive central database of user data is needed. Unless it can be anonymised this data retention will be unpopular with advocates of personal data protection.

If the security of AEE's database cannot be trusted then the major current identity management service providers such as the banks and government will have to continue to rely on PINs and passwords with all their known disadvantages for the consumer and citizen. Moreover, Fraud and ID theft will continue to flourish.

18.2 The main challenges

18.2.1 Realising the promised utility and convenience of biometric authentication despite ageing

To have enduring value to the citizen or consumer future Identity Management Services (IMS) need to incorporate dependable means for biometric authentication across a wide range of non-government applications in both the physical world and online.

Given that said means are ageing tolerant through the deployment of good AEEs, this option offers the attractive prospect of fewer passwords and PINs to remember and less frequent need for password changes.

For most short-term (transient) applications the available ageing compensation means of the AEE need not be invoked. Even for long-term applications complete elimination of PINs, passwords plus 'something else you know' would be unwise. There must always remain the fall-back option of using 'something you know' – a shared secret for emergency use.

18.2.2 Some generic requirements set by the context

The likely emergence of independent multi-application IMS which people will be prepared to pay to use is forecast for online eAuthentication by Kim Cameron of Microsoft [4]. Preferably such an IMS will also be used in the physical world. Hopefully there will emerge a number of competing IMS systems. Each IMS system must be compliant to the relevant personal data privacy and security laws in the territories where it operates. Trans-national IMS systems are highly desirable in the medium term, say within five years of introduction. Ideally the popular IMS systems will be those catering at a reasonable price per ID check for:

1. anybody as subject (no discrimination);
2. elderly subjects (inclusive);
3. difficult subjects, e.g. disabled persons (assistive features needed);
4. multiple applications.

In addition it will be preferable for subject persons to be able to use the same IMS for both applications in the physical world and online (eIMS).

The legal liabilities of the IMS service provider towards individual citizens, the subject of biometric records, must be minimised. This caveat will keep the lawyers busy for years. In the United Kingdom one must take into account the guidance of the Information Commissioner's Office [5]. When combined with guidance and possible directives from the EU the legal issues raised may even prove to be intractable. Closed IMS systems are more likely to appear than open systems.

Disregarding set-up costs for the moment, the operating cost per biometric authentication must be no higher than for today's password- or PIN-based chip card authentication. Moreover it should not take much more time. Here it is important to note that the time problem is largely solved for those biometric checks that can run in parallel with the transaction concerned. Examples include face recognition, handwritten signature recognition, vascular (vein ID) and speaker verification.

18.2.3 Standards compliance

Each IMS must operate in ways compliant with all relevant international standards. Most current biometric modalities are now the subject of ISO Standards from SC37. ISO also develops relevant IT security standards.

18.3 Why is it timely to plan for use of robust biometric authentication on a much wider scale than now?

(a) PINs and passwords are proving to be insufficiently secure [6] and, as already noted, they place a serious burden on the user's memory ('Password Hell').
(b) The progress of Moore's law allows chip design and supply companies like Intel and ARM Holdings to plan for dedicated tamper resistant *hardware* to be incorporated in chip cards and tokens based on smartphones or tablets.

18.4 Choice of biometric sub-systems

These will be expected to meet the above requirements based on:

(a) today's proven biometric modalities with known means for ageing adaptation/compensation as described in the earlier chapters of this book,
(b) tomorrow's promising new biometric methods.

18.5 Guiding principles

The following guiding principles need to be borne in mind:

18.5.1 Laws of identity

Designers of biometric tokens need to bear in mind Kim Cameron's Seven Laws of Identity [4].

18.5.2 Encryption methods

The IMS designers must use appropriate encryption methods for all message traffic needed for biometric authentication. They must also remember 'Encryption only Spreads Security' [7]. So the physical security of tokens remains important. Consequently the microchips employed as crypto-engines must be tamper-evident.

18.5.3 Secure hardware

Secure hardware platforms need to be employed. This is because hardware is generally more difficult for hackers to subvert than software. Within a year or two we can expect to see some of the microchip makers introduce tamper resistant microchip hardware that supports biometric scanning and authentication for use in tokens.

These tokens will be enhanced chip cards and 'smarter' mobile phones and tablets. To enable offline uses these tokens and their authorised reading terminals need to be able to carry out local biometric authentication. Client terminals will also need to support remote sample data capture with central template matching designs. The ACBio protocols devised by Toshiba meet this need. These protocols were standardised in 2009 as ISO 2476 [8]. This matching means could be 'cloud' based. If chip card tokens are used they will preferably be NFC capable. This will enable a smartphone to be used as the card reading terminal. If vascular biometrics is used the camera in the smartphone should be able to capture the palm creases and vein pattern in the front of the subject's hand using visible light (see References 8 and 9).

18.5.4 Location – where am I now? At home or 1000 miles away?

Obviously nobody can be in two places at once. So time/date stamping and the absolute location of persons seeking to authenticate themselves is highly desirable in the IMS. For smart card-based tokens location can be provided by a trusted terminal at the point of authentication. The time reference comes from the terminal's clock.

For a smart mobile phone or tablet used as a trusted token two independent location means are available. These are cell location address and GPS. The time reference comes either from the smart phone's on-board clock or the central server.

18.6 Scope for the independent IMS provider to operate a profitable business – the business case

The business model for providing a token-based IMS for use in cyberspace has hitherto been weak. Instead Internet service providers have each devised their own password- or PIN-based approach – known as the 'walled garden'. Hence the current state of 'password hell' is imposed on consumers and, increasingly, by governments on their citizens.

One established way of developing biometric authentication technology has been to employ it in applications which are viable in a local area and on a small scale. This approach will continue. One such example would be employee access control. This could be for logical access to computer systems or for physical access to restricted premises. Another application area for the introduction of biometrics starting on a small scale could be for detecting some types of fraud and identity theft now affecting UK elderly citizens. In many cases the identity thieves must sign on a document or suitable tablet the signature of the person whose identity they are stealing. So they are committing forgery and can thus be prosecuted under existing laws. Identifying and successfully prosecuting identity thieves will be a popular trend politically. The use of automatic handwritten signature verification need not be carried out on 100% of signed-for documents. So small-scale deployment of signature verification could be used at first. The necessary stylus-operated tablets are already available at reasonable cost. As with other biometric methods the legal issues involved need to be approved by the information commissioner [5].

In a related move focussed on young people Visa has just announced the 'Visa Prepaid Citizen Card'. This is a combined re-chargeable money storage card and proof of age card for young people. The 'proof of age' function is supported by a colour photograph of the authorised card holder on the card surface. The card costs £15 and can store up to £5000. Hopefully parents can provide such 'pocket money' for their older teenage children. For making payments or obtaining cash from an ATM the card operates currently by 'Chip & PIN' [10]. For automatic authentication written or spoken PIN or password version is conceivable.

18.7 The role of the smart phone as a biometric token

As noted above the first biometric token was devised at the National Physical Laboratory (NPL) in the late 1980s [1]. Within a year or two of this work at NPL this author, Bob Carter and Chris Stanford were all forecasting a future role for the mobile phone as a payment token. More recently, David Birch of Consult Hyperion has been pointing out the suitability of the microphone and keypad of the mobile phone for entering data into a remote system. He sees the benefit of most biometrics as giving customer convenience [11].

Smart phones incorporate GPS and cell location. So, as mentioned in section 18.5.4, this makes fraud more difficult.

18.8 Applications requiring enablement by biometrics – a proposed taxonomy

For a given application we need maximum user convenience and minimum time overhead.

Consider the following application-based taxonomy. It assumes the availability of powerful multi-core hardware such as is increasingly found in many smartphones and tablets.

We can have a biometric ID verification replacing the password or PIN and preceding the execution of the application. This is where the steps go in sequence. For example some laptops incorporate a fingerprint scanner for log-on. This is a *serial* use since the biometric verification is preceding and so enabling an application.

Alternatively, the biometric can operate concurrently with the application, e.g. data entry. So we already have, for example, on the market a rearward facing camera on a smartphone or laptop. The camera captures face image data concurrently with the data entry by the subject person. The face data goes to Face matching software. Presumably an iris scan and match could also be made. In these cases this is a *parallel* use of the biometric with the application. Parallel usage saves the subject person time and is therefore quite convenient. On the other hand, posing for the camera is a distraction for many people. It is especially awkward for non-touch typists. Alternative biometrics like Finger Vein (Hitachi) or Finger Vein + Fingerprint (Sagem Sécurité) [12] would fit well with a subject person holding a smart phone with one hand and entering data by touch with the other.

The third case in the proposed taxonomy is parallel operation of the biometric on the same input data as is being entered for the application. This is possible now with the combination of text independent speech recognition with speaker verification. This case can be called *parallel integrated.*

So we have a possible classification of biometric modalities in applications:

(a) serial,
(b) parallel,
(c) series integrated.

In many applications involving a smart phone or tablet as biometric token including the IMS possibility, it seems that class (c) will provide maximum convenience to the subject person. However, there might be problems if the subject has a cough or cold. So it would be as well to have a backup biometric. The Sagem/Hitachi product combining fingerprint and finger vein might be very suitable. The extra cost of two more sensors could however rule out this multi-modal idea for some IMS applications.

A significant advantage for class (c) when using the speaker verification series integrated case would be that updating the template could be done as part of normal spoken telephone calls.

For adapting the template from a remote server the protocols needed will be similar to the digital signature methods used for the current designs for ICAO ePassports. For example there must be mutual authentication between token and the master server. Some details are given by Dr Yamada of Toshiba in Reference 8.

To ensure some utility when 'match on server' is not possible it will be desirable to offer a local 'match on token' feature.

18.9 Conclusions

(i) A generic AAE is proposed for future use. Its operation on a central server seems compatible with the ACBio protocols of ISO 2476.

(ii) Smart mobile telephones and tablets have significant technical potential to function as the citizen/consumer held components in novel IMS exploiting biometric authentication with a central AAE.

(iii) There will be difficulty in introducing such services. This arises from high start-up costs and difficult legal liability issues. Proposed EU legislation in support of personal privacy could hinder or even prevent the introduction of innovative biometric services.

(iv) Handwritten signature verification could be used for early commercial deployment starting on a small scale.

(v) Vascular (vein) biometrics have a promising future. The extent to which human ageing affects the template needs to be further investigated.

References

[1] W. L. Price and B. J. Chorley (1990). 'Secure transactions with an intelligent token' (Chapter 5). In P. Hawkes, D. Davies, & W. L. Price (Eds.), *Integrated Circuit Cards, Tags and Tokens* (pp. 81–91). Oxford: BSP Professional Books, 1990, ISBN 0-632-01935-7.

[2] Veincheck^{TM} patents of Joe Rice (UK) (1984). USP4699259 published 13 October 2987 with priority from a UK application first filed May 20, 1984.

[3] L. Sharpe (2012, March). 'Turkish Bank turns to biometrics for customer ID'. *E & T Magazine*, vol. 7, issue 2. Retrieved from http://www.eandtmagazine.com

[4] Interview with Kim Cameron of Microsoft published on *ZD Net* in March 2012 (Parts I and II). Retrieve by Google search on "Kim Cameron" and "ZD Net".
[5] R. Simon (2012). 'UK data protection obligations and the use of biometrics' (21 slides). Retrieved from http://www.ico.gov.uk [This site also has information on current EU Data protection initiatives.].
[6] D. Everett. (2012, May). 'Passwords – a modern day security folly'. *Smart Card & Identity News*, volume 21, issue 5, pp. 1, 4.
[7] D. Wheeler (1980). Private communication by Professor David Wheeler, Cambridge Computer Laboratory, 1980.
[8] A. Yamada (2012). 'ACBio, Internet enabler for biometrics'. Twenty-two slides presented at the Biometrics Consortium by Dr Ashahiko Yamada of Toshiba in September 2012. Retrieved as a pdf from the Consortium's Website http://www.biometrics.org
[9] 'Biometrics into the NET with smartphones'. Discussion Notes. Retrieved from http://iiw.idcommons.org/Biometrics_into_the_NET_with_Smartphones
[10] *The Times* 23.07.2012. See also www.citizencard.com
[11] Talks by David Birch of Consult Hyperion at various recent conferences. Access by Google search on 'Convenience of Biometrics Birch'.
[12] http://www.hitachi.eu/vein/latestnres.html. Look for Press release of 19 October 2009.

Index